CONGRESSIONAL POLICIES, PRACTICES AND PROCEDURES

CONVENTIONAL WEAPONS SYSTEMS

BACKGROUND AND ISSUES FOR CONGRESS

Congressional Policies, Practices and Procedures

Additional books and e-books in this series can be found on Nova's website under the Series tab.

CONGRESSIONAL POLICIES, PRACTICES AND PROCEDURES

CONVENTIONAL WEAPONS SYSTEMS

BACKGROUND AND ISSUES FOR CONGRESS

STEWART KELLER
EDITOR

snova
New York

Library of Congress Cataloging-in-Publication Data

ISBN: 978-1-53615-414-6

Published by Nova Science Publishers, Inc. † New York

Contents

PREFACE

This book is a compilation of CRS reports on conventional weapons systems. The first report focuses on the background and issues for Congress on cluster munitions, which are air-dropped or ground-launched weapons that release a number of smaller submunitions intended to kill energy personnel or destroy vehicles. The next report focuses on the background and issues for Congress on Navy Aegis Ballistic Missile (BMD) Defense Program. The subsequent report discusses the background and issues for Congress on the Coast Guard's programs for procuring National Security Cutters (NCS), 25 Offshore Patrol Cutters (OPCs) and 58 Fast Response Cutters (FRCs). Finally, the process and procedures that currently apply to congressional consideration of foreign arms sales proposed by the President is reviewed. This includes consideration of proposals to sell major defense equipment, defense articles and services, or the retransfer to third-party states of such military items.

Chapter 1 - Cluster munitions are air-dropped or ground-launched weapons that release a number of smaller submunitions intended to kill enemy personnel or destroy vehicles. Cluster munitions were developed in World War II and are part of many nations' weapons stockpiles. Cluster munitions have been used frequently in combat, including the early phases of the current conflicts in Iraq and Afghanistan. Cluster munitions have been highly criticized internationally for causing a significant number of civilian deaths, and efforts have been undertaken to ban and regulate their use. The Department of Defense (DOD) continues to view cluster munitions as a military necessity but in 2008 instituted a policy to reduce the failure rate of cluster munitions to 1% or less after 2018. In November 2017, a new DOD policy was issued that essentially reversed the 2008 policy. Under the new policy, combatant commanders can use cluster munitions that do not meet the 1% or less unexploded submunitions standard in extreme situations to meet immediate warfighting demands. In addition, the new policy does not establish a deadline to replace cluster munitions exceeding the 1% rate and states that DOD "will retain cluster munitions currently in active inventories until the capabilities they provide are replaced with enhanced and more reliable munitions." Potential issues for Congress include cluster munitions in an era of precision weapons, other weapons in lieu of cluster munitions, and the potential impact of DOD's 2017 revised cluster munitions policy.

Chapter 2 - The Aegis ballistic missile defense (BMD) program, which is carried out by the Missile Defense Agency (MDA) and the Navy, gives Navy Aegis cruisers and destroyers a capability for conducting BMD operations. Under the FY2019 budget submission, the number of BMD-capable Aegis ships is scheduled to be 41 at the end of FY2019 and 57 at the end of FY2023. Two Japan-homeported Navy BMD-capable Aegis destroyers included in the above figures—the *Fitzgerald* (DDG-62) and the *John S McCain* (DDG-56)—were seriously damaged in collisions with merchant ships in waters off the coasts of Japan and Singapore in June 2017 and August 2017, respectively, and are currently being repaired. The temporary loss of these two BMDcapable ships reinforced, at the margin, concerns among some observers about required numbers of BMD-capable Aegis ships versus available numbers of BMD-capable Aegis ships, particularly for performing BMD operations in the Western Pacific. Under the European Phased Adaptive Approach (EPAA) for European BMD operations, BMDcapable Aegis ships are operating in European waters to defend Europe from potential ballistic missile attacks from countries such as Iran. BMD-capable Aegis ships also operate in the Western Pacific and the Persian Gulf to provide regional defense against potential ballistic missile attacks from countries such as North Korea and Iran.

The Aegis BMD program is funded mostly through MDA's budget. The Navy's budget provides additional funding for BMD-related efforts. MDA's proposed FY2019 budget requests a total of $1,711.8 million in procurement and research and development funding for Aegis BMD efforts, including funding for two Aegis Ashore sites in Poland and Romania that are to be part of the EPAA. MDA's budget also includes operations and maintenance (O&M) and military construction (MilCon) funding for the Aegis BMD program. Issues for Congress regarding the Aegis BMD program include the following:

- whether to approve, reject, or modify MDA's FY2019 funding procurement and research and development funding requests for the program;
- required numbers of BMD-capable Aegis ships versus available numbers of BMD-capable Aegis ships;
- the burden that BMD operations may be placing on the Navy's fleet of Aegis ships, and whether there are alternative ways to perform BMD missions now performed by U.S. Navy Aegis ships, such as establishing more Aegis Ashore sites;
- burden sharing—how European naval contributions to European BMD capabilities and operations compare to U.S. naval contributions to European BMD capabilities and operations;

- the potential for ship-based lasers, electromagnetic railguns (EMRGs), and hypervelocity projectiles (HVPs) to contribute in coming years to Navy terminal-phase BMD operations and the impact this might eventually have on required numbers of ship-based BMD interceptor missiles; and
- technical risk and test and evaluation issues in the Aegis BMD program.

Chapter 3 - The Coast Guard Polar Security Cutter (PSC) program, previously known as the polar icebreaker (PIB) program, is a program to acquire three new heavy polar icebreakers, to be followed years from now by the acquisition of up to three new medium polar icebreakers. The Coast Guard wants to begin construction of the first new heavy polar icebreaker in FY2019 and have it enter service in 2023. The PSC program has received about $359.6 million in acquisition funding through FY2018, including $300 million provided through the Navy's shipbuilding account and $59.6 million provided through the Coast Guard's acquisition account. The Coast Guard's proposed FY2019 budget requests $750 million in Coast Guard acquisition funding for the program. The acquisition cost of a new heavy polar icebreaker had earlier been estimated informally at roughly $1 billion, but the Coast Guard and Navy now believe that three heavy polar icebreakers could be acquired for a total cost of about $2.1 billion, or an average of about $700 million per ship. The first ship will cost more than the other two because it will incorporate design costs for the class and be at the start of the production learning curve for the class. When combined with the program's $359.6 million in prior-year funding, the $750 million requested for FY2019 would fully fund the procurement of the first new heavy polar icebreaker and partially fund the procurement of the second. The operational U.S. polar icebreaking fleet currently consists of one heavy polar icebreaker, *Polar Star*, and one medium polar icebreaker, *Healy*. In addition to *Polar Star*, the Coast Guard has a second heavy polar icebreaker, *Polar Sea*. *Polar Sea*, however, suffered an engine casualty in June 2010 and has been nonoperational since then. *Polar Star* and *Polar Sea* entered service in 1976 and 1978, respectively, and are now well beyond their originally intended 30-year service lives. The Coast Guard has used *Polar Sea* as a source of spare parts for keeping *Polar Star* operational. A Department of Homeland Security (DHS) Mission Need Statement (MNS) approved in June 2013 states that "current requirements and future projections ... indicate the Coast Guard will need to expand its icebreaking capacity, potentially requiring a fleet of up to six icebreakers (3 heavy and 3 medium) to adequately meet mission demands in the high latitudes...." The current condition of the U.S. polar icebreaker fleet, the DHS MNS, and concerns among some observers about whether the United States is adequately investing in capabilities to carry out its responsibilities and defend its interests in the Arctic, have focused policymaker attention on the question of whether and when to acquire one or more new heavy polar icebreakers as replacements for *Polar Star* and *Polar Sea*. On March 2, 2018, the U.S. Navy, in collaboration with the U.S. Coast Guard under the polar icebreaker integrated program office, released a request for proposal (RFP) for the advance procurement and detail design for the Coast Guard's heavy polar icebreaker, with options for detail design and construction for up to three heavy polar icebreakers. Issues for Congress for FY2019 for the PSC program include, inter alia, whether to approve, reject, or modify the Coast Guard's FY2019 acquisition funding request; whether to use a contract with options or a block buy contract to acquire the ships; whether to continue providing at least some of the acquisition funding for the PSC program through the Navy's shipbuilding account; technical, schedule, and cost risk in the PSC program; and whether to procure heavy and medium polar icebreakers to a common basic design.

Chapter 4 - The Coast Guard's acquisition program of record (POR) calls for procuring 8 National Security Cutters (NSCs), 25 Offshore Patrol Cutters (OPCs), and 58 Fast Response Cutters (FRCs) as replacements for 90 aging Coast Guard high-endurance cutters, medium-endurance cutters, and patrol craft. The Coast Guard's proposed FY2019 budget requests a total of $705 million in acquisition funding for the NSC, OPC, and FRC programs. NSCs are the Coast Guard's largest and most capable general-purpose cutters; they are intended to replace the Coast Guard's 12 aged Hamilton-class high-endurance cutters. NSCs have an estimated average procurement cost of about $682 million per ship. Although the Coast Guard's POR calls for procuring a total of 8 NSCs to replace the 12 Hamilton-class cutters, Congress through FY2018 has funded 11 NSCs, including two (the 10th and 11th) in FY2018. Six NSCs are now in service. The seventh is scheduled to be commissioned into service in January 2019, and the eighth and ninth are scheduled for delivery in 2019 and 2020, respectively. The Coast Guard's proposed FY2019 budget requests $65 million in acquisition funding for the NSC program; this request does not include additional funding for a 12th NSC. OPCs are to be smaller, less expensive, and in some respects less capable than NSCs; they are intended to replace the Coast Guard's 29 aged medium-endurance cutters. Coast Guard officials describe the OPC program as the service's top acquisition priority. OPCs have an estimated average procurement cost of about $391 million per ship. On September 15, 2016, the Coast Guard announced that it was awarding a contract with options for building up to nine ships in the class to Eastern Shipbuilding Group of Panama City, FL. The first OPC was funded in FY2018 and is to be delivered in 2021. The Coast Guard's proposed FY2019 budget requests $400 million in acquisition funding for the OPC program for the construction of the second OPC (which is scheduled for delivery in 2022) and procurement of long leadtime materials (LLTM) for the third OPC (which is scheduled for delivery in 2023). FRCs are considerably smaller and less expensive than OPCs; they are intended to replace the Coast Guard's 49 aging Island-class patrol boats. FRCs have an estimated average procurement cost of about $58 million per boat. A total of 50 have been funded through FY2018. The 29th was commissioned into service on November 8, 2018. The Coast Guard's proposed FY2019 budget requests $240 million in acquisition funding for the procurement of four more FRCs. The NSC, OPC, and FRC programs pose several issues for Congress, including the following:

- whether to fully or partially fund the acquisition of a 12th NSC in FY2019;
- whether to fund the acquisition of four FRCs in FY2019, as requested, or some other number, such as six, which is the maximum number that has been acquired in some prior fiscal years;
- whether to use annual or multiyear contracting for procuring OPCs;
- the procurement rate for the OPC program;
- the impact of Hurricane Michael on Eastern Shipbuilding of Panama City, FL, the shipyard that is to build the first nine OPCs;
- planned procurement quantities for NSCs, OPCs, and FRCs; and
- initial testing of the NSC.

Congress's decisions on these programs could substantially affect Coast Guard capabilities and funding requirements, and the U.S. shipbuilding industrial base.

Chapter 5 - This report reviews the process and procedures that currently apply to congressional consideration of foreign arms sales proposed by the President. This includes consideration of proposals to sell major defense equipment, defense articles and services, or the retransfer to third-party states of such military items. Under Section 36(b) of the Arms Export Control Act (AECA), Congress must be formally notified *30* calendar days before the Administration can take the final steps to conclude a government-to-government foreign military sale of major defense equipment valued at $14 million or more, defense articles or services valued at $50 million or more, or design and construction services valued at $200 million or more. In the case of such sales to NATO member states, NATO, Japan, Australia, South Korea, Israel, or New Zealand, Congress must be formally notified *15* calendar days before the Administration can proceed with the sale. However, the prior notice threshold values are higher for sales to NATO members, Japan, Australia, South Korea, Israel, or New Zealand. Commercially licensed arms sales also must be formally notified to Congress *30* calendar days before the export license is issued if they involve the sale of major defense equipment valued at $14 million or more, or defense articles or services valued at $50 million or more (Section 36(c) AECA). In the case of such sales to NATO member states, NATO, Japan, Australia, South Korea, Israel, or New Zealand, Congress must be formally notified *15* calendar days before the Administration is authorized to proceed with a given sale. As with government-to-government sales, the prior notice threshold values are higher for sales to NATO members, Japan, Australia, South Korea, Israel, or New Zealand. Furthermore, commercially licensed arms sales cases involving defense articles that are firearms-controlled under category I of the United States Munitions List and valued at $1 million or more must also be formally notified to Congress for review *30* days prior to the license for export being approved. In the case of proposed licenses for such sales to NATO members, Japan, Australia, South Korea, Israel, or New Zealand, *15* days prior notification is required. In general, the executive branch, after complying with the terms of applicable U.S. law, principally contained in the AECA, is free to proceed with an arms sales proposal unless Congress passes legislation prohibiting or modifying the proposed sale. Under current law Congress faces two fundamental obstacles to block or modify a presidential sale of military equipment: it must pass legislation expressing its will on the sale, and it must be capable of overriding a presumptive presidential veto of such legislation. Congress, however, is free to pass legislation to block or modify an arms sale *at any time* up to the point of delivery of the items involved. This report will be updated, if notable changes in these review procedures or applicable law occur.

In: Conventional Weapons Systems
Editor: Stewart Keller
ISBN: 978-1-53615-414-6

Chapter 1

CLUSTER MUNITIONS: BACKGROUND AND ISSUES FOR CONGRESS*

Andrew Feickert and Paul K. Kerr

SUMMARY

Cluster munitions are air-dropped or ground-launched weapons that release a number of smaller submunitions intended to kill enemy personnel or destroy vehicles. Cluster munitions were developed in World War II and are part of many nations' weapons stockpiles. Cluster munitions have been used frequently in combat, including the early phases of the current conflicts in Iraq and Afghanistan. Cluster munitions have been highly criticized internationally for causing a significant number of civilian deaths, and efforts have been undertaken to ban and regulate their use. The Department of Defense (DOD) continues to view cluster munitions as a military necessity but in 2008 instituted a policy to reduce the failure rate of cluster munitions to 1% or less after 2018.

In November 2017, a new DOD policy was issued that essentially reversed the 2008 policy. Under the new policy, combatant commanders can use cluster munitions that do not meet the 1% or less unexploded submunitions standard in extreme situations to meet immediate warfighting demands. In addition, the new policy does not establish a deadline to replace cluster munitions exceeding the 1% rate and states that DOD "will retain cluster munitions currently in active inventories until the capabilities they provide are replaced with enhanced and more reliable munitions."

Potential issues for Congress include cluster munitions in an era of precision weapons, other weapons in lieu of cluster munitions, and the potential impact of DOD's 2017 revised cluster munitions policy.

---

* This is an edited, reformatted and augmented version of Congressional Research Service Publication No. RS22907, Updated January 7, 2019.

WHAT ARE CLUSTER MUNITIONS?[1]

Cluster munitions are weapons that open in mid-air and disperse smaller submunitions— anywhere from a few dozen to hundreds—into an area. They can be delivered by aircraft or from ground systems such as artillery, rockets, and missiles. Cluster munitions are valued militarily because one munition can kill or destroy many targets within its impact area, and fewer weapons systems are needed to deliver fewer munitions to attack multiple targets. Cluster munitions also permit a smaller force to engage a larger adversary and are considered by some an "economy of force" weapon. Many cluster munitions rely on simple mechanical fuzes that arm the submunition based on its rate of spin and explode on impact or after a time delay. A newer generation of sensor-fuzed submunitions is being introduced by a number of nations to improve the munitions' and submunitions' accuracy and to reduce the large number of residual unexploded submunitions. These sensor-fuzed submunitions are designed to sense and destroy vehicles without creating an extensive hazard area of unexploded submunitions.

HISTORY[2]

Cluster bombs were first used in World War II, and inclusive of their debut, cluster munitions have been used in at least 21 states by at least 13 different countries. Cluster munitions were used extensively in Southeast Asia by the United States in the 1960s and 1970s, and the International Committee of the Red Cross (ICRC) estimates that in Laos alone, 9 million to 27 million unexploded submunitions remained after the conflict, resulting in over 10,000 civilian casualties to date. Cluster munitions were used by the Soviets in Afghanistan, by the British in the Falklands, by the Coalition in the Gulf War, and by the warring factions in Yugoslavia. In Kosovo and Yugoslavia in 1999, NATO forces dropped 1,765 cluster bombs containing approximately 295,000 submunitions. From 2001 through 2002, the United States dropped 1,228 cluster bombs containing 248,056 submunitions in Afghanistan, and U.S. and British forces used almost 13,000 cluster munitions containing an estimated 1.8 million to 2 million submunitions during the first three weeks of combat in Iraq in 2003. Senior U.S. government officials have stated that the United States has not used cluster munitions since 2003, during the intervention in Iraq.[3] It is widely believed that confusion over U.S. cluster submunitions (BLU-97/B) that were the same color and size as air-dropped humanitarian food packets

---

[1] Information in this section unless otherwise noted is from Mark Hiznay, "Operational and Technical Aspects of Cluster Munitions," *United Nations Institute for Disarmament Research Disarmament Forum*, 2006, pp. 15-25.

[2] Ibid.

[3] Stephen D. Mull, Acting Assistant Secretary of State for Political-Military Affairs, U.S. Cluster Munitions Policy, On-the-Record Briefing, Washington, DC, May 21, 2008.

played a major role in the U.S. decision to suspend cluster munitions use in Afghanistan but not before using them in Iraq.

In 2006, Israeli use of cluster munitions against Hezbollah forces in Lebanon resulted in widespread international criticism. Israel was said to have fired significant quantities of cluster munitions—primarily during the last 3 days of the 34-day war after a U.N. cease-fire deal had been agreed to[4]—resulting in almost 1 million unexploded cluster bomblets to which the U.N. attributed 14 deaths during the conflict.[5] Reports maintain that Hezbollah fired about 113 "cluster rockets" at northern Israel and, in turn, Israel's use of cluster munitions supposedly affected 26% of southern Lebanon's arable land and contaminated about 13 square miles with unexploded submunitions.[6] One report states that there was a failure rate of upward of 70% of Israel's cluster weapons.[7]

Cluster Munitions Criticisms

The fundamental criticisms of cluster munitions are that they disperse large numbers of submunitions imprecisely over an extended area, that they frequently fail to detonate and are difficult to detect, and that submunitions can remain explosive hazards for decades. Civilian casualties are primarily caused by munitions being fired into areas where soldiers and civilians are intermixed, inaccurate cluster munitions landing in populated areas, or civilians traversing areas where cluster munitions have been employed but failed to explode. Two technical characteristics of submunitions—failure rate and lack of a self-destruct capability—have received a great deal of attention.

Failure Rate[8]

There appear to be significant discrepancies among failure rate estimates. Some manufacturers claim a submunition failure rate of 2% to 5%, whereas mine clearance specialists have frequently reported failure rates of 10% to 30%. A number of factors influence submunition reliability. These include delivery technique, age of the submunition, air temperature, landing in soft or muddy ground, getting caught in trees and vegetation, and submunitions being damaged after dispersal, or landing in such a manner that their impact fuzes fail to initiate.

[4] Scott Peterson, "Cluster Bombs: A War's Perilous Aftermath," *Christian Science Monitor*, February 7, 2007, and "Israel Criticized for Cluster Bombs," *Los Angeles Times*, February 1, 2008.

[5] "Major Violations on Both Sides in Israel-Lebanon Conflict, Say UN Experts," *United Nations Office at Geneva News & Media*, October 4, 2006.

[6] Scott Peterson.

[7] Ibid. Failure rate as described here is either a failure to detonate on impact or a failure to detonate after a predetermined time delay.

[8] Unless otherwise noted, information in this section is from Hiznay. p. 22.

Lack of Self-Destruct Capability

Submunitions lacking a self-destruct capability—referred to as "dumb" munitions—are of particular concern because they can remain a hazard for decades, thereby increasing the potential for civilian casualties. Some nations are developing "smart" or sensor-fuzed weapons with greater reliability and a variety of self-destruct mechanisms intended to address the residual hazard of submunitions.[9] Experts maintain that self-destruct features reduce—but do not eliminate—the unexploded ordnance problem caused by cluster munitions and that the advantage gained by using "smart" cluster munitions is negated when high-failure rate and/or "dumb" cluster munitions are used in the same area.[10] For some nations, replacing "dumb" and high-failure rate cluster munitions may not be an option—China, Russia, and the Republic of Korea maintain that they cannot afford to replace all current submunitions with "smart" submunitions.[11]

International Attempts to Regulate Use

U.N. Convention on Prohibitions or Restrictions on the Use of Certain Conventional Weapons (CCW)

In an effort to restrict or ban specific types of weapons used in armed conflicts, 51 states negotiated the CCW in 1980.[12] When the treaty entered into force in December 1983, it applied only to incendiary weapons, mines and booby-traps, and weapons intended to cause casualties through very small fragments. Since then, some states-parties have added provisions through additional protocols to address other types of weapons. Acting in accordance with the recommendation of a group of experts established during the 2006 CCW review conference, states-parties to the convention decided in 2007 to "negotiate a proposal to address urgently the humanitarian impact of cluster munitions."[13] Negotiations took place in 2008 and 2009, but the parties have not reached agreement on a new proposal. The experts group continued negotiations in 2011 "informed by" a Draft Protocol on Cluster Munitions.[14] However, the CCW states-parties were unable to reach agreement on a protocol during their November 2011 review conference.

---

[9] Gayle S. Putrich, "Cluster-Bomb Ban Exempts Some Smart Munitions," *Defense News,* June 2, 2008, p. 4.

[10] Hiznay, p. 23.

[11] Ibid.

[12] Information in this section is from an Arms Control Association Fact Sheet. "Convention on Certain Conventional Weapons Convention (CCW) at a Glance," Washington, DC, October 2007.

[13] Report from the November 2007 meeting of states-parties to the CCW, December 3, 2007, http://www.unog.ch/80256EDD006B8954/(httpAssets)/029247C7A309EAC2C12573CF005B93B6/$file/CCW+MSP+2007+5+E.pdf.

[14] "*Cluster Munitions Remain the Focus of CCW Ahead of the 2011 Review Conference*," December 3, 2010, http://www.onug.ch/80256EDD006B9C2E/%28httpNewsByYear_

Convention on Cluster Munitions (CCM)[15]

Described as "frustrated with the CCW process," a number of CCW members—led by Norway— initiated negotiations in 2007 outside of the CCW to ban cluster munitions.[16] On May 30, 2008, they reached an agreement to ban cluster munitions.[17] The United States, Russia, China, Israel, Egypt, India, and Pakistan did not participate in the talks or sign the agreement. During the Signing Conference in Oslo on December 3-4, 2008, 94 states signed the convention and 4 of the signatories ratified the convention at the same time.[18] China, Russia, and the United States did not sign the convention, but France, Germany, and the United Kingdom were among the 18 NATO members to do so.[19] The convention was to enter into force six months after the deposit of the 30th ratification. The United Nations received the 30th ratification on February 16, 2010, and the convention entered into force on August 1, 2010. As of January 2, 2019, 105 states were party to the convention.

The Convention on Cluster Munitions (CCM), inter alia, bans the use of cluster munitions, as well as their development, production, acquisition, transfer, and stockpiling.[20] The convention does not prohibit cluster munitions that can detect and engage a single target or explosive submunitions equipped with an electronic self-destruction or self-deactivating feature[21]—an exemption that seemingly permits sensor-fuzed or "smart" cluster submunitions. U.S. officials were concerned that early versions of the CCM would prevent military forces from non-statesparties from providing humanitarian and peacekeeping support and significantly affect NATO military operations, but the version signed May 30, 2008, does permit states-parties to engage in military cooperation and operations with non-states-parties (Article 21, Paragraph 3).

U.S. Policy on Cluster Munitions

Then-Acting Assistant Secretary for Political-Military Affairs Stephen Mull stated in May 2008 that the United States relies on cluster munitions "as an important part of our

---

en%29/33009AD680B169BDC12577EE003F9762? OpenDocument. A copy of the Draft Protocol is available at http://daccess-ods.un.org/TMP/834449.529647827.html.

[15] For detailed information on the Convention on Cluster Munitions, see http://www.cluster muni tionsdublin.ie/ documents.asp.

[16] Arms Control Association Fact Sheet. "Convention on Certain Conventional Weapons Convention (CCW) at a Glance," Washington, DC, October 2007.

[17] Kevin Sullivan and Josh White, "111 Nations, Minus the U.S., Agree to Cluster-Bomb Ban," *Washington Post,* May 29, 2008.

[18] *Convention on Cluster Munitions Homepage* http://www.clusterconvention.org/, January 28, 2009.

[19] Marina Malenic, "Dozens of Nations Sign Cluster Bomb Treaty, U.S. Begins Upgrading Related Technology," *Defense Daily,* December 5, 2008.

[20] Diplomatic Conference for the Adoption of a Convention on Cluster Munitions, *Convention on Cluster Munitions, Dublin, Ireland*, May 30, 2008, http://www.clustermunitions dublin.ie/documents.asp.

[21] Ibid.

own defense strategy," and that Washington's preferred alternative to a ban is "to pursue technological fixes that will make sure that these weapons are no longer viable once the conflict is over."[22] U.S. officials note that

> Cluster munitions are available for use by every combat aircraft in the U.S. inventory, they are integral to every Army or Marine maneuver element and in some cases constitute up to 50 percent of tactical indirect fire support. U.S. forces simply can not fight by design or by doctrine without holding out at least the possibility of using cluster munitions.[23]

The United States also maintains that using cluster munitions reduces the number of aircraft and artillery systems needed to support military operations, and that if cluster munitions were eliminated, significantly more money would need to be spent on new weapons systems, ammunition, and logistical resources. Officials further suggest that if cluster munitions were eliminated, most militaries would increase their use of massed artillery and rocket barrages, which would likely increase destruction of key infrastructure. Then-Department of State Legal Adviser Harold Koh stated November 9, 2009, that the United States has determined that its "national security interests cannot be fully ensured consistent with the terms" of the CCM.[24]

2008 Department of Defense (DOD) Policy on Cluster Munitions[25]

The Barack Obama Administration announced on November 25, 2011, that the United States would continue to implement the DOD policy on cluster munitions issued June 19, 2008, which recognized the need to minimize harm to civilians and infrastructure but also reaffirmed that "cluster munitions are legitimate weapons with clear military utility." The central directive in the Pentagon's policy was the unwaiverable requirement that cluster munitions used after 2018 must leave less than 1% of unexploded submunitions on the battlefield. Prior to that deadline, U.S. use of cluster munitions that did not meet this criterion required combatant commander approval.

---

[22] Mull, 2008.

[23] Richard Kidd, Director of the Office of Weapons Removal and Abatement, U.S. Department of State, "Is There a Strategy for Responsible U.S. Engagement on Cluster Munitions ?" April 28, 2008.

[24] "Opening Statement for the United States Delegation by Harold Hongju Koh, Legal Adviser, United States Department of State, at the Third Conference of the High Contracting Parties to Protocol V on Explosive Remnants of War," November 9, 2009.

[25] Information in this section is from DOD Policy on Cluster Munitions and Unintended Harm to Civilians, June 19, 2008.

Revised 2017 DOD Policy on Cluster Munitions

On November 30, 2017, then-Deputy Secretary of Defense Patrick Shanahan issued a revised policy on cluster munitions.[26] The memorandum describing the policy noted that

> [c]luster munitions provide the Joint Force with an effective and necessary capability to engage area targets, including massed formations of enemy forces, individual targets dispersed over a defined area, targets whose precise location are not known, and time-sensitive or moving targets. Cluster munitions are legitimate weapons with clear military utility, as they provide distinct advantages against a range of threats in the operating environment. Additionally, the use of cluster munitions may result in less collateral damage than the collateral damage that results from use of unitary munitions alone.
>
> Since the inception of the 2008 policy, in the midst of extended combat operations in Iraq and Afghanistan, we have witnessed important changes in the global security environment and experienced several years of budgets that under-invested in replacement systems and the modernization of the Joint Force more broadly. Our adversaries and our potential adversaries have developed advanced capabilities and operational approaches specifically designed to limit our ability to project power.[27]

Both Shanahan and Admiral Harry Harris Jr. have also argued that sustaining the current U.S. cluster munitions arsenal is necessary to prepare for a potential conflict with North Korea.[28]

The revised policy reverses the 2008 policy that established an unwaiverable requirement that cluster munitions used after 2018 must leave less than 1% of unexploded submunitions on the battlefield. Combatant commanders can use cluster munitions that do not meet the 1% or less unexploded submunitions standard in extreme situations to meet immediate warfighting demands. Furthermore, the new policy does not establish a deadline to replace cluster munitions exceeding the 1% rate, and these munitions will be removed only after new munitions that meet the 1% or less unexploded submunitions standard are fielded in sufficient quantities to meet combatant commander requirements. However, the new DOD policy stipulates that the department "will only procure cluster munitions containing submunitions or submunition warheads" meeting the 2008 UXO requirement or possessing "advanced features to minimize the risks posed by unexploded submunitions."

Specifically, DOD's revised policy stipulates the following:[29]

---

[26] Memorandum from the Deputy Secretary of Defense, *Subject: DoD Policy on Cluster Munitions*, November 30, 2017.

[27] Ibid.

[28] "Statement of Admiral Harry B. Harris Jr,. U.S. Navy Commander, U.S. Pacific Command, Senate Armed Services Committee," March 15, 2018; "Remarks by Deputy Secretary of Defense Patrick M. Shanahan at the 2018 Military Reporters & Editors Conference, Navy League Headquarters of the U.S.," October 26, 2018.

[29] Memorandum from the Deputy Secretary of Defense, November 30, 2017.

- Continuing or beginning with their respective FY2019 budgets, the military departments will program for capabilities to replace cluster munitions currently in active inventories that do not meet the above-described standards for procuring new cluster munitions. The department's annual Program and Budget Review will be used to assess the sufficiency of the replacement efforts.
- The department's operational planners should plan for the availability of cluster munitions. The approval authority to employ cluster munitions that do not meet the standards prescribed by this policy for procuring new cluster munitions, however, rests with the combatant commanders. In accordance with their existing authorities, commanders may use cluster munitions that meet the standards prescribed by this policy for procuring new cluster munitions.
- The military departments and combatant commands, in keeping with U.S. legal obligations under CCW Protocol V on Explosive Remnants of War and consistent with past practices, will continue to record and retain information on the use of cluster munitions and provide relevant information to facilitate the removal or destruction of unexploded submunitions.
- The military departments and combatant commands will maintain sufficient inventories and a robust stockpile surveillance program to ensure operational quality and reliability of cluster munitions. In extremis, to meet immediate warfighting demand, combatant commanders may accept transfers of cluster munitions that do not meet the above-described cluster-munition procurement standards.
- Cluster munitions that do not meet the standards prescribed by this policy for procuring new cluster munitions will be removed from active inventories and demilitarized after their capabilities have been replaced by sufficient quantities of munitions that meet the standards in this policy.
- The department will not transfer cluster munitions except as provided for under U.S. law. The operational use of cluster munitions that include Anti-Personnel Landmines (APL) submunitions shall comply with presidential policy.

Furthermore, the Deputy Secretary of Defense

> Expect(s) the Department to achieve the goals in this policy as rapidly as industry can support. Combatant Commanders will continue to ensure that the employment of cluster munitions is consistent with the law of war and applicable international agreements in order to minimize their harmful effects on civilian populations and infrastructure.[30]

[30] Ibid.

In developing a new generation of cluster munitions less dangerous to civilians, DOD will need to determine whether such a high level of performance is achievable under both controlled laboratory conditions and real-world conditions. Factors such as delivery technique, landing in soft or muddy ground, getting caught in trees and vegetation, and submunitions being damaged after dispersal or landing could result in an appreciable number of dud submunitions, even if they have a self-deactivation feature.

DOD Efforts to Reduce Unexploded Ordnance Rates for Its Cluster Munitions

DOD and the services have been and are currently involved in efforts to reduce cluster munitions failure rates. The Army's Alternative Warhead Program (AWP) is intended to assess and recommend new technologies to reduce or eliminate cluster munitions failure rates.[31] The AWP program is viewed as particularly relevant, as the Pentagon estimates that "upward of 80 percent of U.S. cluster munitions reside in the Army artillery stockpile."[32] In December 2008, the Army decided to cease procurement of a Guided Multiple Launch Rocket System (GMLRS) warhead— the Dual-Purpose Improved Conventional Munition (DPICM) warhead—because its submunitions had a dud rate up to 5%.[33] The Air Force has also acquired cluster munitions that comply with the less than 1% failure rate—the CBU-97 Sensor Fuzed Weapon (SFW) and the CBU-105 WCMD/SFW.[34] While DOD's new 2017 cluster munitions policy calls for DOD to continue its efforts to meet the 1% or less unexploded submunitions standard "as rapidly as industry can support,"[35] it is not yet known how this policy will affect the aforementioned programs or how it could result in the establishment of new programs.

Potential Issues for Congress

Cluster Munitions in an Era of Precision Weapons

It may be argued that even with advances in "sensor-fuzed" type submunitions that seek out and destroy certain targets, cluster munitions are still essentially an

[31] Daniel Wasserbly, "Army Warhead Program to Reduce Dud Rate for Cluster Munitions," *InsideDefense.com*, July21, 2008.

[32] Ibid.

[33] Kate Brannen, "Army Will Stop Procurement of GMLRS-DPICM After Final Deliveries," *InsideDefense.com*, December 1, 2008.

[34] Marina Malenic, "Dozens of Nations Sign Cluster Bomb Treaty, U.S. Begins Upgrading Related Technology," *Defense Daily*, December 5, 2008.

[35] Memorandum from the Deputy Secretary of Defense, Subject: DoD Policy on Cluster Munitions, November 30, 2017.

indiscriminate area weapon in an era where precision weapons are increasingly becoming the military norm. In Operation Desert Storm in 1991, only about 10% of ordnance used were precision-guided, but by the time of the Iraq invasion in 2003, "the ratio of 'smart' to dumb weapons was nearly reversed."[36] Since then, this trend toward greater precision has continued, if not accelerated with the development of precision rocket, artillery, mortar munitions, and smaller precision aerial bombs designed to reduce collateral damage.[37] Given current and predicted future precision weaponry trends, cluster munitions might be losing their military relevance—much as chemical weapons did between World War I and World War II.

Other Weapons in Lieu of Cluster Munitions

According to the State Department, the U.S. military suspended its use of cluster munitions in Iraq and Afghanistan in 2003. For subsequent military operations, where cluster munitions would otherwise have been the weapon of choice, Congress might review what types of weapons were substituted in place of cluster munitions and how effective they were in achieving the desired tactical results. Also worth considering are effects-based weapons systems and operations, which seek to achieve the same or similar effect against a potential target without applying a "kinetic solution" such as a cluster munition. Such insights could prove valuable in analyzing U.S. policy options on the future of cluster munitions.

What is the Impact of DOD's 2017 Revised Cluster Munitions Policy?

DOD's November 2017 revised policy on cluster munitions potentially raises a number of issues for possible congressional consideration. With limits on cluster munition use after 2018 rescinded, how does this affect combatant commanders' operational plans in their respective theaters? Does this mean a lesser degree of military risk because combatant commanders can employ cluster munitions to meet warfighting demands, possibly translating into fewer forces needed to achieve the same result when the 2008 policy was in effect? Despite DOD emphasis on achieving a 1% or less unexploded submunitions standard "as rapidly as industry can support," will DOD funding restrictions slow or stall programs previously intended to replace those systems that exceeded 1% because there no longer is an urgent operational need to replace those systems? In a similar manner, will defense industry view this as a renewed opportunity

[36] J.R. Wilson, "Pinpoint Accuracy: Bombs, Rockets, and Mortars Get New Brains," *Armed Forces Journal*, January 2005, pp. 32-34.

[37] Jeffrey Kofman, "Introducing the Small Diameter Bomb," *ABC News*, February 16, 2008.

to develop systems with a 1% or less unexploded submunitions standard or take a more sanguine view that since DOD is no longer time constrained to develop and field 1% or less weapons that funding these programs will be less of a priority and, therefore, an unprofitable venture? Another possible issue for consideration is how this U.S. policy reversal on the military use of cluster munitions will be perceived by the international community and how this might affect future U.S. and international military treaty initiatives.

SELECTED LEGISLATION

Consolidated Appropriations Acts

The Consolidated Appropriations Act, 2010 (P.L. 111-117), which the President signed into law December 16, 2009, prohibits the provision of military assistance for cluster munitions, the issuing of defense export licenses for cluster munitions, or the sale or transfer of cluster munitions or cluster munitions technology unless "the submunitions of the cluster munitions, after arming, do not result in more than 1 percent unexploded ordnance across the range of intended operational environments." Moreover, any agreement "applicable to the assistance, transfer, or sale of such cluster munitions or cluster munitions technology" must specify that the munitions "will only be used against clearly defined military targets and will not be used where civilians are known to be present or in areas normally inhabited by civilians."

Subsequent appropriations laws have included similar provisions; the most recent is the Consolidated Appropriations Act, 2018 (P.L. 115-141), which the President signed into law on March 23, 2018. The Department of Defense and Labor, Health and Human Services, and Education Appropriations Act, 2019, and Continuing Appropriations Act, 2019 (P.L. 115-245), which the President signed into law on September 28, 2018, and P.L. 115-298, which the President signed into law on December 7, 2018, have continued these provisions.

In: Conventional Weapons Systems
Editor: Stewart Keller
ISBN: 978-1-53615-414-6

Chapter 2

NAVY AEGIS BALLISTIC MISSILE DEFENSE (BMD) PROGRAM: BACKGROUND AND ISSUES FOR CONGRESS (UPDATED)*

Ronald O'Rourke

SUMMARY

The Aegis ballistic missile defense (BMD) program, which is carried out by the Missile Defense Agency (MDA) and the Navy, gives Navy Aegis cruisers and destroyers a capability for conducting BMD operations. Under the FY2019 budget submission, the number of BMD-capable Aegis ships is scheduled to be 41 at the end of FY2019 and 57 at the end of FY2023.

Two Japan-homeported Navy BMD-capable Aegis destroyers included in the above figures—the *Fitzgerald* (DDG-62) and the *John S McCain* (DDG-56)—were seriously damaged in collisions with merchant ships in waters off the coasts of Japan and Singapore in June 2017 and August 2017, respectively, and are currently being repaired. The temporary loss of these two BMD capable ships reinforced, at the margin, concerns among some observers about required numbers of BMD-capable Aegis ships versus available numbers of BMD-capable Aegis ships, particularly for performing BMD operations in the Western Pacific.

Under the European Phased Adaptive Approach (EPAA) for European BMD operations, BMD capable Aegis ships are operating in European waters to defend Europe from potential ballistic missile attacks from countries such as Iran. BMD-capable Aegis ships also operate in the Western Pacific and the Persian Gulf to provide regional defense against potential ballistic missile attacks from countries such as North Korea and Iran.

The Aegis BMD program is funded mostly through MDA's budget. The Navy's budget provides additional funding for BMD-related efforts. MDA's proposed FY2019

* This is an edited, reformatted and augmented version of Congressional Research Service, Publication No. RL33745, dated December 13, 2018.

budget requests a total of $1,711.8 million in procurement and research and development funding for Aegis BMD efforts, including funding for two Aegis Ashore sites in Poland and Romania that are to be part of the EPAA. MDA's budget also includes operations and maintenance (O&M) and military construction (MilCon) funding for the Aegis BMD program.

Issues for Congress regarding the Aegis BMD program include the following:

- whether to approve, reject, or modify MDA's FY2019 funding procurement and research and development funding requests for the program;
- required numbers of BMD-capable Aegis ships versus available numbers of BMD-capable Aegis ships;
- the burden that BMD operations may be placing on the Navy's fleet of Aegis ships, and whether there are alternative ways to perform BMD missions now performed by U.S. Navy Aegis ships, such as establishing more Aegis Ashore sites;
- burden sharing—how European naval contributions to European BMD capabilities and operations compare to U.S. naval contributions to European BMD capabilities and operations;
- the potential for ship-based lasers, electromagnetic railguns (EMRGs), and hypervelocity projectiles (HVPs) to contribute in coming years to Navy terminal-phase BMD operations and the impact this might eventually have on required numbers of ship-based BMD interceptor missiles; and
- technical risk and test and evaluation issues in the Aegis BMD program.

INTRODUCTION

This report provides background information and issues for Congress on the Aegis ballistic missile defense (BMD) program, which is carried out by the Missile Defense Agency (MDA) and the Navy, and gives Navy Aegis cruisers and destroyers a capability for conducting BMD operations. The issue for Congress is whether to approve, reject, or modify Department of Defense (DOD) acquisition strategies and proposed funding levels for the Aegis BMD program. Congress's decisions on the Aegis BMD program could significantly affect U.S. BMD capabilities and funding requirements, and the BMD-related industrial base.

BACKGROUND

Aegis Ships

The Navy's cruisers and destroyers are called Aegis ships because they are equipped with the Aegis ship combat system—an integrated collection of sensors, computers, software, displays, weapon launchers, and weapons named for the mythological shield that defended Zeus. The Aegis system was originally developed in the 1970s for defending ships against aircraft, anti-ship cruise missiles (ASCMs), surface threats, and subsurface threats. The system was first deployed by the Navy in 1983, and it has been updated many times since. The Navy's Aegis ships include Ticonderoga (CG-47) class cruisers and Arleigh Burke (DDG-51) class destroyers.

Ticonderoga (CG-47) Class Aegis Cruisers

Overview

A total of 27 CG-47s (CGs 47 through 73) were procured for the Navy between FY1978 and FY1988; the ships entered service between 1983 and 1994. The first five ships in the class (CGs 47 through 51), which were built to an earlier technical standard in certain respects, were judged by the Navy to be too expensive to modernize and were removed from service in 2004-2005, leaving 22 ships in operation (CGs 52 through 73).

"2-4-6" Program for Modernizing 11 Existing Aegis Cruisers

Congress has directed the Navy to implement the so-called "2-4-6" program for modernizing the 11 youngest Aegis cruisers. Under the 2-4-6 program, no more than two of the cruisers are to enter the modernization program each year, none of the cruisers is to remain in reduced status for modernization for more than four years, and no more than six of the cruisers are to be in the program at any given time. Among the 11 Aegis cruisers that are to be modernized under this program are four that are BMD-capable—CG-67 (*Shiloh*), CG-70 (*Lake Erie*), CG-72 (*Vella Gulf*), and CG-73 (*Port Royal*).

Arleigh Burke (DDG-51) Class Aegis Destroyers[1]

Flight I/II and Flight IIA DDG-51s Procured in FY1985-FY2005

A total of 62 DDG-51s were procured for the Navy between FY1985 and FY2005; the first entered service in 1991 and the 62nd entered service in FY2012. The first 28 ships are known as Flight I/II DDG-51s. The next 34 ships, known as Flight IIA DDG-51s, incorporate some design changes.

No DDG-51s Procured in FY2006-FY2009

No DDG-51s were procured in FY2006-FY2009. The Navy during this period instead procured three Zumwalt (DDG-1000) class destroyers. The Navy does not plan to procure any additional DDG-1000s. The DDG-1000 design does not use the Aegis system and does not include a capability for conducting BMD operations. Navy plans do not call for modifying the three DDG1000s to make them BMD-capable.

---

[1] For more on the DDG-51 program, see CRS Report RL32109, *Navy DDG-51 and DDG-1000 Destroyer Programs: Background and Issues for Congress*, by Ronald O'Rourke.

Procurement of DDG-51s Resumed in FY2010

Procurement of DDG-51s resumed in FY2010, following procurement of the three DDG-1000s. A total of 17 DDG-51s were procured in FY2010-FY2018.[2]

Transition to Flight III DDG-51 Design in FY2017

DDG-51s procured in FY2017 and subsequent years are being built to a new version of the DDG51 design called the Flight III version. The Flight III version is to be equipped with a new radar, called the Air and Missile Defense Radar (AMDR) or the SPY-6 radar, that is more capable than the SPY-1 radar installed on all previous Aegis cruisers and destroyers.

Aegis Ships in Allied Navies

Sales of the Aegis system to allied countries began in the late 1980s. Allied countries that now operate, are building, or are planning to build Aegis-equipped ships include Japan, South Korea, Australia, Spain, and Norway.[3]

Aegis BMD System[4]

Aegis ships are given a capability for conducting BMD operations by incorporating changes to the Aegis system's computers and software, and by arming the ships with BMD interceptor missiles. In-service Aegis ships can be modified to become BMD-capable ships, and DDG-51s procured in FY2010 and subsequent years are being built from the start with a BMD capability.

Versions and Capabilities of Aegis BMD System

The Aegis BMD system exists in several variants. Listed in order of increasing capability, these are the 3.6.X variant, the 4.0.3 variant, the 4.1 variant, the 5.0 CU (Capability Upgrade) variant (also known as the Baseline [BL] 9.C1 variant), the 5.1 variant (also known as the BL 9.C2 variant), and the 6.X variant (also known as the BL 10 variant).

Figure 1 summarizes the capabilities of most of these variants and correlates them with the phases of the European Phased Adaptive Approach (or EPAA; see below) for European BMD operations. As shown in Figure 1, the Aegis BMD system is designed to intercept theater-range ballistic missiles, meaning short-, medium-, and intermediate-

---

[2] The 15 DDG-51s procured in FY2010-FY2017 include one in FY2010, two in FY2011, one in FY2012, three in FY2013, one in FY2014, two in FY2015, three in FY2016, two in FY2017, and two in FY2018.

[3] The Norwegian ships are somewhat smaller than the other Aegis ships, and consequently carry a reduced-size version of the Aegis system that includes a smaller, less-powerful version of the SPY-1 radar.

[4] Unless stated otherwise, information in this section is taken from MDA briefings on the Aegis BMD program given to CRS and CBO analysts on the MDA's FY2019 and prior-year budget submissions.

range ballistic missiles (SRBMs, MRBMs, and IRBMs, respectively). The Aegis BMD system is not designed, at least at present, to intercept longer-ranged intercontinental ballistic missiles (ICBMs). Detection and tracking data collected by the Aegis BMD system's radar, however, might be passed to other U.S. BMD systems that are designed to intercept ICBMs, which might support intercepts of ICBMs that are conducted by those other U.S. BMD systems. See also "Reported Cooperative Development of New Radar Technology" in the section below entitled "Allied Participation and Interest in Aegis BMD Program."

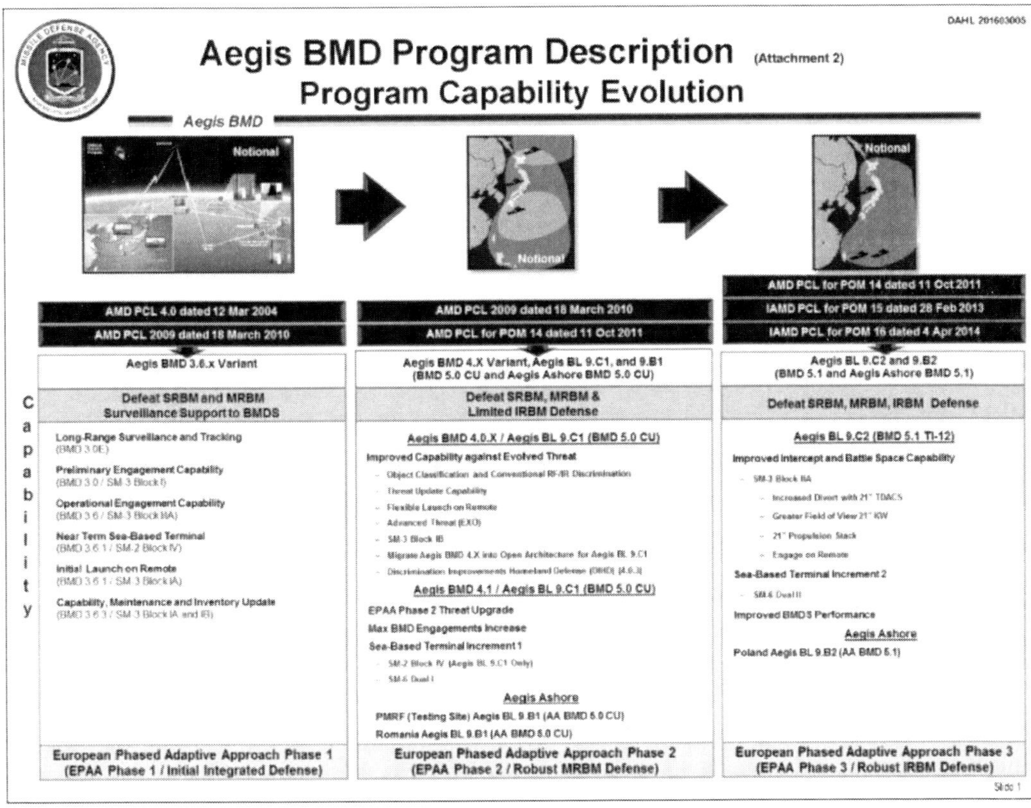

Source: MDA briefing slide provided to CRS on March 25, 2016.

Figure 1. Aegis BMD System Variants; (Summary of capabilities).

Aegis BMD Interceptor Missiles

The BMD interceptor missiles used by Aegis ships are the Standard Missile-3 (SM-3), the SM-2 Block IV, and the SM-6.

SM-3 Midcourse Interceptor

The SM-3 is designed to intercept ballistic missiles above the atmosphere (i.e., exo-atmospheric intercept), in the midcourse phase of an enemy ballistic missile's flight. It is equipped with a "hitto-kill" warhead, called a kinetic vehicle, that is designed to destroy a ballistic missile's warhead by colliding with it. MDA and Navy plans call for fielding increasingly capable versions of the SM-3 in coming years. The current versions, called the SM-3 Block IA and SM-3 Block IB, are to be supplemented in coming years by SM-3 Block IIA.

Compared to the Block IA version, the Block IB version has an improved (two-color) target seeker, an advanced signal processor, and an improved divert/attitude control system for adjusting its course. Compared to the Block IA and 1B versions, which have a 21-inch-diameter booster stage at the bottom but are 13.5 inches in diameter along the remainder of their lengths, the Block IIA version has a 21-inch diameter along its entire length. The increase in diameter to a uniform 21 inches provides more room for rocket fuel, permitting the Block IIA version to have a burnout velocity (a maximum velocity, reached at the time the propulsion stack burns out) that is greater than that of the Block IA and IB versions,[5] as well as a larger-diameter kinetic warhead. The United States and Japan have cooperated in developing certain technologies for the Block IIA version, with Japan funding a significant share of the effort.[6]

A March 26, 2018, press report states the following:

> [MDA] Director Lt. Gen. Sam Greaves said MDA "is evaluating the technical feasibility of the capability of the SM-3 Block IIA missile, currently under development, against an ICBM-class target."
>
> "If proven to be effective against an ICBM, this missile could add a layer of protection, augmenting the currently deployed GMD system," Greaves said in written testimony submitted March 22 to the Senate Armed Services strategic forces

---

[5] Some press reports and journal articles, all of which are now more than a decade old, report unconfirmed figures on the burnout velocities of various SM-3 missile configurations (some of which were proposed but ultimately not pursued). See, for example, J. D. Williams, *The Future Of Aegis Ballistic Missile Defense*, point paper dated October 15, 2004, accessed online at http://marshall.org/wp-content/uploads/2013/08/Williams-The-Future-of-Aegis-BallisticMissile-Defense.pdf; "STANDARD Missile-3 Destroys a Ballistic Missile Target in Test of Sea-based Missile Defense System," Raytheon news release circa January 26, 2002; Gopal Ratnam, "U.S. Navy To Play Larger Role In Missile Defense, *Defense News*, January 21-27, 2002: 10; Hans Mark, "A White Paper on the Defense Against Ballistic Missiles," *The Bridge*, Summer 2001, pp. 17-26, accessed online at https://www.nae.edu/File.aspx?id=7315; Michael C. Sirak, "White House Decision May Move Sea-Based NMD Into Spotlight," *Inside Missile Defense*, September 6, 2000: 1; Henry F. Cooper and J.D. Williams, "The Earliest Deployment Option—Sea-Based Defenses," *Inside Missile Defense*, September 6, 2000 (guest perspective; including graphic on page 21); Robert Holzer, "DoD Weighs Navy Interceptor Options, *Defense News*, July 24, 2000: 1, 60 (graphic on page 1); and Robert Holzer, "U.S. Navy Gathers Strength, Allies in NMD Showdown," *Defense News*, March 15, 1999: 1, 42 (graphic on page 1).

[6] The cooperative research effort has been carried out under a U.S.-Japan memorandum of agreement signed in 1999. The effort has focused on risk reduction for four parts of the missile: the sensor, an advanced kinetic warhead, the second-stage propulsion, and a lightweight nose cone. The Block IIA development effort includes the development of a missile, called the Block II, as a stepping stone to the Block IIA. As a result, the Block IIA development effort has sometimes been called the Block II/IIA development effort. The Block II missile is not planned as a fielded capability.

> subcommittee. [Greaves] said MDA will conduct a demonstration of the SM-3 Block IIA against an ICBM-like target by the end of 2020....
>
> [The FY2018 national defense authorization act] called on MDA to "conduct a test to evaluate and demonstrate, if technologically feasible, the capability to defeat a simple intercontinental ballistic missile threat using the Standard Missile-3 Block IIA missile interceptor."[7]

MDA and Navy plans at one point called for the SM-3 Block IIA to be succeeded by a still-more capable interceptor called the SM-3 Block IIB. The effort to develop that missile, however, was ended, and MDA reportedly is not pursuing any follow-on capabilities to the SM-3 Block IIA.[8]

SM-2 and SM-6 Terminal Interceptors

The SM-2 Block IV is designed to intercept ballistic missiles inside the atmosphere (i.e., endoatmospheric intercept), during the terminal phase of an enemy ballistic missile's flight. It is equipped with a blast fragmentation warhead. The existing inventory of SM-2 Block IVs—72 as of February 2012—was created by modifying SM-2s that were originally built to intercept aircraft and ASCMs. A total of 75 SM-2 Block IVs were modified, and at least 3 were used in BMD flight tests.

MDA and the Navy are now procuring a more capable terminal-phase (endo-atmospheric intercept) BMD interceptor based on the SM-6 air defense missile (the successor to the SM-2 air defense missile). The SM-6 is a dual-capability missile that can be used for either air defense (i.e., countering aircraft and anti-ship cruise missiles) or ballistic missile defense.

A July 23, 2018, press report states:

> The Defense Department has launched a prototype project that aims to dramatically increase the speed and range of the Navy's Standard Missile-6 by adding a larger rocket motor to the ship-launched weapon, a move that aims to improve both the offensive and defensive reach of the Raytheon-built system.
>
> On Jan. 17, the Navy approved plans to develop a Dual Thrust Rocket Motor with a 21- inch diameter for the SM-6, which is currently fielded with a 13.5-inch propulsion package. The new rocket motor would sit atop the current 21-inch booster, producing a new variant of the missile: the SM-6 Block IB.[9]

---

[7] Jason Sherman, "MDA Exploring Potential of SM-3 Block IIA Against ICBM Threat," *Inside the Navy*, March 26, 2018.

[8] See, for example, Justin Doubleday, "Missile Defense Agency Not Pursuing Follow-On to SM-3 Block IIA Interceptor," *Inside the Navy*, October 24, 2016.

[9] Jason Sherman, "Navy Looking to Increase Range, Speed of SM-6 with Larger Rocket Motor," *Inside the Navy*, July 23, 2018.

European Phased Adaptive Approach (EPAA) for European BMD

On September 17, 2009, the Obama Administration announced a new approach for regional BMD operations called the Phased Adaptive Approach (PAA). The first application of the approach is in Europe, and is called the European PAA (EPAA). EPAA calls for using BMD-capable Aegis ships, a land-based radar in Europe, and two Aegis Ashore sites in Romania and Poland to defend Europe against ballistic missile threats from countries such as Iran.

Phase I of EPAA involved deploying Aegis BMD ships and a land-based radar in Europe by the end of 2011. Phase II involved establishing the Aegis Ashore site in Romania with SM-3 IB interceptors in 2016.[10] Phase 3 involves establishing the Aegis Ashore site in Poland with SM-3 IIA interceptors by perhaps FY2020. The completion of construction of the Poland site has been delayed by at least a year, MDA says, due to contractor performance issues.[11] Each Aegis Ashore site in the EPAA is to include a structure housing an Aegis system similar to the deckhouse on an Aegis ship and 24 SM-3 missiles launched from a re-locatable Vertical Launch System (VLS) based on the VLS that is installed in Navy Aegis ships.[12]

Although BMD-capable Aegis ships were deployed to European waters before 2011, the first BMD-capable Aegis ship officially deployed to European waters as part of the EPAA departed its home port of Norfolk, VA, on March 7, 2011, for a deployment to the Mediterranean that lasted several months.[13]

Numbers of BMD-Capable Aegis Ships and SM-3 Interceptors

Table 1 shows numbers of BMD-capable Aegis ships and SM-3 interceptor deliveries under DOD's FY2019 budget submission. Two Japan-homeported BMD-capable Aegis destroyers included in the figures shown in Table 1—the *Fitzgerald* (DDG-62) and the *John S McCain* (DDG-56)—were seriously damaged in collisions with merchant ships in waters off the coasts of Japan and Singapore in June 2017 and August 2017, respectively,

---

[10] The Aegis Ashore site in Romania was operationally certified on May 12, 2016. (See "Aegis Ashore Missile Defense System-Romania Operationally Certified, *Navy News Service*, May 12, 2016; Sam LaGrone, "Aegis Ashore Site in Romania Declared Operational," *USNI News*, May 12, 2016).

[11] See, for example, Jen Judson, "Construction Issues Still Plague Polish Aegis Ashore Site," *Defense News*, August 14, 2018.

[12] For additional discussion of the Aegis Ashore sites, see Edward Lundquist, "Aegis Ashore Adapts Sea-Based Missile Defense System to Protect Europe," *National Defense*, September 2016.

[13] Karen Parrish, "Milestone nears for European Missile Defense Plan," *American Forces Press Service*, March 2, 2011 (http://archive.defense.gov/news/newsarticle.aspx?id=62997); Untitled "Eye On The Fleet" news item, *Navy News Service*, March 7, 2011 (accessed online at http://www.navy.mil/view_single.asp?id=98184); "Warship With Radar Going To Mediterranean," *Washington Post*, March 2, 2011; Brock Vergakis, "US Warship Deploys to Mediterranean to Protect Europe Form Ballistic Missiles, *Canadian Press*, March 7, 2011.

and are now being repaired. Of the 35 BMD-capable ships in operation as of February 2018, 19 were homeported in the Pacific fleet, and 16 in the Atlantic fleet. The inventories of SM-3 interceptors are lower than the delivery figures shown in the table due to the use of SM-3s in tests.

Homeporting of BMD-Capable DDG-51s in Spain

On October 5, 2011, the United States, Spain, and NATO jointly announced that, as part of the EPAA, four BMD-capable Aegis ships were to be forward-homeported (i.e., based) at the naval base at Rota, Spain.[14] The four ships are the destroyers *Ross* (DDG-71) and *Donald Cook* (DDG75), which moved to Rota in FY2014, and the destroyers *Carney* (DDG-64) and *Porter* (DDG78), which moved to Rota in FY2015. The moves involved an estimated 1,239 military billets (including 1,204 crew members for the four ships and 35 shore-based support personnel),[15] and about 2,100 family members.[16] The Navy estimated the up-front costs of transferring the four ships at $92 million in FY2013, and the recurring costs of basing the four ships in Spain rather than in the United States at roughly $100 million per year.[17]

Rota is on the southwestern Atlantic coast of Spain, a few miles northwest of Cadiz, and about 65 miles northwest of the Strait of Gibraltar leading into the Mediterranean. U.S. Navy ships have been homeported at Rota at various points in the past, most recently (prior to the current arrangement) in 1979.[18] For additional background information on the Navy's plan to homeport four BMD-capable Aegis destroyers at Rota, Spain, see Appendix B.

---

[14] "Announcement on missile defence cooperation by NATO Secretary General Anders Fogh Rasmussen, the Prime Minister of Spain, Jose Luis Rodriguez Zapatero and US Defense Secretary Leon Panetta," October 5, 2011, accessed October 6, 2011, at http://www.nato.int/cps/en/SID-107ADE55-FF83A6B8/natolive/opinions_78838.htm. See also "SECDEF Announces Stationing of Aegis Ships at Rota, Spain," accessed October 6, 2011, at http://www.navy.mil/ search/display.asp?story_id=63109.

[15] Source: Navy information paper dated March 8, 2012, provided by Navy Office of Legislative Affairs to CRS on March 9, 2012.

[16] Source: Navy briefing slides dated February 27, 2012, provided by the Navy to CRS on March 9, 2012.

[17] Source: Navy briefing slides dated February 27, 2012, provided by the Navy to CRS on March 9, 2012. The briefing slides state that the estimated up-front cost of $92 million includes $13.5 million for constructing a new weapon magazine, $0.8 million for constructing a pier laydown area, $3.4 million for constructing a warehouse, $5.0 million for repairing an existing facility that is to be used as an administrative/operations space, and $69.3 million for conducting maintenance work on the four ships in the United States prior to moving them to Rota. The briefing states that the estimated recurring cost of $100 million per year includes costs for base operating support, annual PCS (personnel change of station) costs, a pay and allowances delta, annual mobile training team costs, ship maintenance work, the operation of a Ship Support Activity, and higher fuel costs associated with a higher operating tempo that is maintained by ships that are homeported in foreign countries.

[18] Source: Sam Fellman, "U.S. To Base Anti-Missile Ships in Spain," *Defense News*, October 10, 2011: 76.

Table 1. Numbers of BMD-Capable Aegis Ships and SM-3 Missiles under FY2019 Budget Submission

	FY17	FY18	FY19 (req.)	FY20 (proj.)	FY21 (proj.)	FY22 (proj.)	FY23 (proj.)
BMD-capable Aegis ships							
3.6 version	17	15	10	6	5	4	4
4.0.X version	9	2	0	0	0	0	0
4.1 version	1	9	16	20	21	22	22
BL 9C.1 version	8	10	7	4	1	0	0
BL 9.C2 version	0	2	8	16	22	29	31
Total	35	38	41	46	49	55	57
Aegis Ashore sites	1	1	2	2	2	2	2
SM-3 missile cumulative deliveries							
Block I/IA	150	150	150	150	150	150	150
Block IB	147	182	218	253	287	324	362
Block IIA	0	4	15	17	27	42	48
TOTAL	297	336	383	420	464	516	560

Source: FY2019 MDA budget submission. The FY2019 quantity of 2 for Aegis Ashore sites in FY2019 may on reflect the delay in the construction of the second (Poland) site to FY2020.

Aegis BMD Flight Tests

Since January 2002, the Aegis BMD system has achieved 33 successful exo-atmospheric intercepts in 42 attempts using the SM-3 missile (including 4 successful intercepts in 5 attempts by Japanese Aegis ships, and two successful intercepts in three attempts attempt using the Aegis Ashore system), and 7 successful endo-atmospheric intercepts in 7 attempts using the SM-2 Block IV and SM-6 missiles, making for a combined total of 40 successful intercepts in 49 attempts.

In addition, on February 20, 2008, a BMD-capable Aegis cruiser operating northwest of Hawaii used a modified version of the Aegis BMD system with the SM-3 missile to shoot down an inoperable U.S. surveillance satellite that was in a deteriorating orbit.[19]

---

[19] The modifications to the ship's Aegis BMD midcourse system reportedly involved primarily making changes to software. DOD stated that the modifications were of a temporary, one-time nature. Three SM-3 missiles reportedly were modified for the operation. The first modified SM-3 fired by the cruiser successfully intercepted the satellite at an altitude of about 133 nautical miles (some sources provide differing altitudes). The other two modified SM-3s (one carried by the cruiser, another carried by an engage-capable Aegis destroyer) were not fired, and the Navy stated it would reverse the modifications to these two missiles. (For additional information, see the MDA discussion available online at http://www.mda.mil/system/aegis_one_time_mission.html, and also Peter Spiegel, "Navy Missile Hits Falling Spy Satellite," *Los Angeles Times*, February 21, 2008; Marc Kaufman and Josh White, "Navy Missile Hits Satellite, Pentagon Says," *Washington Post*, February 21, 2008; Thom Shanker, "Missile Strikes A Spy Satellite Falling From Its Orbit," *New York Times*, February 21, 2008; Bryan Bender, "US Missile Hits Crippled Satellite," *Boston Globe*, February 21, 2008; Zachary M. Peterson, "Navy Hits Wayward Satellite On First Attempt," *NavyTimes.com*, February 21, 2008; Dan Nakaso, "Satellite Smasher Back At Pearl," *Honolulu Advertiser*, February 23, 2008; Zachary M. Peterson, "Lake Erie CO Describes Anti-Satellite Shot," *NavyTimes.com*, February 25, 2008; Anne Mulrine, "The Satellite Shootdown: Behind the Scenes," *U.S. News & World Report*, February 25, 2008;

Including this intercept in the count increases the totals to 34 successful exo-atmospheric intercepts in 43 attempts using the SM-3 missile, and 41 successful exo- and endo-atmospheric intercepts in 50 attempts using SM3, SM-2 Block IV, and SM-6 missiles.

The Aegis BMD development effort, including Aegis BMD flight tests, is often described as following a development philosophy long held within the Aegis program office of "build a little, test a little, learn a lot," meaning that development is done in manageable steps, then tested and validated before moving on to the next step.[20]

For further background information on Aegis BMD flight tests, see Appendix A.

Allied Participation and Interest in Aegis BMD Program

Japan[21]

Overview

Japan's interest in BMD, and in cooperating with the United States on BMD matters, was heightened in August 1998, when North Korea test-fired a Taepo Dong-1 ballistic missile that flew over Japan before falling into the Pacific.[22] Japan's interest has been periodically reinforced since then by subsequent North Korean ballistic missile test flights.

BMD-Capable Aegis Destroyers

Japan is modifying all six of its Aegis destroyers to include the Aegis BMD capability. As of August 2017, four of the six ships reportedly had been modified, and

---

Nick Brown, "US Modified Aegis and SM-3 to Carry Out Satellite Interception Shot," *Jane's International Defence Review*, April 2008: 35).

MDA states that the incremental cost of the shoot-down operation was $112.4 million when all costs are included. MDA states that this cost is to be paid by MDA and the Pacific Command (PACOM), and that if MDA is directed to absorb the entire cost, "some realignment or reprogramming from other MDA [program] Elements may be necessary to lessen significant adverse impact on [the] AEGIS [BMD program's] cost and schedule." (MDA information paper dated March 7, 2008, provided to CRS on June 6, 2008. See also Jason Sherman, "Total Cost for Shoot-Down of Failed NRO Satellite Climbs Higher," *InsideDefense.com*, May 12, 2008).

[20] See, for example, "Aegis BMD: "Build a Little, Test a Little, Learn a Lot," *USNI blog*, March 15, 2010, accessed September 11, 2013, at http://blog.usni.org/2010/03/15/aegis-bmd-build-a-little-test-a-little-learn-a-lot, and "Aegis Ballistic Missile Defense, Aegis Ballistic Missile Defense Overview for the George C. Marshall Institute, RADM Alan B. Hicks, USN, Aegis BMD Program Director, August 3, 2009, slide 16 of 20, entitled "Some of our Philosophies In a Nutshell (1 of 2)," accessed September 11, 2011, at http://www.marshall.org/pdf/materials/743.pdf.

[21] For a research paper providing additional background information U.S.-Japan cooperation in ballistic missile defense, see Rachel Hoff, "U.S.-Japan Missile Defense Cooperation: Increasing Security and Cutting Costs," American Action Forum, December 2, 2015.

[22] For a discussion, see CRS Report RL31337, *Japan-U.S. Cooperation on Ballistic Missile Defense: Issues and Prospects*, by Richard P. Cronin. This archived report was last updated on March 19, 2002. See also CRS Report RL33436, *Japan-U.S. Relations: Issues for Congress*, coordinated by Emma Chanlett-Avery.

Japan planned to modify a fifth by March 2018, or perhaps sooner than that.[23] In November 2013, Japan announced plans to procure two additional Aegis destroyers and equip them as well with the Aegis BMD capability, which will produce an eventual Japanese force of eight BMD-capable Aegis destroyers. As of 2016, the two additional ships were expected to enter service in 2020 and 2021. Japanese BMD capable Aegis ships have participated in some of the flight tests of the Aegis BMD system using the SM-3 interceptor (see Table A-1 in Appendix A).

Cooperative Development of SM-3 Block IIA Missile

Japan has cooperated with the United States on development the SM-3 Block IIA missile. Japan developed certain technologies for the missile, and paid for the development of those technologies, reducing the missile's development costs for the United States.

Aegis Ashore Sites

In May 2017, it was reported that Japan was considering purchasing an Aegis Ashore capability to further bolster Japan's BMD capabilities for defending against North Korean ballistic missiles.[24] In August 2017, it was reported that the Japanese government planned to deploy an Aegis Ashore system and would seek funding in the budget for Japan's next fiscal year to cover Aegis Ashore system design costs.[25] In October 2017, it was reported that Japan was interested in purchasing SM-6 interceptors for its desired Aegis Ashore sites, so that the sites would employ both SM-3 midcourse-interceptors and SM-6 terminal-phase interceptors.[26]

In November 2017, it was reported that the United States was providing Japan initial pricing and technical data for both the existing Aegis Ashore system and a version equipped with the AMDR. The report stated that Japan was interested in purchasing two Aegis Ashore systems, and that the systems, if purchased, would go into operation by 2023.[27] It was also reported in November 2017 that the two Aegis Ashore systems might be located at Ground Self-Defense Force (GSDF) facilities in Akita Prefecture in eastern Japan and Yamaguchi Prefecture in western Japan, and would be operated mainly by the

[23] Richard Abott, "Japan Plans To Expand Missile Defense Systems, Includes Aegis Ashore," *Defense Daily*, August 21, 2017: 7-9.

[24] Sam LaGrone, "Defense Minister: Japan Considering Purchasing Aegis Ashore Following North Korean ICBM Test," *USNI News*, May 16, 2017.

[25] Richard Abott, "Japan Plans To Expand Missile Defense Systems, Includes Aegis Ashore," *Defense Daily*, August 21, 2017: 7-9.

[26] Richard Tomkins, "Report: Japan Eyeing SM-6 Missiles for Defense Program," *United Press International*, October 20, 2017.

[27] Anthony Capaccio, "Japan in Talks With U.S. on Buying Aegis Missile Defense," *Bloomberg*, November 7, 2017.

GSDF (i.e., Japan's army).[28] In December 2017, it was reported that that the Japanese cabinet had approved the purchase of two Aegis Ashore systems.[29] In July 2018, it was reported that Japan had decided to equip its two Aegis Ashore systems with a new Lockheed-made radar called the Long Range Discrimination Radar (LRDR) rather than the Raytheon-made SPY-6 AMDR that is being installed on U.S. Navy Flight III DDG-51s.[30]

Reported Cooperative Development of New Radar Technology

A July 6, 2018, press report states:

> The U.S. and Japan are looking to jointly develop next-generation radar technology that would use Japanese semiconductors to more than double the detection range of the Aegis missile defense system....
>
> Japanese vessels equipped with the system currently employ Lockheed Martin's SPY-1 radar. The U.S. Navy plans to upgrade its ships with Raytheon's SPY-6, which can detect missiles more than 1,000 kilometers away, or more than twice as far as the SPY-1.
>
> The U.S. and Japan now aim to develop an even longer-range system that is also more compact, letting ships stationed in the Sea of Japan watch for missiles across all of the Korean Peninsula and part of eastern China.
>
> Washington apparently broached the topic of collaborating on this next-generation system at high-level military talks in June. It hopes to incorporate technologies from companies such as Mitsubishi Electric that use gallium nitride chips, which make radar systems much more powerful at little additional energy cost. While American contractors have similar technology, Japan is thought to be the leader in the field.
>
> The countries are expected to agree on the joint project as early as this year. Studies would come first, preparing for mass production five or 10 years down the road. Japan plans to begin budgeting research expenses for the project as early as in fiscal 2019.[31]

---

[28] Yomiuri Shimbun, "Akita, Yamaguchi to Get Aegis Ashore/GSDF Involvement Expected to Strengthen Missile Defense," *The Japan News*, November 11, 2017. See also Kyodo, "Japan Mulling News Missile Interceptor Deployment to Guard Against North Korea," *South China Morning Post*, November 11, 2017.

[29] "Japan to Expand Ballistic Missile Defense with Ground-Based Aegis Batteries," *Reuters*, December 18, 2017; Daisuke Kikuchi, "Japan Approves Introduction of Aegis Ashore Missile Defense System amid North Korea Threat," *Japan Times*, December 19, 2017; Mari Yamaguchi, "Japan to Buy Aegis Ashore Missile Defense Systems," *Defense News*, December 19, 2017; Megan Eckstein, "Japan Cabinet Approves Aegis Ashore Buy to Supplement DDGs In Ballistic Missile Defense," *USNI News*, December 22, 2017.

[30] Nobuhiro Kubo, "Japan Picks Lockheed Martin Radar for Missile Defence System: Ministry Official," *Reuters*, July 2, 2018; Reuters Staff, "Japan Picks $1.2 Billion Lockheed Radar for Aegis Ashore Batteries," *Reuters*, July 30, 2018; Ben Werner, "Japan Selects Lockheed Martin to Supply Radar for Aegis Ashore System," *USNI News*, July 30, 2018; Dan Leaf, "Japan's Risky Aegis Ashore Radar Choice," *Japan Times*, July 30, 2018; Rich Abott, "Japan Chooses Lockheed Martin For Aegis Ashore Radar," *Defense Daily*: July 31, 2018: 1-2.

[31] Nikkei staff writers, "US Taps Japan Radar Tech to Double Missile Defense Range," *Nikkei Asian Review*, July 6, 2018.

Other Countries

Other countries that MDA views as potential naval BMD operators (using either the Aegis BMD system or some other system of their own design) include the United Kingdom, the Netherlands, Spain, Germany, Denmark, South Korea, and Australia. Spain, South Korea, and Australia either operate, are building, or are planning to build Aegis ships. The other countries operate destroyers and frigates with different combat systems that may have potential for contributing to BMD operations. For additional background information on allied participation and interest in the Aegis BMD program for countries other than Japan, see Appendix C.

FY2019 MDA Funding Request

The Aegis BMD program is funded mostly through MDA's budget. The Navy's budget provides additional funding for BMD-related efforts. Table 2 shows MDA procurement and research and development funding for the Aegis BMD program.

Table 2. MDA Funding for Aegis BMD Efforts, FY2019-FY2023 (In millions of dollars, rounded to nearest tenth; totals may not add due to rounding)

	FY19 (req.)	FY20 (proj.)	FY21 (proj.)	FY22 (proj.)	FY23 (proj.)
Procurement					
Aegis BMD (line 29)	593.5	576.2	538.6	565.7	767.0
Aegis BMD Advance Procurement (line 30)	115.2	97.0	44.9	17.5	0
Aegis Ashore Phase III (line 34)	15.0	0	0	0	0
Aegis BMD hardware and software (line 36)	97.1	125.8	60.4	87.2	85.0
SUBTOTAL Procurement	**820.8**	**799.0**	**643.9**	**670.4**	**852.0**
Research and development					
Aegis BMD (PE 0603892C) (line 78)	767.5	780.1	707.9	693.3	562.7
Aegis BMD Test (PE 0604878C) (line 107)	95.8	80.7	94.1	146.9	136.6
Land-based SM-3 (PE 0604880C) (line 109)	27.7	29.3	28.4	27.2	28.2
Aegis SM-3 IIA (PE 0604881C) (line 110)	0	0	0	0	0
SUBTOTAL RDT&E	**891.0**	**890.1**	**830.4**	**867.4**	**727.5**
TOTAL	**1,711.8**	**1,689.1**	**1,474.3**	**1,537.8**	**1,579.5**

Source: Table prepared by CRS based on FY2019 MDA budget submission.

As shown in Table 2, which shows MDA funding only, MDA's proposed FY2019 budget requests a total of $1,711.8 million in procurement and research and development funding for Aegis BMD efforts, including funding for the two Aegis Ashore sites that are part of the EPAA, which are referred to in the table as funding for the land-based SM-3. MDA's budget also includes additional funding not shown in the table for operations and maintenance (O&M) and military construction (MilCon) for the Aegis BMD program.

ISSUES FOR CONGRESS

FY19 Funding Request

One issue for Congress is whether to approve, reject, or modify MDA's FY2019 procurement and research and development funding requests for the program. In considering this issue, Congress may consider various factors, including whether the work that MDA is proposing to fund for FY2019 is properly scheduled for FY2019, and whether this work is accurately priced.

Required vs. Available Numbers of BMD-Capable Aegis Ships

Overview

Another potential issue for Congress concerns required numbers of BMD-capable Aegis ships versus available numbers of BMD-capable Aegis ships. Some observers are concerned about the potential operational implications of a shortfall in the available number of BMD-capable relative to the required number. Regarding the required number of BMD-capable Aegis ships, an August 15, 2018, Navy information paper states the following:

> The [Navy's] 2016 Force Structure Assessment [FSA][32] sets the requirement [for BMD capable ships] at 54 BMD-capable ships, as part of the 104 large surface combatant requirement, to meet Navy unique requirements to support defense of the sea base and limited expeditionary land base sites....
>
> The minimum requirement for 54 BMD ships is based on the Navy unique requirement as follows. It accepts risk in the sourcing of combatant commander (CCDR) requests for defense of land.
>
> - 30 to meet CVN escort demand for rotational deployment of the carrier strike groups
> - 11 INCONUS for independent BMD deployment demand
> - 9 in forward deployed naval forces (FDNF) Japan to meet operational timelines in USINDOPACOM
> - 4 in FDNF Europe for rotational deployment in EUCOM[33]

[32] The FSA is the Navy's analysis, performed every few years, that establishes the Navy's ship force structure requirements. For further discussion, see CRS Report RL32665, *Navy Force Structure and Shipbuilding Plans: Background and Issues for Congress*, by Ronald O'Rourke.

[33] Navy information paper dated August 15, 2018, entitled "Ballistic Missile Defense (BMD) Capable Ships requirement," provided by Navy Office of Legislative Affairs to CBO and CRS on August 15, 2018. The information paper was requested by CBO.

BMD-Capable Destroyers Fitzgerald and John S McCain Seriously Damaged

Two Japan-homeported Navy BMD-capable Aegis destroyers—*Fitzgerald* (DDG-62) and *John S McCain* (DDG-56)—were seriously damaged in collisions with merchant ships in waters off the coasts of Japan and Singapore in June 2017 and August 2017, respectively, and are currently being repaired. Reportedly, *Fitzgerald* will remain nonoperational for more than a year, and *John S McCain* for at least several months, while repairs on the two ships are completed.[34] The temporary loss of these two BMD-capable ships reinforced, at the margin, concerns among some observers about required numbers of BMD-capable Aegis ships versus available numbers of BMD-capable Aegis ships, particularly for performing BMD operations in the Western Pacific.[35]

An October 12, 2017, press report states the following:

> The Navy is surging a guided-missile destroyer and altering the deployment of a guided-missile destroyer to cover gaps left by two ballistic missile defense-capable destroyers that were damaged in collisions with merchant ships, USNI News has learned.
>
> The Norfolk-based cruiser USS Monterey (CG-61) and Pearl Harbor-based destroyer USS O'Kane (DDG-77) will be deployed to assist in BMD missions, two Navy officials confirmed to USNI News....
>
> O'Kane was scheduled for an independent patrol to an unspecified location before it was tasked to U.S. 7th Fleet for BMD operations in the Western Pacific, a Navy official confirmed to USNI News.
>
> Monterey will conduct an independent BMD deployment in the U.S. 6th and 5th Fleet areas of operation in Europe and the Middle East to ease to overall BMD burden for the service, a Navy official told USNI News.
>
> Both deployments will last about six months.
>
> Monterey's surge deployment follows a seven-month Middle East deployment as part of the Eisenhower Carrier Strike Group, which ended in late December.
>
> Both ships will share the U.S. BMD burden left after collisions with merchant ships sidelined guided-missile destroyers USS Fitzgerald (DDG-62) and USS John S. McCain (DDG-56).[36]

---

[34] Sam LaGrone, "USS Fitzgerald Repair Will Take More Than a Year; USS John S. McCain Fix Could Be Shorter," *USNI News*, September 2017.

[35] See, for example, Chandan Prasad, "USS John McCain: Destroyer Collision Opens Gap In Missile Defense Against North Korea," *International Business Times*, August 21, 2107; Kristen Doerer, "Do U.S. Navy Collisions Weaken Our Defense Against A North Korean Missile Attack?" *PBS News Hour*, August 22, 2017; Aaron Barruga, "The USS McCain Tragedy Has A Dire Impact On US Missile Defenses," *Task and Purpose*, August 24, 2017; Saadia M. Pekkanen, "To Contend With North Korea, Can Japan Rely On The U.S.?" *Forbes*, August 29, 2017; Mike Fabey "North Korean Missiles Are Testing A Stressed U.S. Defense Net," *Space News*, August 31, 2017.

[36] Sam LaGrone, "Navy to Surge USS Monterey, Deploy USS O'Kane for BMD Operations to Cover Ships Sidelined in Collisions," *USNI News*, October 12, 2017. See also Mark D. Faram, "Navy to Send Destroyer to Far East to Boost Ballistic Missile Defense," *Navy Times*, October 12, 2017.

Burden of BMD Mission on U.S. Navy Aegis Ships

A related potential issue for Congress is the burden that BMD operations may be placing on the Navy's fleet of Aegis ships, and whether there are alternative ways to perform BMD missions now performed by U.S. Navy Aegis ships, such as establishing more Aegis Ashore sites. A June 16, 2018, press report states

> The U.S. Navy's top officer wants to end standing ballistic missile defense patrols and transfer the mission to shore-based assets.
>
> Chief of Naval Operations Adm. John Richardson said in no uncertain terms on June 12 that he wants the Navy off the tether of ballistic missile defense patrols, a mission that has put a growing strain on the Navy's hard-worn surface combatants, and the duty shifted towards more shore-based infrastructure.
>
> "Right now, as we speak, I have six multi-mission, very sophisticated, dynamic cruisers and destroyers—six of them are on ballistic missile defense duty at sea," Richardson said during his address at the U.S. Naval War College's Current Strategy Forum. "And if you know a little bit about this business you know that geometry is a tyrant.
>
> "You have to be in a tiny little box to have a chance at intercepting that incoming missile. So, we have six ships that could go anywhere in the world, at flank speed, in a tiny little box, defending land."
>
> Richardson continued, saying the Navy could be used in emergencies but that in the long term the problem demands a different solution.
>
> "It's a pretty good capability and if there is an emergent need to provide ballistic missile defense, we're there," he said. "But 10 years down the road, it's time to build something on land to defend the land. Whether that's AEGIS ashore or whatever, I want to get out of the long-term missile defense business and move to dynamic missile defense."
>
> The unusually direct comments from the CNO come amid growing frustration among the surface warfare community that the mission, which requires ships to stay in a steaming box doing figure-eights for weeks on end, is eating up assets and operational availability that could be better used confronting growing high-end threats from China and Russia.
>
> The BMD mission was also a factor in degraded readiness in the surface fleet. Amid the nuclear threat from North Korea, the BMD mission began eating more and more of the readiness generated in the Japan-based U.S. 7th Fleet, which created a pressurized situation that caused leaders in the Pacific to cut corners and sacrifice training time for their crews, an environment described in the Navy's comprehensive review into the two collisions that claimed the lives of 17 sailors in the disastrous summer of 2017.
>
> Richardson said that as potential enemies double down on anti-access technologies designed to keep the U.S. Navy at bay, the Navy needed to focus on missile defense for its own assets.

"We're going to need missile defense at sea as we kind of fight our way now into the battle spaces we need to get into," he said. "And so restoring dynamic maneuver has something to do with missile defense.[37]

A June 23, 2018 press report states:

The threats from a resurgent Russia and rising China—which is cranking out ships like it's preparing for war—have put enormous pressure on the now-aging [U.S. Navy Aegis destroyer] fleet. Standing requirements for BMD patrols have put increasing strain on the U.S. Navy's surface ships.

The Navy now stands at a crossroads. BMD, while a burden, has also been a cash cow that has pushed the capabilities of the fleet exponentially forward over the past decade. The game-changing SPY-6 air and missile defense radar destined for DDG Flight III, for example, is a direct response to the need for more advanced BMD shooters. But a smaller fleet, needed for everything from anti-submarine patrols to freedom-of-navigation missions in the South China Sea, routinely has a large chunk tethered to BMD missions.

"Right now, as we speak, I have six multimission, very sophisticated, dynamic cruisers and destroyers—six of them are on ballistic missile defense duty at sea," Chief of Naval Operations Adm. John Richardson said during an address at the recent U.S. Naval War College's Current Strategy Forum. "You have to be in a tiny little box to have a chance at intercepting that incoming missile. So we have six ships that could go anywhere in the world, at flank speed, in a tiny little box, defending land."

And for every six ships the Navy has deployed in a standing mission, it means 18 ships are in various stages of the deployment cycle preparing to relieve them.

The Pentagon, led by Defense Secretary Jim Mattis, wants the Navy to be more flexible and less predictable—"dynamic" is the buzzword of moment in Navy circles. What Richardson is proposing is moving standing requirements for BMD patrols away from ships underway and all the associated costs that incurs, and toward fixed, shore-based sites, and also surging the Navy's at-sea BMD capabilities when there is an active threat....

In a follow-up response to questions posed on the CNO's comments, Navy spokesman Cmdr. William Speaks said the Navy's position is that BMD is an integral part of the service's mission, but where long-term threats exist, the Navy should "consider a more persistent, land-based solution as an option."

"This idea is not about the nation's or the Navy's commitment to BMD for the U.S. and our allies and partners—the Navy's commitment to ballistic missile defense is rock-solid," Speaks said. "In fact, the Navy will grow the number of BMD-capable ships from 38 to 60 by 2023, in response to the growing demand for this capability.

"The idea is about how to best meet that commitment. In alignment with our national strategic documents, we have shifted our focus in an era of great power competition—this calls us to think innovatively about how best to meet the demands of this mission and optimize the power of the joint force."...

---

[37] David B. Larter, "The US Navy Is Fed Up with Ballistic Missile Defense Patrols," *Defense News*, June 16, 2018.

While the idea of saving money by having fixed BMD sites and freeing up multimission ships is sensible, it may have unintended consequences, said Bryan McGrath, a retired destroyer skipper and owner of the defense consultancy The Ferry Bridge Group.

"The BMD mission is part of what creates the force structure requirement for large surface combatants," McGrath said on Twitter after Defense News reported the CNO's comments. "Absent it, the number of CG's and DDG's would necessarily decline. This may in fact be desirable, depending on the emerging fleet architecture and the roles and missions debate underway. Perhaps we need more smaller, multi-mission ships than larger, more expensive ones.

"But it cannot be forgotten that while the mission is somewhat wasteful of a capable, multi-mission ship, the fact that we have built the ships that (among other things) do this mission is an incredibly good thing. If there is a penalty to be paid in peacetime sub-optimization in order to have wartime capacity--should this not be considered a positive thing?"

McGrath went on to say that the suite of combat systems that have been built into Aegis have been in response to the BMD threat. And indeed, the crown jewels of the surface fleet—Aegis Baseline 9 software, which allows a ship to do both air defense and BMD simultaneously; the Aegis common-source library; the forthcoming SPY-6; cooperative engagement—have come about either in part or entirely driven by the BMD mission....

A Navy official who spoke on condition of anonymity, to discuss the Navy's shifting language on BMD, acknowledged the tone had shifted since the 2000s when the Navy latched onto the mission. But the official added that the situation more than a decade later has dramatically shifted.

"The strategic environment has changed significantly since the early 2000s—particularly in the western Pacific. We have never before faced multiple peer rivals in a world as interconnected and interdependent as we do today," the official said. "Nor have we ever seen technologies that could alter the character of war as dramatically as those we see emerging around us. China and Russia have observed our way of war and are on the move to reshape the environment to their favor."

In response to the threat and Defense Secretary Jim Mattis' desire to use the force more dynamically, the Navy is looking at its options, the official said. "This includes taking a look at how we employ BMD ships through the lens of great power competition to compete, deter and win against those who threaten us."[38]

Burden Sharing: U.S. vs. European Naval Contributions to European BMD

Another related potential issue for Congress concerns burden sharing—how European naval contributions to European BMD capabilities and operations compare to

[38] David B. Larter, "As Threats Mount, US Navy Grapples with Costly Ballistic Missile Defense Mission," *Defense News*, June 23, 2018.

U.S. naval contributions to European BMD capabilities and operations, particularly in light of constraints on U.S. defense spending, worldwide operational demands for U.S. Navy Aegis ships, and calls by some U.S. and European observers for increased defense efforts by NATO countries in Europe. Potential oversight issues for Congress include the following:

- How does the total value of European naval contributions to European BMD capabilities and operations compare to the total value of the U.S. naval contributions (including the Aegis Ashore sites) to European BMD capabilities and operations?
- Given constraints on U.S. defense spending, worldwide operational demands for U.S. Navy Aegis ships,[39] and calls by some U.S. and European observers for increased defense efforts by NATO countries in Europe—as well as the potential for European countries to purchase or build BMD-capable Aegis ships, upgrade existing ships with BMD capabilities, or purchase Aegis ashore systems—should the United States seek increased investment by European countries in their regional BMD capabilities so as to reduce the need for assigning BMD-capable U.S. Navy Aegis ships to the EPAA? Why should European countries not pay a greater share of the cost of the EPAA, since the primary purpose of the EPAA is to defend Europe against theater-range missiles?

Potential Future BMD Contribution from Lasers, Railguns, and Gun-Launched Guided Projectiles

Another potential issue for Congress concerns the potential for ship-based lasers, electromagnetic railguns (EMRGs), and gun-launched guided projectiles (GLGPs, previously known as hypervelocity projectiles [HVPs]) to contribute in coming years to Navy terminal-phase BMD operations and the impact this might eventually have on required numbers of ship-based BMD interceptor missiles. Another CRS report discusses the potential value of ship-lasers, EMRGs, and GLGPs for performing various missions, including, potentially, terminal-phase BMD operations.[40]

Technical Risk and Test and Evaluation Issues

Another potential oversight issue for Congress is technical risk and test and evaluation issues in the Aegis BMD program.

[39] See, for example, Lance M. Bacon, "Missile Defense Ships Face Arms Race, High Op Tempo," *Navy Times*, January 31, 2015.

[40] See CRS Report R44175, *Navy Lasers, Railgun, and Gun-Launched Guided Projectile: Background and Issues for Congress*, by Ronald O'Rourke.

January 2018 DOT&E Report

A January 2018 report from DOD's Director, Operational Test and Evaluation (DOT&E) — DOT&E's annual report for FY2017—stated the following in its section on the Aegis BMD program:

Assessment

- With one exception, the MDA completed its planned flight testing with the SM-3 Block IB TU [Threat Update] missile as documented in the Integrated Master Test Plan. The lone exception is FTM-24, a planned engagement against a complex MRBM target that the MDA delayed until FY20. The legacy SM-3 Block IB missile (i.e., without the TU) completed its flight testing in November 2014.
- DOT&E has lower confidence in SM-3 missile reliability due to recent in-flight failures, coupled with MDA shortfalls in simulating the in-flight environment in its SM-3 ground test program, addressing failures and anomalies identified during flight testing; and implementing a rigorous configuration management and control process for SM-3 production.
- The MDA missile ground test program may not adequately simulate the in-flight environment:
- Contractors introduced a software design flaw into the SM-3 Block IB that was not present in the SM-3 Block IA. The MDA did not discover this flaw during ground testing, but instead discovered this flaw during a failed SM CTV [Controlled Test Vehicle]-01 launch in 2016 and subsequent investigation after the EPAA Phase 2 capability declaration.
- During the course of routine production testing, Raytheon discovered a rare condition that could cause the SM-3 Block IB Kinetic Warhead Guidance Unit Guidance Unit to fail. The MDA halted deliveries of SM-3 Block IB missiles for approximately 5 months while it identified a root cause. The MDA corrected the problem with Block IB software build 6.404, released in August 2016.
- The SM-3 Block IB electromagnetic interference test and subsequent ground tests have not been compliant with Military Standard 461F, did not evaluate the self-compatibility of SM-3 Block IB electrical and software systems, and did not reflect in-flight electrical grounding, including electrical isolation and grounding shifts due to stage separations.
- The MDA did not thoroughly address, prior to flight testing, the software flaws that were present during recent flight testing:
- The MDA did not correct the software design flaw that led to the SM CTV-01 failure before conducting the test. The MDA did not correct this problem before retesting the SM3 during SM CTV-01a, but rather employed patches in a non-tactical software build to conduct the test.
- Another software design flaw that caused kinetic warhead guidance units to be unresponsive was observed during contractor acceptance testing, but was not addressed prior to conducting five subsequent flight tests. Although the flaw did not

adversely affect the flight tests, it represented an unmitigated risk to SM-3 reliability. The root cause of this flaw appears to be the MDA configuration and control process for SM-3, discussed below.

- The SM-3 program may need to improve configuration management and control:
- The software design flaw that caused the failed SM CTV-01 launch was associated with a change to the software boot-up processes and not related to capability upgrades. The MDA's continuing efforts to improve the SM-3 Block IB could introduce other unintended consequences.
- The MDA discovered the software design flaw associated with kinetic warhead guidance units (also discussed above) when it observed a performance difference in one of the circuit cards in 2016. This performance difference resulted from an approved manufacturing tooling change made in 2011. The MDA did not evaluate the potential for software performance problems caused by the tooling change until it conducted the SM CTV-01 failure investigation 5 years later.
- The MDA did not discover an unapproved manufacturing process change in 2014 associated with wiring harnesses until one failed a hardware inspection over a year later. Failures associated with this change had the potential to prevent stage separation during SM-3 Block IB missile operational use.
- Results from flight testing, high-fidelity M&S [modeling and simulation], and HWIL [hardware in the loop] and distributed ground testing demonstrate that Aegis BMD 4.0 and Baseline 9.1 firing assets can engage and intercept non-separating, simple-separating, and complex-separating ballistic missiles in the midcourse phase with SM-3 Block IB and Block IB TU guided missiles. However, flight testing and M&S are not yet sufficient to assess the full range of expected threat types, ground ranges, and raid sizes.
- The SM-3 Block IIA guided missile has flown in two developmental intercept flight tests, the first achieving a successful intercept. The second attempt, during SFTM-02, was unsuccessful because a sailor onboard the firing ship inadvertently pushed a button that caused the Aegis Weapon System to break engagement and initiate a message commanding the SM-3 Block IIA missile to destruct, destroying the missile in flight. DOT&E attributes this flight test failure to a design deficiency that allows an operator to break a ballistic missile engagement with the push of a button, without having to confirm the action. After conducting a Failure Review Board (FRB), the MDA provided a number of recommendations to the Navy that, if implemented, would preclude this type of failure from reoccurring.
- Two intercept flight tests in previous fiscal years and accredited high-fidelity M&S demonstrated that the Aegis Baseline 9.C1 system's SBT capability can successfully engage select SRBMs with SM-6 Dual I and SM-2 Block IV missiles. The SBT flight tests in FY17 demonstrated the ability to engage select MRBMs in the terminal phase of flight with SM-6 Dual I missiles, but the MDA has not yet performed M&S analyses with accredited models. The MDA plans to conduct M&S studies for select MRBM threats in FY19 and COMOPTEVFOR plans to accredit the M&S in the same timeframe.

- SM CTV-03 in October 2017 demonstrated the capability of the Aegis BMD 4.1 upgrade to fire an SM-6 Dual I missile. The BMD 4.1 build incorporates Baseline 9.C1 capabilities into the BMD 4.0 baseline.
- SM-6 Dual I and SM-2 Block IV missiles have been reliable in SBT flight tests. Missile reliability estimates for these missiles meet the specification, but not with statistical confidence due to the limited number of firings. To date, the MDA and Navy have conducted nine firings of the SM-6 Dual I or SM-6 Processor Replacement Program missile, and five firings of the SM-2 Block IV missile after modification for the SBT mission.
- Reliability, maintainability, availability, and supportability (RMA&S) data that the MDA collected during Aegis Baseline 9.1 BMD-related testing through FY17 show that the system's availability is less than desired due to large repair and logistics delay times. However, the DOT&E estimate of availability is consistent with the specification.
- The MDA demonstrated the Aegis Baseline 9.C1 system IAMD capabilities to a limited degree in flight testing. IAMD flight test engagements to date have included at most two cruise missile surrogates and a single ballistic missile target.
- MDA ground test events routinely demonstrated that inter-element coordination and interoperability need improvement to increase situational awareness. The tests also highlighted an Aegis BMD problem related to track management when it operates with other elements of the BMDS.
- The FS-17 fleet exercise demonstrated the ability of Aegis BMD 4.0.3 to interoperate with NATO partners over operational communication architectures during cruise missile and ballistic missile engagements, and to use remote data provided by NATO partners to prosecute remote engagements.

Recommendations

- Status of Previous Recommendations. The MDA:
  1) Partially addressed the second recommendation from FY13 to conduct operationally realistic testing that exercises Aegis BMD 4.0's improved engagement coordination with Terminal High-Altitude Area Defense (THAAD) and Patriot, when it conducted Flight Test Operational-02 (FTO-02) Event 2a (FY16) using an Aegis Baseline 9.C1 destroyer and THAAD firing assets. This flight test did not include Patriot. The MDA plans to include Patriot in FTO-03 Event 2 in FY18.
  2) Partially addressed the third recommendation from FY14 to ensure that the Aegis Baseline 9.C1 system conducts sufficient flight testing to allow for verification, validation, and accreditation (VV&A) of the M&S suite to cover the full design to Aegis BMD battlespace. The MDA has collected sufficient flight test data to allow the BMDS Operational Test Agency (OTA) to accredit the high fidelity M&S suite over a portion of the engagement battlespace for

Aegis Baseline 9.B1. The MDA and the BMDS OTA plan to conduct VV&A over the remaining battlespace for Baseline 9.C1 in FY18.

3) Has not addressed the second recommendation from FY15 to conduct stressing simultaneous air and ballistic missile defense engagements with the Aegis Baseline 9.C1 system operating in IAMD radar priority mode, with simultaneous engagement of multiple ballistic missile and anti-ship cruise missile threats.
4) Has not addressed the first recommendation from FY16 to conduct high-fidelity M&S runs-for-the-record for Aegis Baseline 9.B2 and 9.C2 to assess performance across the expected engagement battlespace in all Combatant Command areas of responsibility and develop an appropriate M&S VV&A plan to support that effort. The MDA developed a VV&A plan, but it will not perform Aegis Baseline 9.2 runs-for-the-record until FY20.
5) Has not addressed the second recommendation from FY16 to conduct a live-flight test demonstration of a fully remote engagement. The MDA plans to conduct this type of engagement in FY18 during FTM-29.
6) Partially addressed the third recommendation from FY16 to include BMDS OTA RMA&S data collectors in all flight test missions to improve the accuracy and statistical confidence of future suitability assessments. COMOPTEVFOR [Commander, Operational Test and Evaluation Force] works with the program to have data collectors present at each flight test event. However, the MDA has not always funded data collectors for follow-on system-level flight tests like FTO-02 Event 1a and FTO-02 Event 2a.

- FY17 Recommendations. The MDA should:
  1) Conduct an in-depth review of SM-3 missile reliability to ensure ground testing is adequately simulating the in-flight environment as observed during recent test failures.
  2) Implement processes to fix failures and anomalies identified during SM-3 ground testing prior to flight testing.
  3) Ensure that SM-3 production configuration management, manufacturing control processes, and reporting requirements are adequate.
  4) Conduct high-fidelity M&S analysis of the performance of an Aegis Baseline 9 variant ship operating in IAMD radar priority mode when simultaneously engaging multiple ballistic missile and AAW threats.
  5) Work with the Navy to implement recommendations from the SFTM-02 FRB report, including the implementation of fail-safe software designs, to preclude future inadvertent operator actions from breaking engagements against hostile ballistic missile tracks.[41]

Regarding the SM-6 missile, the January 2018 DOT&E report also stated the following:

---

[41] Department of Defense, Director, Operational Test & Evaluation, *FY2017 Annual Report*, January 2018, pp. 293-295.

Assessment

- As reported in DOT&E's memorandum, "Post Initial Operational Test and Evaluation Observations and Assessment of Standard Missile-6 Block I Suitability," dated December 2016, DOT&E considers the previously reported uplink/downlink antenna shroud reliability deficiency resolved.
- The Navy developed specific software improvements to SM -6 BLK I to mitigate the classified performance deficiency discovered during IOT&E and in DOT&E's classified IOT& E report. V C D FOT& E events conducted by the Navy demonstrated that the software improvements work as intended and lessen the severity of the deficiency, but the improvements did not resolve the deficiency in all instances. The testing identified two concerns that contributed to the deficiency not being fully resolved.
    - Testing revealed a classified concern with the missile's Target Detection Device.
    - Testing revealed a classified concern with the missile's active seeker.
- NIFC-CA FTS event L FT-5 further demonstrates the NIFC-CA FTS capability, but – as with previous NIFC-CA FTS tests – the Navy did not conduct the test under operationally realistic conditions. Moreover, the Navy's test scenarios are not sufficiently challenging to demonstrate the N IFC-CA FTS requirements defined in the Navy's September 2012 NIFCCA FTS Testing Capability Definition Letter. Nevertheless, the Navy has deployed the NIFC-CA FTS capability as a tactical option in fleet air defense. DOT&E reported on NIFC-CA FTS in the classified "Aegis Weapon System Baseline 9A Early Fielding Report" issued in July 2015, and will continue to report on NIFC-CA FTS in future Aegis Weapon System assessments.
- During SM-6 BLK 1A event GTV -3a, the SM -6 BLK IA experienced an inflight failure that prevented the target from intercepting its intended target. The failure delayed the start of SM -6 BLK IA operational testing.
- The Launch Availability Key Performance Parameter was unresolved in SM -6 BLK I IOT&E. During SM -6 BLK I FOT&E, the Navy fired, without failure, seven missiles that met the required storage requirements. While these results were not sufficient to state that BLK I meets its required Launch Availability with high statistical confidence, the results were sufficient to indicate no significant problem exists with storage reliability.

  Recommendations
- Status of Previous Recommendations.
    - The Navy is addressing the previous recommendations from FY14 to 1) complete corrective actions of the classified performance deficiency discovered during IOT&E and 2) develop a flight test program to test those corrective actions.
    - The Navy has not addressed the FY15 recommendation to provide DOT&E an operational test concept and operational test plan for NIFC-CA FTS Increment 2; DOT&E rescinded this recommendation as the Navy integrated NIFC-CA

FTS as a tactical option in fleet air defense. DOT&E removed the NIFC-CA FTS program from T&E oversight and it will be tested as a normal tactic in future Aegis/SM -6 testing.

- FY17 Recommendation. The Navy should continue investigating the classified performance deficiency discovered during IOT&E, perform corrective actions, and verify corrective actions with flight tests. This includes correcting the two new problems encountered during FY17 SM6 BLK I VCD tests.[42]

May 2017 GAO Report

An May 2017 Government Accountability Office (GAO) report on missile defense programs stated the following for the Aegis BMD program:

> The Aegis BMD program supported the European Phased Adaptive Approach (EPAA) Phase 2 delivery in December 2015 with the delivery of AWS 5.0CU. However, another spiral that was part of the original concept was delayed. EPAA Phase 2 capability was supported, in part, by two BMDS level operational flight tests called Flight Test Operational (FTO)-2 Event 1a and FTO-02 Event 2a. While MDA experienced challenges in fiscal year 2015 with the initial attempts at both tests, once conducted, they demonstrated AWS capabilities, allowing EPAA Phase 2 declared delivered. Specifically:
>
> - MDA successfully conducted FTO-02 Event 1a in December 2015. In the test, the Aegis Ashore test installation in Hawaii engaged an air-launched medium range target with the upgraded Aegis BMD Standard Missile (SM)-3 Block IB interceptor using data from the collocated off board radar. The initial test attempt in June 2015—FT0-02 Event 1—failed after a new intermediate-range target malfunctioned. As we previously reported, the delay between the initial test and the retest reduced the time available to assess all aspects of performance prior to the Romania site delivery. In addition, although the system intercepted the target, the Aegis Ashore installation was not equipped with operational version of the planned software known as AWS Baseline 9.B1, which reduced the extent the test reflected the operational architecture.
> - MDA conducted FTO-02 Event 2a, in November 2015. The test was designed to demonstrate a layered BMDS with Aegis BMD and Terminal High-Altitude Area Defense (THAAD) sharing common defended areas and shot opportunities against two threat-representative ballistic missile targets. The primary Aegis BMD test objective was to conduct a ballistic missile engagement in the presence of debris generated by a THAAD intercept, while simultaneously conducting anti-air warfare against an anti-ship cruise missile target. Although the Aegis ship successfully engaged the cruise missile, the Aegis BMD SM-3 Block IB failed in flight, preventing an intercept of the ballistic missile target. Moreover, according to the

---

[42] Department of Defense, Director, Operational Test & Evaluation, *FY2017 Annual Report*, January 2018, pp. 220-221.

Director, Operational Test and Evaluation, the scenario with two Aegis targets was less stressing than prior test of similar capabilities. However, despite the Aegis BMD SM-3 Block IB failure, THAAD intercepted the target, MDA was able to collect important data on AWS tracking and engagement processing performance. For further details on the Aegis BMD SM-3 Block IB and THAAD programs, see appendix IV and IX.

Both upgraded software packages— Baseline 9.B1 and 9.C1— offer advanced defense capabilities and integration capability with other systems external to the Aegis ships. However, according to MDA officials both versions required updates, after the EPAA Phase 2 delivery, which were certified July 2016. Additionally, according to DOT&E, testing of Baseline 9.B1 indicated that the weapons system has software issues, which lowers its reliability and availability.

In fiscal year 2016, Aegis BMD continued to assess options for developing AWS 4.1, which was initially planned to provide ballistic missile defense capabilities for additional ships. Specifically, the upgrade was initially planned to be retrofitted on ships, especially those planned for EPAA Phase 2 as ships equipped with Baseline 9.C1 were slated for other regions. However, the effort was put on hold last year, following development challenges and program funding issues. For example, as we reported in May 2015, technical assessments revealed challenges with matching Baseline 9.C1 performance characteristics on ships which utilize older hardware. This year however, MDA continued to explore options to deliver this capability, but not in support of EPAA Phase 2. Rather, current plans indicate full delivery in December 2020.

Aegis BMD also participated in two key BMD system-level assessments for the delivery of discrimination upgrades for Homeland Defense. The tests, called Ground Test Integrated -06 Part 2 and Ground Test Distributed -06 Part 2, employed models and simulations to assess upgraded AWS software— including AWS 4.0.3 and AWS 3.6.3. The new software is designed to provide upgrades for discrimination and interoperability with other BMDS elements. According to MDA, the tests successfully demonstrated the upgrades but analysis delayed the delivery of the associated capability planned for March 2017.

Aegis BMD made progress in development of AWS 5.1 for Aegis ships (Baseline 9.C2) and Aegis Ashore (Baseline 9.B2). AWS 5.1 improves AWS 5.0CU capability against longer range and more complex threats in the middle and terminal phases of flight. It also extends defended areas by engaging threats based on tracks from forward-based sensors. These AWS versions, expected to be delivered in December 2018, are slated to support EPAA Phase 3 and defend against intermediate range ballistic missile attacks.

However, the program schedule to meet EPAA Phase 3 lacks margins and has risk. For example, the deliveries of AWS Baseline 9.C2 and 9.B2 are now planned at the beginning of EPAA Phase 3 integration activities, leaving no time to rectify challenges that could still arise during development. Moreover, communication upgrades for this AWS version, are now behind schedule. Program documentation indicates that the lag in development could result in compatibility issues between these upgrades with the rest of the weapons system. This, in turn, could require retrofits and reduce performance. In addition, C2BMC delays deferred completion of Aegis BMD's capability to intercept threats based on tracks from forward-based sensors until fiscal year 2021.[43]

Regarding Aegis Ashore systems, the report stated:

MDA delivered the Aegis Ashore facility in Romania to support the European Phased Adaptive Approach Phase 2 declaration in December 2015, after demonstrating performance with only one intercept test. As we have previously reported, insufficient testing while fielding assets increases the risk of performance shortfalls and increased costs if issues are discovered as a result of flight testing. In December 2015, MDA conducted a BMDS level flight test—FTO-02 Event 1a—intended to demonstrate the operational capability of EPAA Phase 2 and Aegis Ashore's ability to defend Europe against medium-range ballistic threats. In the test, the Aegis Ashore Missile Defense test facility successfully engaged an air-launched medium-range target with an upgraded Aegis BMD SM-3 Block IB interceptor.

FT0-02 Event 1a was Aegis Ashore's first and only intercept test prior to declaring the site operational. Moreover, since 2013, Aegis Ashore has reduced the number of planned intercept tests prior to EPAA Phase 3 from four to two. According to program officials, they are leveraging data from sea-based Aegis BMD tests, however, conditions at sea are different than on land, as are the system configurations. While MDA delivered Aegis Ashore in Romania, incomplete test data delayed the evaluation of Aegis Ashore's performance against all expected engagement scenarios to determine its capabilities and limitations. This analysis is not expected to be completed until at least fiscal year 2018.

Construction of the Aegis Ashore site in Poland, the schedule for which has already been compressed, has been further complicated due to delays in completing work at the site in Romania. The Romanian site's late finish and the Poland site's slow start has resulted in the Aegis Ashore program concurrently working on both sites, leading to the increased risk of schedule delays or reduced testing. Aegis Ashore program documentation indicated that all work on the Aegis Ashore site in Romania was to be complete by December 2015; however work will be ongoing until at least fiscal year 2017 due to, among other things, the necessity to resolve power system issues required to complete system verification and validation. Overseeing work at two sites simultaneously has placed additional burdens on the program and the contractor's managers. According to program documentation, these issues place the program at risk of not being able to meet its testing and delivery milestones.

MDA has taken some steps to mitigate the project's schedule risks. According to Aegis Ashore officials; they believe "lessons learned" from the construction of the Romania site can be utilized in Poland to offset some schedule compression. Specifically, to reduce schedule risks, the program plans to add additional personnel and is working with the contractor to prevent the delays from further impacting the project schedule.

---

[43] Government Accountability Office, *Missile Defense[:] Some Progress Delivering Capabilities, but Challenges with Testing Transparency and Requirements Development Need to Be Addressed*, GAO-17-381, May 2017, pp. 81-83.

Consequently, according to program documentation, the program expects to complete the installation and delivery of Aegis Ashore Poland on schedule in 2018 and meet the EPAA Phase 3 deadline. However, further delays could result in either delaying the planned delivery or not having sufficient time to conduct all planned testing limiting the warfighter's understanding of the system's capabilities and limitations.[44]

Regarding the SM-3 Block IB interceptor, the report stated:

MDA successfully tested and delivered the Standard Missile-3 Block IB for operational use in fiscal year 2016, but the program still faces several technical issues, some of which have implications for performance or reliability. According to the Department of Operational Test and Evaluation, these reliability issues could negatively affect the interceptor's operational effectiveness due to the chance of the missile failing in flight. MDA assessed the interceptor's reliability as being within its requirements, but is taking steps to address the risks. These steps include redesigning certain components and working with Raytheon to address quality and production issues that have been discovered during recent reviews.

Addressing reliability concerns discovered during testing introduced delays and additional costs. The SM-3 Block IB program experienced two separate test failures in fiscal year 2016 that required convening a Failure Review Board to identify root causes for the failure and implement corrective actions. As a result of the failures, MDA suspended deliveries of additional interceptors, and as a result MDA missed its target for interceptor delivery. The program has identified the components responsible for the failures and will incorporate fixes during the recertification process.

In part in response to one of our prior recommendations, MDA postponed putting into production a significant design change to the interceptor's third-stage rocket motor until properly tested. MDA further delayed its decision to enter full production, from the 2nd quarter of fiscal year 2016 to the 2nd quarter of fiscal year 2017 while attempting to address issues identified in the most recent test failure. MDA has delayed full production multiple times over the life of the SM-3 Block IB, which was initially scheduled for fourth quarter, fiscal year 2012.

MDA successfully tested the new third-stage rocket motor design with two non-intercept flight tests– Standard Missile Controlled Test Vehicle (SM CTV)-01a and SM CTV-02. The redesign is intended to increase interceptor reliability, and was necessitated by a test failure in October 2013. The program initially planned to execute the tests in February 2016, but delayed the tests after the SM-3 Block IB failed a diagnostic test before the initial attempt. The tests were successfully completed in May 2016 and the redesign was approved for production in July 2016. The first missiles with the redesigned rocket motor are expected for delivery in the second quarter of fiscal year 2017. MDA plans to retrofit existing SM-3 Block IB interceptors during the periodic recertification

---

[44] Government Accountability Office, *Missile Defense[:] Some Progress Delivering Capabilities, but Challenges with Testing Transparency and Requirements Development Need to Be Addressed*, GAO-17-381, May 2017, pp. 85-86.

process. The cost to fix each interceptor is expected to be about $545 thousand per interceptor.[45]

Regarding the SM-3 Block IIA interceptor, the report stated:

The Aegis BMD SM-3 Block IIA conducted one test in fiscal year 2016, which revealed some technical challenges. Although delayed due to developmental challenges with the guidance system, the SM-3 Block IIA program conducted a non-intercept test named Standard Missile-3 Cooperative Development Controlled Test Vehicle (SCD CTV)-01 in June 2015, as well as the second non-intercept test–SCD CTV-02–in December 2015. Both tests demonstrated key capabilities including the ability to control the interceptor through the final rocket stage, separation of the kinetic warhead, and operation of the warhead after separation. The tests were successful by these measures, but still exposed some technical problems that could affect its schedule and result in further cost overruns. These challenges include design issues with missile guidance systems, which steer the interceptor to the target, and missile communication with sensors.

In order to assess these issues and incorporate lessons learned, the next test, which was also the first intercept test – Standard Missile-3 Block IIA Flight Test Mission (SFTM)-01 – was delayed from the end of fiscal year 2016 to February, 2017. MDA previously stated that any further delay in SFTM-01 could impact the schedule for SFTM-02, and with it, MDA's initial production decision scheduled for the fourth quarter of fiscal year 2017. Program documentation also stated that additional schedule delays could affect future planned testing and increase overall program cost. Despite these risks, MDA does not believe these issues will impact the schedule for EPAA Phase 3.

The development of the SM-3 Block IIA continues to deal with cost growth, in addition to schedule problems and technical issues, all of which threaten the ability to deliver an effective interceptor on time and within budget. The Missile Defense Agency reports that the contractor's estimated cost at program completion increased by around $61 million. While the program has implemented some mitigation measures, according to program documentation, additional growth threatens to result in funding shortfalls.

The program has also experienced a delay in awarding the contract for the materials needed to build interceptor test rounds, which has had further impact on the program schedule. In particular, the procurement delay has required adjustments to Flight Test Mission (FTM) - 29, the first flight test designed to fire the interceptor against an intermediate-range target while relying on remote sensor data. While the program is taking steps to mitigate the delay, risk remains that the test will be delayed which could affect the scheduled EPAA Phase 3 declaration date.[46]

---

[45] Government Accountability Office, *Missile Defense[:] Some Progress Delivering Capabilities, but Challenges with Testing Transparency and Requirements Development Need to Be Addressed*, GAO-17-381, May 2017, pp. 88-89.

[46] Government Accountability Office, *Missile Defense[:] Some Progress Delivering Capabilities, but Challenges with Testing Transparency and Requirements Development Need to Be Addressed*, GAO-17-381, May 2017, pp. 91-92.

Legislative Activity for FY2019

Summary of Action on FY2019 MDA Funding Request

Table 3. Summary of Congressional Action on FY2019 Request for MDA Procurement and RDT&E Funding for Aegis BMD Program (In millions of dollars, rounded to nearest tenth; totals may not add due to rounding)

	Request	Authorization			Appropriation		
		HASC	SASC	Conf.	HAC	SAC	Conf.
Procurement							
Aegis BMD (line 29)	593.5	593.5	593.5	593.5	679.6	708.7	700.5
Aegis BMD advance procurement (line 30)	115.2	115.2	115.2	115.2	0	0	0
Aegis Ashore Phase III (line 34)	15.0	15.0	15.0	15.0	15.0	35.0	15.0
Aegis BMD hardware and software (line 36)	97.1	97.1	97.1	97.1	97.1	97.1	97.1
Subtotal Procurement	**820.8**	**820.8**	**820.8**	**820.8**	**791.7**	**840.8**	**812.6**
Research, development, test, and evaluation (RDT&E)							
Aegis BMD (PE 0603892C) (line 78)	767.5	767.5	767.5	767.5	726.1	773.5	741.1
Aegis BMD test (PE 0604878C) (line 107)	95.8	95.8	95.8	95.8	95.8	95.8	95.8
Land-based SM-3 (PE 0604880C) (line 109)	27.7	27.8	27.7	27.7	27.7	27.7	27.7
Aegis SM-3 IIA (PE 0604881C) (line 110)	0	0	0	0	0	0	0
Subtotal RDT&E	**891.0**	**891.1**	**891.0**	**891.0**	**849.6**	**897.0**	**864.6**
TOTAL	**1,711.8**	**1,711.9**	**1,711.8**	**1,711.8**	**1,641.3**	**1,737.8**	**1,677.2**

Source: Table prepared by CRS based on DOD's original FY2019 budget submission, committee and conference reports, and explanatory statements on FY2018 National Defense Authorization Act and FY2019 DOD Appropriations Act.

Notes: HASC is House Armed Services Committee; SASC is Senate Armed Services Committee; HAC is House Appropriations Committee; SAC is Senate Appropriations Committee; Conf. is conference agreement.

Table 3 summarizes congressional action on the FY2018 request for MDA procurement and research and development funding for the Aegis BMD program.

National Defense Authorization Act for Fiscal Year 2019/ John S. McCain National Defense Authorization Act for Fiscal Year 2019 (H.R. 5515/S. 2987/P.L. 115-232)

House

The House Armed Services Committee, in its report (H.Rept. 115-676 of May 15, 2018) on H.R. 5515, recommended the funding levels for the Aegis BMD program shown in the HASC column of Table 3. The recommended increase of $150,000 [i.e., $0.15 million] for line 109 is for "Retain Poland CHUs." (Page 418)

Section 125 of H.R. 5515 as reported by the committee states the following:

> SEC. 125. Multiyear procurement authority for standard missile–6.
>
> (a) Authority for multiyear procurement.—Subject to section 2306b of title 10, United States Code, the Secretary of the Navy may enter into one or more multiyear contracts, beginning with the fiscal year 2019 program year, for the procurement of up to 625 standard missile–6 missiles at a rate of not more than 125 missiles per year during the covered period.
>
> (b) Condition for out-year contract payments.—A contract entered into under subsection (a) shall provide that any obligation of the United States to make a payment under the contract for a fiscal year after fiscal year 2019 is subject to the availability of appropriations or funds for that purpose for such later fiscal year.
>
> (c) Covered period defined.—In this section, the term "covered period" means the 5-year period beginning with the fiscal year 2019 program year and ending with the fiscal year 2023 program year.

Section 1255 of H.R. 5515 as reported by the committee states the following (emphasis added):

> SEC. 1255. Missile defense exercises in the Indo-Pacific region with United States regional allies and partners.
>
> (a) Findings.—Congress finds the following:
>
> The Democratic People's Republic of Korea (North Korea) continues to develop, test, and threaten the use of intercontinental ballistic missiles and nuclear weapons that threaten the United States and United States allies and partners.
>
> The People's Republic of China and the Russian Federation continue to develop and deploy advanced counter-intervention technologies, including fielding and testing highly maneuverable reentry vehicles and warheads (such as hypersonic weapons), and cruise missiles and small-unmanned aircraft systems (UAS) that challenge United States strategic, operational, and tactical freedom of movement and maneuver.
>
> (b) Sense of Congress.—It is the sense of Congress that the United States should—
>
> (1) continue to develop and deploy a robust missile defense in the Indo-Pacific region;
>
> (2) increase the capacity of interceptors, sensors, and operational concepts in the region;
>
> (3) continue bilateral and multilateral operationally realistic missile defense exercises in the region;
>
> (4) increase coordination with United States regional allies and partners, including Japan, South Korea, Australia, India, and other countries, as appropriate;
>
> (5) begin planning for military exercises in 2020 with United States regional allies and partners that is specifically focused on interoperability;

(6) integrate radar information from United States and allied Patriot, Terminal High Altitude Area Defense, Aegis, and other systems for region-wide command and control capabilities;

(7) increase the capacity of United States allies and partners through foreign military sales;

(8) seek increased areas of co-production for components of missile defense systems; and

(9) develop new capabilities to address threats to the region.

(c) Missile defense exercises in the Indo-Pacific region.—The Secretary of Defense may conduct missile defense exercises in the Indo-Pacific region with United States regional allies and partners to improve interoperability.

(d) Briefing.—Not later than 120 days after the date of the enactment of this Act, the Secretary of Defense shall provide to the appropriate congressional committees a briefing on plans for missile defense exercises as described in subsection (c).

(e) Appropriate congressional committees defined.—In this section, the term "appropriate congressional committees" means—

(1) the congressional defense committees; and

(2) the Committee on Foreign Relations of the Senate and the Committee on Foreign Affairs of the House of Representatives.

Section 1666 of H.R. 5515 as reported by the committee states the following:

SEC. 1666. Requirements for ballistic missile defense capable ships.

(a) Force structure assessment.—The Secretary of the Navy, in consultation with the Director of the Missile Defense Agency, shall include in the first force structure assessment conducted following the date of the enactment of this Act the following:

(1) An assessment of the requirements for ballistic missile defense capable ships.

(2) The force structure requirements associated with advanced ballistic missile defense capabilities.

(b) Force structure assessment defined.—The term "force structure assessment" has the meaning given the term in Chief of Naval Operations Instruction 3050.27.

Section 1667 of H.R. 5515 as reported states the following:

SEC. 1667. Multiyear procurement authority for standard missile–3 block IB missiles.

(a) Authority for multiyear procurement.—Subject to section 2306b of title 10, United States Code, the Director of the Missile Defense Agency may enter into one or more multiyear contracts, beginning with the 2019 program year, for the procurement of standard missile–3 block IB missiles.

(b) Condition for out-year contract payments.—A contract entered into under subsection (a) shall provide that any obligation of the United States to make a payment under the contract for a fiscal year after fiscal year 2019 is subject to the availability of appropriations or funds for that purpose for such later fiscal year.

H.Rept. 115-676 states the following:

Surface Fleet Live Fire Training

The committee recognizes the Navy's desire to increase fleet readiness training and exercise ship systems before deployment by including live-firing of missiles in pre-deployment training exercises. The committee also notes the Navy's Standard Missile-3 Block IA inventory is approaching the end of service life. Furthermore, the committee is aware that in lieu of demilitarization, the Navy intends to assess repurposing these missiles to conduct live-fire readiness training using shipboard ballistic missile defense systems. The committee encourages this initiative and directs the Secretary of the Navy to provide a briefing to the House Committee on Armed Services not later than November 5, 2018, on the Navy's progress in making SM–3 Block IA missiles approaching the end of their service life available for live-fire readiness training for ships and crews. (Page 91)

H.Rept. 115-676 also states the following:

Options to Supplement Missile Defense of Hawaii

The committee notes that Hawaii is currently defended against missile threats from North Korea by the deployed ground-based interceptors located at Fort Greely, Alaska, and Vandenberg Air Force Base, California. Mindful of potential costs and untested capability of Standard Missile–3 (SM–3) interceptors against long-range missile threats, the committee directs the Secretary of the Navy, in consultation with the Director of the Missile Defense Agency, to provide a briefing to the Committee on Armed Services of the House of Representatives, not later than September 15, 2018, on the potential to supplement this defense by assigning a permanent Aegis ship patrol to increase a layered ballistic missile defense of Hawaii, with the assumption that SM–3 missiles might be effective against long-range threats. The briefing should address the technical capability, feasibility, benefits, risks, cost, and tradeoffs of this option for the purpose of defending Hawaii.

In addition, mindful of the high demand for Terminal High Altitude Area Defense (THAAD) batteries and the untested capability of the THAAD weapon system against long-range threats, the committee also directs the Director of the Missile Defense Agency, in coordination with the Secretary of the Army, to provide a briefing to the Committee on Armed Services of the House of Representatives, not later than September 15, 2018, on the feasibility of stationing a permanent THAAD battery in Hawaii, and the technical capability, costs, benefits, and risks of testing a THAAD interceptor against an intercontinental ballistic missile. (Page 233)

H.Rept. 115-676 also states the following:

Protection of Ballistic Missile Defense System Components

The committee notes an increase to land-based ballistic missile defense system (BMDS) components with the development and delivery of the Long Range Discriminating Radar, Homeland Defense Radar-Hawaii, Pacific Radar, and completion of the Aegis Ashore site in Poland. These new sites are in addition to already deployed terrestrial weapon system sites and radars. Responsibility for protection of these sites against threats such as cruise missiles, unmanned aerial vehicles, and electronic warfare falls under the combatant commander for which they are located.

The committee directs the Secretary of Defense, in coordination with the Commander, U.S. Strategic Command, and appropriate regional combatant commands, to provide a briefing to the congressional defense committees by November 30, 2018, detailing the current protections of deployed BMDS assets from cruise missile, unmanned aerial vehicle, and electronic warfare threats. The briefing should also include the requirements for protection of the future assets that are in the program of record, as well as any plans to increase protection of current and future assets, including costs and any mitigating measures in the event that a system is degraded or unavailable. (Page 234)

H.Rept. 115-676 also states the following:

Standard Missile–3 Testing and Reliability

The committee is aware of the role and importance of the Standard Missile–3 (SM–3) interceptors in providing missile defense capability to the warfighter. The committee notes that failures of the SM–3 IB and SM–3 IIA revealed issues that may have been avoided with additional system engineering focus, and these recent challenges could have impacts on reliability assessments of these interceptors by the Director, Operational Test and Evaluation.

The committee also notes that section 1680 of the National Defense Authorization Act for Fiscal Year 2018 (Public Law 115–91) included a requirement to test the SM–3 IIA capability against a longer range threat. The committee directs the Director of the Missile Defense Agency to provide a briefing to the Committees on Armed Services of the House of Representatives and the Senate, not later than August 1, 2018, on how the recent SM–3 IIA test failure affects the planned test of this missile against an intercontinental ballistic missile-range target. This briefing should include implications such as changes to timeline of planned tests, requirements for additional tests, and changes in funding requirements.

The committee also directs the Director of the Missile Defense Agency, in coordination with the Director of the Office of Test and Evaluation, to provide a briefing to the Committees on Armed Services of the Senate and the House of Representatives, not later than December 15, 2018, detailing how the Missile Defense Agency will ensure the contractor's systems engineering and ground testing procedures are adequate to support production of SM–3 IB and SM–3 IIA interceptors. The briefing should describe how ground test data from production interceptors supports SM–3 reliability estimates from the Missile Defense Agency and the Office of Test and Evaluation. (Pages 234-235)

H.Rept. 115-676 also states the following:

Aegis Ashore Poland Austere Housing

The committee notes that the U.S. Navy has made the decision to maintain austere housing accommodations for the Aegis Ashore site in Redzikowo, Republic of Poland. This decision was made despite the committee's concerns about the impact that these conditions could have on the quality of life for the sailors manning the site.

Aegis Ashore Poland will provide critical missile defense capability to defend our deployed forces, allies, partners, and friends from missile defense threats. The site will be manned 24/7 by sailors on rotating, unaccompanied tours. The Commander of Naval Installations Command determined that the Aegis Ashore site located in Redzikowo, Poland, warranted "austere" housing, and the Chief of Naval Operations approved this determination. Under this determination, the housing accommodation guidelines will place up to 4 persons in each berthing room.

The committee is concerned that the austere housing may have a negative impact on quality of life for the sailors manning the site as they execute a critical missile defense mission. Therefore, the committee directs the Secretary of the Navy to provide a briefing to the House Committee on Armed Services by November 30, 2018, on options to improve housing standards for sailors at the Aegis Ashore Poland site, including estimated costs and schedule for completing the possible improvements. (Pages 267-268)

Senate

The Senate Armed Services Committee, in its report (S.Rept. 115-262 of June 5, 2018) on S. 2987, recommended the funding levels for the Aegis BMD program shown in the SASC column of Table 3.

Section 125 of S. 2987 as reported states:

SEC. 125. Multiyear procurement authority for Standard Missile-6.

(a) Authority for multiyear procurement.—Subject to section 2306b of title 10, United States Code, the Secretary of the Navy may enter into one or more multiyear contracts, beginning with the fiscal year 2019 program year, for the procurement of up to 625 Standard Missile–6 guided missiles.

(b) Authority for advance procurement and economic order quantity.—The Secretary may enter into one or more contracts for advance procurement associated with the missiles (including economic order quantity) for which authorization to enter into a multiyear procurement contract is provided under subsection (a).

(c) Cost analysis requirement.—The Secretary may not exercise the authority provided under subsection (a) or (b) until the Secretary of Defense submits to the congressional defense committees the report and confirmation required under subparagraphs (A) and (B), respectively, of section 2306b(i)(2) of title 10, United States Code.

(d) Condition for out-year contract payments.—A contract entered into under subsection (a) shall provide that any obligation of the United States to make a payment under the contract for a fiscal year after fiscal year 2019 is subject to the availability of appropriations for that purpose for such later fiscal year.

Section 1652 of S. 2987 as reported states:

SEC. 1652. Multiyear procurement authority for Standard Missile–3 IB guided missiles.

(a) Authority for multiyear procurement.—Subject to section 2306b of title 10, United States Code, the Secretary of Defense may enter into one or more multiyear contracts, beginning with the fiscal year 2019 program year, for the procurement of Standard Missile– 3 Block IB guided missiles.

(b) Authority for advance procurement.—The Secretary may enter into one or more contracts for advance procurement associated with the missiles for which authorization to enter into a multiyear procurement contract is provided under subsection (a).

(c) Cost analysis requirement.—The Secretary may not exercise the authority provided under subsection (a) or (b) until the Secretary submits to the congressional defense committees the report and confirmation required under subparagraphs (A) and (B), respectively, of section 2306b(i)(2) of title 10, United States Code.

(d) Condition for out-year contract payments.—A contract entered into under subsection (a) shall provide that any obligation of the United States to make a payment under the contract for a fiscal year after fiscal year 2019 is subject to the availability of appropriations for that purpose for such later fiscal year.

Regarding Section 1652, S.Rept. 115-262 states:

Multiyear procurement authority for Standard Missile–3 IB guided missiles (sec. 1652)

The committee recommends a provision that would provide authority for the Secretary of Defense to enter into a multiyear contract for the procurement of up to 204 Standard Missile–3 (SM–3) Block IB guided missiles for fiscal year 2019 through fiscal year 2023 program years, with advance procurement for economic order quantities also beginning in fiscal year 2019 pending the Director of Cost Assessment and Program Evaluation confirmation of the Secretary of the Navy's preliminary findings as required in subsection a of section 2306b of title 10, United States Code.

The SM–3 Block IB program is a core element of the Aegis Ballistic Missile Defense program and is approved through the current future years defense program. This provision would provide the following benefits: (1) Generate cost savings compared to annual procurement cost estimates; (2) Provide stable production of SM–3 Block IBs; (3) Provide a long-term commitment to the low-density aerospace industrial base that stabilizes aerospace employment levels; (4) Provide an incentive for industry capital investment for productivity improvements that would benefit several Department of Defense missile programs; and (5) Reduce disruptions in vendor delivery schedules. (Page 344)

Conference

The conference report (H.Rept. 115-874 of July 25, 2018) on H.R. 5515/P.L. 115-232 of August 13, 2018, recommends the funding levels shown in the authorization conference column of Table 3.

Section 124 of H.R. 5515 states:

SEC. 124. MULTIYEAR PROCUREMENT AUTHORITY FOR STANDARD MISSILE–6.

(a) AUTHORITY FOR MULTIYEAR PROCUREMENT.—Subject to section 2306b of title 10, United States Code, the Secretary of the Navy may enter into one or more multiyear contracts, beginning with the fiscal year 2019 program year, for the procurement of up to 625 standard missile–6 missiles at a rate of not more than 125 missiles per year during the covered period.

(b) AUTHORITY FOR ADVANCE PROCUREMENT AND ECONOMIC ORDER QUANTITY.—The Secretary may enter into one or more contracts for advance procurement associated with the missiles (including economic order quantity) for which authorization to enter into a multiyear procurement contract is provided under subsection (a).

(c) CONDITION FOR OUT-YEAR CONTRACT PAYMENTS.—A contract entered into under subsection (a) shall provide that any obligation of the United States to make a payment under the contract for a fiscal year after fiscal year 2019 is subject to the availability of appropriations or funds for that purpose for such later fiscal year.

(d) COVERED PERIOD DEFINED.—In this section, the term "covered period" means the 5-year period beginning with the fiscal year 2019 program year and ending with the fiscal year 2023 program year.

Section 1684 of H.R. 5515 states:

SEC. 1684. REQUIREMENTS FOR BALLISTIC MISSILE DEFENSE CAPABLE SHIPS.

(a) FORCE STRUCTURE ASSESSMENT.—The Secretary of the Navy, in consultation with the Director of the Missile Defense Agency, shall include in the first force structure assessment conducted following the date of the enactment of this Act the following:

(1) An assessment of the requirements for ballistic missile defense capable ships.

(2) The force structure requirements associated with advanced ballistic missile defense capabilities.

(b) FORCE STRUCTURE ASSESSMENT DEFINED.—The term "force structure assessment" has the meaning given the term in Chief of Naval Operations Instruction 3050.27.

Section 1685 of H.R. 5515 states:

SEC. 1685. MULTIYEAR PROCUREMENT AUTHORITY FOR STANDARD MISSILE–3 IB GUIDED MISSILES.

(a) AUTHORITY FOR MULTIYEAR PROCUREMENT.—Subject to section 2306b of title 10, United States Code, the Secretary of Defense may enter into one or more multiyear contracts, beginning with the fiscal year 2019 program year, for the procurement of standard missile–3 block IB guided missiles.

(b) AUTHORITY FOR ADVANCE PROCUREMENT.—The Secretary may enter into one or more contracts for advance procurement associated with the missiles for which authorization to enter into a multiyear procurement contract is provided under subsection (a).

(c) CONDITION FOR OUT-YEAR CONTRACT PAYMENTS.—A contract entered into under subsection (a) shall provide that any obligation of the United States to make a payment under the contract for a fiscal year after fiscal year 2019 is subject to the availability of appropriations for that purpose for such later fiscal year.

FY2019 DOD Appropriations Act (H.R. 6157/S. 3159/Division A of H.R. 6157/P.L. 115-245)

House

The House Appropriations Committee, in its report (H.Rept. 115-769 of June 20, 2018) on H.R. 6157, recommended the funding levels for the Aegis BMD program shown in the HAC column of Table 3.

The recommended net increase of $86.151 million for line 29 includes a recommended reduction of $4.925 million for "Spares excess growth," a recommended reduction of $5.946 million for "SM-3 IIA unit cost growth," a recommended reduction of $2.258 million for "SM-3 IIA obsolescence early to need," and a recommended increase of $99.28 million for "SM-3 IIA production—transfer from line 30." The recommended reduction of $115.206 million (the entire requested amount) for line 30 is for "Excess to need" ($15.926 million) and "Lack of justification—transfer to line 29 for SM-3 IIA production ($99.28 million). (Page 208)

The recommended net reduction of $41.463 million for line 78 includes a recommended reduction of $30.463 million for "Weapon system capability insertion early to need," a recommended reduction of $15.0 million for "Aegis BMD SM-3 development unjustified growth," and a recommended increase of $4.0 million for "Program increase—improved discrimination capabilities." (Page 283)

H.Rept. 115-769 states

> AEGIS ASHORE POLAND
>
> The Committee recognizes that the Aegis Ashore in Redzikowo, Poland will provide critical missile defense capability to defend deployed forces, allies, and partners from ballistic missile threats. The site will be manned and operated by sailors on rotating, unaccompanied tours. The Committee provides for the use of up to $150,000 of Operation and Maintenance, Navy funding to maintain the current containerized handling units on site in Poland for housing.
>
> The Committee directs the Secretary of the Navy to provide a report to the congressional defense committees not later than 60 days after the enactment of this Act on options to improve long-term housing for sailors at the Aegis Ashore Poland site, including estimated costs and schedule for completing the possible improvements. (Pages 71-72)

Senate

The Senate Appropriations Committee, in its report (S.Rept. 115-290 of June 28, 2018) on S. 3159, recommended the funding levels for the Aegis BMD program shown in the SAC column of Table 3.

The recommended increase of $115.206 million for line 29 is for "Transfer: Transfer from line 30 for additional SM–3 Block IIA interceptors." The recommended reduction of $115.206 million (the entire requested amount) for line 30 is for "Restoring acquisition accountability: Lack of justification—transfer to line 29 for additional SM–3 Block IIA interceptors." The recommended increase of $20.0 million for line 34 is for "Program increase: AEGIS Ashore Poland." (Page 142)

The recommended increase of $5.981 million for line 78 is for "Program increase: Discrimination capabilities" ($4.0 million) and "Program increase: Facilities, sustainment, restoration and modernization" ($1.981 million). (Page 197)

S.Rept. 115-290 states

> *Aegis Ashore Poland.*—The Committee understands that completion of the Aegis Ashore site in Poland will slip by at least 1 year, delaying installation of the weapon system and transition of the capability to U.S. European Command. The Committee is concerned with this delay and recommends an additional $20,000,000 to continue combat system and combat structure adaptation, integration, installation, and testing. The Committee directs the Director, MDA, to submit to the congressional defense committees, with the fiscal year 2020 President's budget request, an updated program baseline for Aegis Ashore Poland, to include revised cost estimates. The Director, CAPE, is directed to provide, with the fiscal year 2020 President's budget request, an Independent Cost Estimate for Aegis Ashore Poland. (Page 10)

S.Rept. 115-290 also states

> *AEGIS Modernization.*—The fiscal year 2019 President's budget request includes $396,403,000 to develop modifications to the A EGIS Weapon system and integrate combat capabilities developed by the Navy and the Missile Defense Agency into the AEGIS Combat System, an increase of $44,874,000 over amounts appropriated in fiscal year 2018. The Committee notes the improved joint briefing materials provided by the Navy and the Missile Defense Agency in support of the budget request and directs the Program Executive Officer, Integrated Warfare Systems, and the Director, Missile Defense Agency to continue to provide these materials, as subsequently modified per congressional request, in future budget briefings. (Page 176)

Conference

In final action, the FY2019 DOD Appropriations Act became Division A of the Department of Defense and Labor, Health and Human Services, and Education Appropriations Act, 2019, and Continuing Appropriations Act, 2019 (H.R. 6157/P.L. 115-245 of September 28, 2018).

The joint explanatory statement for H.R. 6157/P.L. 115-245 specified the funding levels shown in the appropriations conference column of Table 3. The net increase of $107.0 million for Line 29 includes a reduction of $5.946 million for "SM-3 IIA unit cost growth," a reduction of $2.258 million for "SM-3 IIA obsolescence early to need," and an increase of $115.206 million for "SM3 IIA additional interceptors - transfer from line 30." (PDF page 229 of 559) The reduction of $115.206 million (the entire requested amount) for Line 30 is for "Lack of justification - transfer to line 29 for SM-3 IIA additional interceptors." (PDF page 229 of 559) The net reduction of $26.463 million for Line 78 includes a reduction of $30.463 million for "Weapon system capability insertion early to need," and an increase of $4.0 million for "Program increase - improved discrimination capabilities." (PDF page 321 of 559)

APPENDIX A. AEGIS BMD FLIGHT TESTS

This appendix presents additional background information on Aegis BMD flight tests.

Summary of Test Flights

Table A-1 presents a summary of Aegis BMD flight tests since January 2002. As shown in the table, since January 2002, the Aegis BMD system has achieved 33 successful exo-atmospheric intercepts in 42 attempts using the SM-3 missile (including 4 successful intercepts in 5 attempts by Japanese Aegis ships, and two successful intercepts in three attempts attempt using the Aegis Ashore system), and 7 successful endo-atmospheric intercepts in 7 attempts using the SM-2 Block IV and SM-6 missiles, making for a combined total of 40 successful intercepts in 49 attempts.

Table A-1. Aegis BMD Flight Tests From January 2002 to the Present

Date	Country	Name of flight test of exercise	Ballistic Missile Target	Successful?	Cumulative successes	Cumulative attempts
Exo-atmospheric (using SM-3 missile)						
1/25/02	US	FM-2	Unitary short-range (TTV)	Yes	1	1
6/13/02	US	FM-3	Unitary short-range (TTV)	Yes	2	2
11/21/02	US	FM-4	Unitary short-range (TTV)	Yes	3	3
6/18/03	US	FM-5	Unitary short-range (TTV)	No	3	4
12/11/03	US	FM-6	Unitary short-range (TTV)	Yes	4	5
2/24/05	US	FTM 04-1 (FM-7)	Unitary short-range (TTV)	Yes	5	6
11/17/05	US	FTM 04-2 (FM-8)	Separating short-range (MRT)	Yes	6	7
6/22/06	US	FTM 10	Separating short-range (TTV)	Yes	7	8
12/7/06	US	FTM 11	Unitary short-range (TTV)	No	7	9
4/26/07	US	FTM 11 Event 4	Unitary short-range (ARAV-A)	Yes	8	10
6/22/07	US	FTM 12	Separating short-range (MRT)	Yes	9	11
8/31/07	US	FTM-11a	Classified	Yes	10	12
11/6/07	US	FTM 13	Unitary short-range (ARAV-A)	Yes	11	13
			Unitary short-range (ARAV-A)	Yes	12	14
12/17/07	Japan	JFTM-1	Separating short-range (MRT)	Yes	13	15
11/1/08	US	Pacific Blitz	Unitary short-range (ARAV-A)	Yes	14	16
			Unitary short-range (ARAV-A)			
				No	14	17
11/19/08	Japan	JFTM-2	Separating short-range (MRT)	No	14	18
7/30/09	US	FTM-17	Unitary short-range (ARAV-A)	Yes	15	19

Date	**Country**	**Name of flight test of exercise**	**Ballistic Missile Target**	**Successful?**	**Cumulative successes**	**Cumulative attempts**
10/27/09	Japan	JFTM-3	Separating short-range (MRT)	Yes	16	20
10/28/10	Japan	JFTM-4	Separating short-range (MRT)	Yes	17	21
4/14/11	US	FTM-15	Separating intermediate range (LV-2)	Yes	18	22
9/1/11	US	FTM-16 E2	Separating short-range (ARAV-B)	No	18	23
5/9/12	US	FTM-16 E2a	Unitary short-range (ARAV-A)	Yes	19	24
6/26/12	US	FTM-18	Separating short-range (MRT)	Yes	20	25
10/25/12	US	FTI-01	Separating short-range (ARAV-B)	No	20	26
2/12/13	US	FTM-20	Separating medium-range (MRBM-T3)	Yes	21	27
5/15/13	US	FTM-19	Separating short-range (ARAV-C)	Yes	22	28
9/10/13	US	FTO-01	Separating medium-range (eMRBM-T1)	Yes	23	29
9/18/13	US	FTM-21	Separating short-range (ARAV-C++)	Yes	24	30
10/3/13	US	FTM-22	Separating medium-range (ARAV-TTO-E)	Yes	25	31
11/6/14	US	FTM-25	Separating short-range (ARAV-B)	Yes	26	32
6/25/15	US	FTO-02 E1	Separating medium-range (IRBM T1)	n/aa	26	32
10/4/15	US	FTO-02 E2	Separating medium-range (eMRBM)	n/ab	26	32
10/20/15	US	ASD-15 E2	Separating short-range (Terrier Orion)	Yes	27	33
11/1/15	US	FTO-02 E2a	Separating medium-range (eMRBM)	No	27	34
12/10/15	US (Aegis Ashore)	FTO02 E1a	Separating medium-range (IRBM T1)	Yes	28	35
2/3/17	US-Japan	SFTM-01	Separating medium-range (MRT)	Yes	29	36
6/21/17	US-Japan	SFTM-02	Medium-range	No	29	37
10/15/17	US	FS17	Medium-range target	Yes	30	38
1/31/18	US (Aegis Ashore)	FTM-29	Intermediate-range target	No	30	39

Table A-1. (Continued)

Date	Country	Name of flight test of exercise	Ballistic Missile Target	Successful?	Cumulative successes	Cumulative attempts
9/11/18	Japan	JFTM-05	Simple separating target	Yes	31	40
10/26/18	US	FTM-45	Medium range	Yes	32	41
12/10/18	US (Aegis Ashore)	FTI-03	Intermediate-range target	Yes	33	42
Endo-atmospheric (using SM-2 missile Block IV missile and [for MMW Event 1] SM-6 Dual 1 missile)						
5/24/06	US	Pacific Phoenix	Unitary short-range target (Lance)	Yes	1	1
6/5/08	US	FTM-14	Unitary short-range target (FMA)	Yes	2	2
3/26/09	US	Stellar Daggers	Unitary short-range target (Lance)	Yes	3	3
7/28/15	US	MMW E1	Unitary short-range target (Lance)	Yes	4	4
7/29/15	US	MMW E2	Unitary short-range target (Lance)	Yes	5	5
12/14/16	US	FTM-27	Unitary short-range target (Lance)	Yes	6	6
8/29/17	US	FTM-27 E2	Medium-range target (MRBM)	Yes	7	7
Combined total for exo- and endo-atmospheric above tests					40	49

Sources: Table presented in MDA fact sheet, "Aegis Ballistic Missile Defense Testing," February 2017, accessed on October 16, 2017, at https://www.mda.mil/global/documents/pdf/aegis_tests.pdf, and (for flight tests subsequent to February 2017) MDA news releases.

Notes: TTV is target test vehicle; ARAV is Aegis Readiness Assessment Vehicle. In addition to the flight tests shown above, there was a successful use of an SM-3 on February 20, 2008, to intercept an inoperative U.S. satellite—an operation called Burnt Frost. Including this intercept in the count increases the totals to 31 successful exo-atmospheric intercepts in 40 attempts using the SM-3 missile, and 38 successful exo- and endoatmospheric intercepts in 47 attempts using SM-3, SM-2 Block IV, and SM-6 missiles.

[a] MDA's table shows this as a test that did not result in the launch of an SM-3. MDA as of August 3, 2015, had not issued a news release discussing this event. MDA's count of 31 successful intercepts in 37 launches through July 29, 2015, does not appear to include this test, suggesting that this was considered a "no test" event—a test in which there was a failure that was not related to the Aegis BMD system or the SM-3 interceptor. News reports state that the test was aborted due to a failure of the target missile. (Andrea Shalal, "U.S. Skips Aegis Ashore Missile Test After Target Malfunction," *Reuters*, June 26, 2015.) MDA's table similarly shows the test of December 7, 2006, as a test that did not result in the launch of an SM-3. MDA issued a news release on this test, which stated that an SM-3 was not launched "due to an incorrect system setting aboard the Aegis-class cruiser USS *Lake Erie* prior to the launch of two interceptor missiles from the ship. The incorrect configuration prevented the fire control system aboard the ship from launching the first of the two [SM-3] interceptor missiles. Since a primary test objective was a near-simultaneous launch of two missiles against two different targets, the second interceptor missile was intentionally not launched." MDA counts the test of December 7, 2006, as an unsuccessful intercept in its count of 31 successful intercepts in 37 launches through July 29, 2015.

[b] MDA's table shows this as a test that did not result in the launch of an SM-3. MDA as of November 10, 2015, had not issued a news release discussing this event. MDA's count of 32 successful intercepts in 39 launches through November 1, 2015, does not appear to include this test, suggesting that this was considered a "no test" event—a test in which there was a failure that was not related to the Aegis BMD system or the SM-3 interceptor.

In addition, on February 20, 2008, a BMD-capable Aegis cruiser operating northwest of Hawaii used a modified version of the Aegis BMD system with the SM-3 missile to shoot down an inoperable U.S. surveillance satellite that was in a deteriorating orbit. Including this intercept in the count increases the totals to 34 successful exo-atmospheric intercepts in 43 attempts using the SM-3 missile, and 41 successful exo- and endo-atmospheric intercepts in 50 attempts using SM3, SM-2 Block IV, and SM-6 missiles.

May 2010 Criticism of Claimed Successes in Flight Tests

In a May 2010 magazine article and supplementary white paper, two professors with scientific backgrounds—George Lewis and Theodore Postol—criticized DOD claims of successes in Aegis (and other DOD) BMD flight tests, arguing that

> the Defense Department's own test data show that, in combat, the vast majority of "successful" SM-3 experiments would have failed to destroy attacking warheads. The data also show potential adversaries how to defeat both the SM-3 and the GMD [ground-based missile defense] systems, which share the same serious flaws that can be readily exploited by adversaries.[47]

The criticisms made by Lewis and Postol were reported in a May 18, 2010, *New York Times* article.[48] In response to the criticisms and the *New York Times* article, MDA issued a press release and other information defending the flight tests and arguing that the criticisms are based on inaccurate or incomplete information.[49]

Details on Selected Exo-Atmospheric (SM-3) Flight Tests Since June 2006

June 22, 2006, Test

This was the first test to use the 3.6 version of the Aegis BMD system.[50]

---

[47] George N. Lewis and Theodore A. Postol, "A Flawed and Dangerous U.S. Missile Defense Plan," *Arms Control Today*, May 2010: 24-32. The quoted passage appears on p. 26. The associated white paper is George N. Lewis and Theodore A. Postol, *A Technically Detailed Description of Flaws in the SM-3 and GMD Missile Defense Systems Revealed by the Defense Department's Ballistic Missile Test Data*, May 3, 2010, 13 pp.

[48] William J. Broad and David E. Sanger, "Review Cites Flaws In U.S. Antimissile Program," *New York Times*, May 18, 2010: 1.

[49] Missile Defense Agency, "Missile Defense Agency Responds to New York Times Article," May 18, 2010 (10-NEWS-0005); Missile Defense Agency, *Missile Defense Agency Response to Request for Information, Standard Missile – 3 Interceptor Testing*, May 18, 2010, 2 pp.; Missile Defense Agency, *Missile Defense Agency Response to Request for Information, Response to New York Times May 18, 2010, Article Regarding SM-3 Testing*, May 18, 2010, 3 pp.; Richard Lehner, "Missile Defense Agerncy Responds to New York Times Article," *DOD Live* (http://www.dodlive.mil), May 18, 2010; Transcript of Department of Defense Bloggers Roundtable With Richard Lehner, Spokesman, Missile Defense Agency (MDA), Subject: Standard Missile 3 Test Program, May 18, 2010.

[50] Missile Defense Agency, "Missile Defense Test Results in Successful 'Hit To Kill' Intercept," June 22, 2006 (06-NEWS-0018).

December 7, 2006, Test

This was the first unsuccessful flight test since June 2003. MDA stated that the ninth test

> was not completed due to an incorrect system setting aboard the Aegis-class cruiser USS Lake Erie prior to the launch of two interceptor missiles from the ship. The incorrect configuration prevented the fire control system aboard the ship from launching the first of the two interceptor missiles. Since a primary test objective was a near-simultaneous launch of two missiles against two different targets, the second interceptor missile was intentionally not launched.
>
> The planned test was to involve the launch of a Standard Missile 3 against a ballistic missile target and a Standard Missile 2 against a surrogate aircraft target. The ballistic missile target was launched from the Pacific Missile Range Facility, Kauai, Hawaii and the aircraft target was launched from a Navy aircraft. The USS Lake Erie (CG 70), USS Hopper (DDG 70) and the Royal Netherlands Navy frigate TROMP were all successful in detecting and tracking their respective targets. Both targets fell into the ocean as planned.
>
> After a thorough review, the Missile Defense Agency and the U.S. Navy will determine a new test date.[51]

A news article about the ninth test stated the following:

> "You can say it's seven of nine, rather than eight of nine," Missile Defense Agency spokesman Chris Taylor said of the second failure in tests of the system by the agency and the Navy....
>
> The drill was planned to demonstrate the Navy's ability to knock down two incoming missiles at once from the same ship.
>
> "In a real world situation it is possible, maybe even probable, that in addition to engaging a ballistic missile threat that was launched, you may be engaging a surface action," said Joe Rappisi before the test. He is director for the Aegis Ballistic Missile Defense system at Lockheed Martin, the primary contractor for the program.
>
> The test would have marked the first time a ship has shot down one target in space and another target in the air at the same time.
>
> The test presented a greater challenge to the ship's crew and the ballistic missile defense system than previous tests, Rappisi said. The multiple target scenario is also closer to what sailors might actually face in battle.
>
> The U.S. Pacific Fleet has been gradually installing missile surveillance and tracking technology on many of its destroyers and cruisers amid concerns about North Korea's long-range missile program.
>
> It is also installing interceptor missiles on many of its ships, even as the technology to track and shoot down incoming missiles is being developed and perfected.

---

[51] Untitled Missile Defense Agency "For Your Information" statement dated December 7, 2006 (06-FYI-0090).

The Royal Netherlands Navy joined the tracking and monitoring off Kauai to see how its equipment works. The Dutch presence marked the first time a European ally has sent one of its vessels to participate in a U.S. ballistic missile defense test.[52]

A subsequent news article stated the following:

the test abort of the Aegis Ballistic Missile Defense system Dec. 7 resulted from human error, [MDA Director USAF Lt. Gen. Henry] Obering says.... Both the ballistic missile and aircraft targets launched as planned, but the first interceptor failed to fire because an operator had selected an incorrect setting for the test. Officials then aborted before the second could boost.

Aegis missile defense system tests are at a standstill until officials are able to identify an appropriate ballistic missile target. The one used Dec. 7 was the last of its kind, Obering says, leaving them empty handed in the near future.[53]

Another article stated the following:

Philip Coyle, a former head of the Pentagon's testing directorate, gives the Navy credit for "discipline and successes so far" in its sea-based ballistic missile defense testing program. Coyle is now a senior adviser at the Center for Defense Information.

"The U.S. Navy has an enviable track record of successful flight intercept tests, and is making the most of its current, limited Aegis missile defense capabilities in these tests," Coyle told [*Inside the Navy*] Dec. 7.

"Difficulties such as those that delayed the latest flight intercept attempt illustrate the complexity of the system, and how everything must be carefully orchestrated to achieve success," Coyle added. "Nevertheless, this particular setback won't take the Navy long to correct."[54]

April 26, 2007, Test

MDA states that this test:

involved the simultaneous engagements of a ballistic missile "unitary" target (meaning that the target warhead and booster remain attached) and a surrogate hostile air target....

The test demonstrated the [Aegis ship's] ability to engage a ballistic missile threat and defend itself from attack at the same time. The test also demonstrated the effectiveness of engineering, manufacturing, and mission assurance changes in the solid divert and attitude control system (SDACS) in the kinetic kill weapon. This was the first

---

[52] David Briscoe, "Test Interceptor Missile Fails To Launch," *NavyTimes.com*, December 8, 2006.

[53] Amy Butler, "GMD Trial Delayed Until Spring; Aegis Failure Human Error," *Aerospace Daily & Defense Report*, December 19, 2006.

[54] Zachary M. Peterson, "Sea-Based Missile Defense Test Fails Due To 'Incorrect Configuration,'" *Inside the Navy*, December 11, 2006.

flight test of all the SM-3 Block IA's upgrades, previously demonstrated in ground tests.[55]

A press report on the test stated that the hostile air target was an anti-ship cruise missile. The article stated that the scenario for the test

called for the [Aegis ship] to come under attack from a cruise missile fired by an enemy plane.... A Navy plane fired the cruise missile target used in the test.[56]

June 22, 2007, Test

MDA states that this test

was the third intercept involving a separating target and the first time an Aegis BMD-equipped destroyer was used to launch the interceptor missile. The USS Decatur (DDG 73), using the operationally-certified Aegis Ballistic Missile Defense Weapon System (BMD 3.6) and the Standard Missile-3 (SM-3) Block IA missile successfully intercepted the target during its midcourse phase of flight....

An Aegis cruiser, USS Port Royal (CG 73), a Spanish frigate, MÉNDEZ NÚÑEZ (F-104), and MDA's Terminal High Altitude Area Defense (THAAD) mobile ground-based radar also participated in the flight test. USS Port Royal used the flight test to support development of the new Aegis BMD SPY-1B radar signal processor, collecting performance data on its increased target detection and discrimination capabilities. MÉNDEZ NÚÑEZ, stationed off Kauai, performed long-range surveillance and track operations as a training event to assess the future capabilities of the F-100 Class. The THAAD radar tracked the target and exchanged tracking data with the Aegis BMD cruiser.

This event marked the third time that an allied military unit participated in a U.S. Aegis BMD test, with warships from Japan and the Netherlands participating in earlier tests.[57]

August 31, 2007, Test

MDA has publicly noted the occurrence of this test and the fact that it resulted in a successful intercept,[58] but states that the details about the test are classified.[59] MDA does

---

[55] Missile Defense Agency, "Successful Sea-Based Missile Defense 'Hit to Kill' Intercept," April 26, 2007 (07-NEWS0032).

[56] Audrey McAvoy, "Aegis Missile Test Successful," *NavyTimes.com*, April 27, 2007.

[57] Missile Defense Agency, "Sea-Based Missile Defense 'Hit to Kill' Intercept Achieved," June 22, 2007 (07-NEWS0037).

[58] See for example, slide 8 in the 20-slide briefing entitled "Ballistic Missile Defense Program Overview For The Congressional Breakfast Seminar Series," dated June 20, 2008, presented by Lieutenant General Trey Obering, USAF, Director, Missile Defense Agency. Source for briefing: *InsideDefense.com* (subscription required). Each slide in the briefing includes a note indicating that it was approved by MDA for public release on June 13, 2008. Slide 8 lists Aegis BMD midcourse flight tests conducted since September 2005, including a test on August 31, 2007. The slide indicates with a check mark that the flight test was successful. A success in this test is also needed to for the total number of successful intercepts to match the reported figure.

not appear to have issued a news release about this flight test following the completion of the test, as it has for other Aegis BMD flight tests.[60]

November 6, 2007, Test

MDA states that this test involved:

> a multiple simultaneous engagement involving two ballistic missile targets.... For the first time, the operationally realistic test involved two unitary "non-separating" targets, meaning that the target's warheads did not separate from their booster rockets....
>
> At approximately 6:12 p.m. Hawaii Standard Time (11:12 p.m. EST), a target was launched from the Pacific Missile Range Facility (PMRF), Barking Sands, Kauai, Hawaii. Moments later, a second, identical target was launched from the PMRF. The USS Lake Erie's Aegis BMD Weapon System detected and tracked the targets and developed fire control solutions.
>
> Approximately two minutes later, the USS Lake Erie's crew fired two SM-3 missiles, and two minutes later they successfully intercepted the targets outside the earth's atmosphere more than 100 miles above the Pacific Ocean and 250 miles northwest of Kauai....
>
> A Japanese destroyer also participated in the flight test. Stationed off Kauai and equipped with the certified 3.6 Aegis BMD weapon system, the guided missile destroyer JS Kongo performed long-range surveillance and tracking exercises. The Kongo used the test as a training exercise in preparation for the first ballistic missile intercept test by a Japanese ship planned for later this year. This event marked the fourth time an allied military unit participated in a U.S. Aegis BMDS test.[61]

December 17, 2007, Test

In this flight test, a BMD-capable Japanese Aegis destroyer used an SM-3 Block IA missile to successfully intercept a ballistic missile target in a flight test off the coast of Hawaii. It was the first time that a non-U.S. ship had intercepted a ballistic missile using the Aegis BMD system.[62]

---

[59] An email from MDA to CRS dated June 30, 2008, states that the flight test "was a hit to kill intercept test but details about the test are classified."

[60] MDA's website, when accessed on June 30, 2008, did not show a news release issued on or soon after August 31, 2007, that discusses this test.

[61] Missile Defense Agency, "Sea-Based Missile Defense "Hit to Kill" Intercept Achieved," November 6, 2007 (07-NEWS-0051).

[62] John Liang, "Japanese Destroyer Shoots Down Ballistic Missile Test Target," *Inside Missile Defense*, December 19, 2007; "Japanese Aegis Destroyer Wins Test By Killing Target Missile With SM-3 Interceptor," *Defense Daily*, December 18, 2007; Reuters, "Japanese Ship Downs Missile In Pacific Test," *New York Times*, December 18, 2007: 8; Audrey McAvoy, "Japan Intercepts Missile In Test Off Hawaii," NavyTimes.com, December 17, 2007.

November 1, 2008, Test

This flight test was reportedly the first U.S. Navy Aegis BMD flight test conducted by the Navy, without oversight by MDA. The test involved two Aegis ships, each attempting to intercept a ballistic missile. The SM-3 fired by the first Aegis ship successfully intercepted its target, but the SM-3 fired by the second Aegis ship did not intercept its target. A press release from the U.S. Third Fleet (the Navy's fleet for the Eastern Pacific) states that

> Vice Adm. Samuel J. Locklear, Commander, U.S. Third Fleet announced today the successful Navy intercept of a ballistic missile target over the Pacific Ocean during Fleet Exercise Pacific Blitz. This was the first Fleet operational firing to employ the Standard Missile-3 (SM-3) against a ballistic missile target. Command and control of this mission resided with Commander, U.S. Third Fleet, based in San Diego, Calif.
>
> Pearl Harbor-based Aegis destroyers, USS Paul Hamilton (DDG 60) and USS Hopper (DDG 70), which have been upgraded to engage ballistic missiles, fired SM-3 missiles at separate targets. During this event, a short-range ballistic missile target was launched from the Pacific Missile Range Facility (PMRF), Barking Sands, Kauai, Hawaii. Upon detecting and tracking the target, USS Paul Hamilton, launched a SM-3 missile, resulting in a direct-hit intercept. Following USS Paul Hamilton's engagement, PMRF launched another target. USS Hopper successfully detected, tracked and engaged the target. The SM-3 followed a nominal trajectory, however intercept was not achieved. Extensive analysis of the flight mission will be used to improve the deployed Aegis BMD system.[63]

November 19, 2008, Test

This was the second Japanese flight test, and involved a single ballistic missile target. The test did not result in a successful intercept. MDA states that

> Rear Admiral Tomohisa Takei, Director General of Operations and Plans, for the Japanese Maritime Staff Office (MSO), Japan Maritime Self Defense Force (JMSDF), and Lt. General Henry "Trey" Obering, United States Missile Defense Agency director, announced the completion today of a cooperative sea-based Aegis Ballistic Missile Defense intercept flight test off the coast of Kauai in Hawaii. The event, designated Japan Flight Test Mission 2 (JFTM-2), marked the second attempt by an Allied naval ship to intercept a ballistic missile target with the sea-based midcourse engagement capability provided by Aegis Ballistic Missile Defense. Target performance, interceptor missile launch and fly out, and operation of the Aegis Weapon System by the crew were successful, but an intercept was not achieved.

---

[63] Commander, U.S. Third Fleet, Public Affairs Office, press release 23-08, dated November 1, 2008, entitled "Navy Intercepts Ballistic Missile Target in Fleet Exercise Pacific Blitz." See also Dave Ahearn, "One of Two Missiles Hit In Aegis Test; Navy For First Time Runs Test Instead of MDA," *Defense Daily*, November 4, 2008: 1-2.

The JFTM-2 was a test of the newest engagement capability of the Aegis Ballistic Missile Defense configuration of the recently upgraded Japanese destroyer, JS CHOKAI (DDG176). At approximately 4:21 pm (HST), 11:21 am (Tokyo time) a ballistic missile target was launched from the Pacific Missile Range Facility, Barking Sands, Kauai, Hawaii. JS CHOKAI crew members detected and tracked the target using an advanced on-board radar. The Aegis Weapon System then developed a fire control solution, and at approximately 4:24 pm (HST), 11:24 am (Tokyo time) on Nov 20, a single Standard Missile -3 (SM-3) Block IA was launched. Approximately two minutes later, the SM-3 failed to intercept the target. There is no immediate explanation for the failed intercept attempt. More information will be available after a thorough investigation. The JS CHOKAI crew performance was excellent in executing the mission. JFTM-2 was the second time that a Japanese ship was designated to launch the interceptor missile, a major milestone in the growing cooperation between Japan and the U.S.[64]

A November 21, 2008, press report states that

An Aegis ballistic missile defense (BMD) test by the Japanese destroyer Chokai (DDG176) ended in failure when the Standard Missile-3 Block 1A interceptor lost track of the target missile in the final seconds before a planned hit-to-kill.

The Chokai and its crew performed well throughout the test, and the SM-3 also performed flawlessly through its first three stages, according to Rear Adm. Brad Hicks, the U.S. Navy Aegis ballistic missile defense program director. He spoke with several reporters in a teleconference around midnight ET Wednesday-Thursday, after the test in the area of the Pacific Missile Range Facility, Barking Sands, Kauai, Hawaii.

This was the second Aegis BMD test failure in less than a month.

These latest two failures come as some Democrats in Congress are poised to cut spending on missile defense programs when they convene next year to consider the Missile Defense Agency budget for the fiscal year ending Sept. 30, 2010....

Still, in the coming money debates next year, missile defense advocates will be able to point out that even including the Hopper and Chokai failures, the record for the Aegis tests is an overwhelming 16 successful hits demolishing target missiles out of 20 attempts.

Those successes included the first Japanese attempt. The Japanese destroyer Kongo (DDG173) successfully used its SM-3 interceptor to kill a target missile. The difference in tests is that the Kongo crew was advised beforehand when the target missile would be launched, while the Chokai crew wasn't....

[Hicks] said a board will be convened to examine why the latest test failed. Hicks declined to speculate on why the SM-3 interceptor missed the target. "I'm confident we'll find out the root cause" of the Chokai interceptor failure to score a hit, he said.

However, he was asked by *Space & Missile Defense Report* whether the prior SM-3 successes make it unlikely the Chokai failure stems from some basic design flaw in all

---

[64] Missile Defense Agency, "Japan/U.S. Missile Defense Flight Test Completed," November 19, 2008 (08-NEWS-0087).

SM3s, and whether it is more likely that the Chokai SM-3 failed because of some flaw or glitch in just that one interceptor.

Hicks said that is likely.

"Obviously, we believe this is hopefully related to this one interceptor," and doesn't reflect any basic design flaw in the SM-3 interceptors, he said.

The Chokai test failure cost Japan a $55 million loss, he said, adding, "It wasn't cheap."...

In the Chokai test, the target missile was launched from Barking Sands, and about three minutes later the Chokai crew had spotted the target, the Aegis system had developed a tracking and hit solution, and the SM-3 interceptor was launched.

The first, second and third stages of the interceptor performed nominally, without problems, but then came the fourth stage. The nosecone components opened to expose the kill vehicle area, and somehow the program to track the target missile failed.

"It lost track," Hicks said, only seconds before the hit would have been achieved.

If the kill had occurred, it would have been about 100 nautical miles (roughly 115 statute miles) above Earth, and some 250 miles away from Barking Sands, Hicks said.

It took the interceptor about two minutes flight time to reach the near miss with the target missile.

Meanwhile, the Hamilton was nearby watching the test. The Hamilton Aegis system successfully spotted and tracked the target, and developed a simulated solution and simulated interceptor launch that, if it had been real, would have resulted in a successful hit on the target, Hicks said. The Hamilton didn't cue the Chokai, however. "It was strictly Chokai's engagement," Hicks said.[65]

July 30, 2009, Test

MDA states that

In conjunction with the Missile Defense Agency (MDA), U.S. Pacific Fleet ships and crews successfully conducted the latest Aegis Ballistic Missile Defense (BMD) at-sea firing event on July 30. During this event, entitled Stellar Avenger, the Aegis BMD-equipped ship, USS Hopper (DDG 70), detected, tracked, fired and guided a Standard Missile -3 (SM-3) Block (Blk) IA to intercept a sub-scale short range ballistic missile. The target was launched from the Kauai Test Facility, co-located on the Pacific Missile Range Facility (PMRF), Barking Sands, Kauai. It was the 19th successful intercept in 23 at-sea firings, for the Aegis BMD Program, including the February 2008 destruction of the malfunctioning satellite above the earth's atmosphere. Stellar Avenger was part of the continual evaluation of the certified and fielded Aegis BMD system at-sea today.

At approximately 5:40 pm (HST), 11:40 pm (EDT), a target was launched from PMRF. Three U.S. Navy Aegis BMD-equipped ships, the cruiser, USS Lake Erie (CG 70) and destroyers USS Hopper (DDG 70) and USS O'Kane (DDG 77) detected and tracked the target with their SPY radars. Each developed fire control solutions. At 5:42

[65] Dave Ahearn, "Japanese Aegis Missile Defense Test Fails, But Aegis Record Is 16 Hits in 20 Tries," *Defense Daily*, November 21, 2008: 5-6.

pm (HST), 11:42 pm (EDT) the crew of USS Hopper fired one SM-3 Blk IA missile. The USS Hopper's Aegis BMD Weapon System successfully guided the SM-3 to a direct body to body hit, approximately two minutes after leaving the ship. The intercept occurred about 100 miles above the Pacific Ocean. USS O'Kane conducted a simulated engagement of the target. USS Lake Erie, with its recently installed upgraded Aegis BMD 4.0.1 Weapons System, detected and tracked the same target.[66]

A July 31, 2009, press report states the following:

> The test was the first Aegis BMD exercise to feature two versions of the software in a single event, according to Lisa Callahan, Lockheed's vice president for ballistic missile defense programs.
>
> A goal of the exercises was to test the Aegis system's ability to discern all the different parts and pieces of a ballistic missile, Nick Bucci, Lockheed's director for Aegis BMD development programs, told reporters July 29 during a pre-exercise conference call.
>
> Three more flight tests this fall will further test the system's discrimination capabilities, Bucci added, with each test becoming more complex. The last test will "be against a pretty darn complex target," he said.
>
> The July 30 tests also validated fixes put in place after a BMD test last November involving a missile launched from the Aegis BMD Japanese destroyer Chokai failed to intercept its target, according to MDA spokesman Chris Taylor. The improvements—which were successful in the most recent test—involved fixes to the Solid Divert Attitude Control System.
>
> The Chokai is the second of four Japanese Aegis ships being upgraded with BMD capability. A third ship, the Myoko, is scheduled to carry out a BMD test this fall.[67]

An August 3, 2009, press report states the following:

> This test was added to the schedule to evaluate changes made after last year's failed attempt to intercept a target with an SM-3 Block IA launched by a Japanese Aegis-equipped ship .... After the Nov. 19 test, MDA officials said, "Target performance, interceptor missile launch and flyout, and operation of the Aegis Weapon System by the crew were successful, but an intercept was not achieved."
>
> A root cause has not been identified, and an MDA spokesman did not say whether fixes have been made to hardware or operational procedures resulting from the failure review. It is also unclear why a subscale target was used in the July 30 trial.[68]

---

[66] Missile Defense Agency, "Aegis Ballistic Missile Defense Test Successful," July 31, 2009 (09-News-0015).

[67] Christopher P. Cavas, "Aegis BMD Test Successful," *DefenseNews.com*, July 31, 2009.

[68] Amy Butler, "SM-3 Scores Hit After Fixes Implemented," *Aerospace Daily & Defense Report*, August 3, 2009: 5.

An August 4, 2009, press report states the following:

> [Rear Admiral Alan "Brad" Hicks, Aegis/SM-3 program manager for MDA], said that a November [2008] failure of an SM-3 Block IA... during a flight-test was attributable to poor adherence to processes on Raytheon's assembly line in Tucson, Ariz.
>
> This was isolated to that missile, and it was the result of perturbations to the build process encountered when shifting from development to production operations.
>
> During the November test, a Japanese Aegis-equipped ship fired the interceptor and it flew "perfectly," Hicks said. In the endgame, a failure of the divert and attitude control system on the unitary kill vehicle led to a miss.
>
> The July 30 demonstration using a U.S. ship "restored confidence" for the Japanese that the miss last fall was an isolated incident, he says.[69]

October 27, 2009, Test

This was the third Japanese flight test, and it involved a single ballistic missile target. MDA states that

> The Japan Maritime Self-Defense Force (JMSDF) and the United States Missile Defense Agency (MDA) announced the successful completion of an Aegis Ballistic Missile Defense (BMD) intercept flight test, in cooperation with the U.S. Navy, off the coast of Kauai in Hawaii. The event, designated Japan Flight Test Mission 3 (JFTM-3), marked the third time that a JMSDF ship has successfully engaged a ballistic missile target, including two successful intercepts, with the sea-based midcourse engagement capability provided by Aegis BMD.
>
> The JFTM-3 test event verified the newest engagement capability of the Japan Aegis BMD configuration of the recently upgraded Japanese destroyer, JS MYOKO (DDG-175). At approximately 6:00pm (HST), 1:00 pm Tokyo time on Oct 28, a separating, medium-range ballistic missile target was launched from the Pacific Missile Range Facility, Barking Sands, Kauai, Hawaii. JS MYOKO crew members detected and tracked the target. The Aegis Weapon System then developed a fire control solution and, at approximately 6:04pm (HST), 1:04 pm Tokyo time a Standard Missile-3 (SM-3) Block IA interceptor missile was launched. Approximately 3 minutes later, the SM-3 successfully intercepted the target approximately 100 miles above the Pacific Ocean. JFTM-3 is a significant milestone in the growing cooperation between Japan and the U.S. in the area of missile defense.
>
> Also participating in the test, were the Pearl Harbor-based USS Lake Erie (CG 70) and USS Paul Hamilton (DDG 60) which detected and tracked the target and conducted a simulated engagement.[70]

---

[69] Amy Butler, "SM-3 Upgrade Program Cost Increases," *Aerospace Daily & Defense Report*, August 4, 2009: 1-2. See also Dan Taylor, "Navy Conducts Aegis BMD Test, New Baseline System Participates," *Inside the Navy*, August 3, 2009; Daniel Wasserbly, "US Aegis BMD System Achieves Trial Success," *Jane's Defence Weekly*, August 5, 2009: 8.

[70] Missile Defense Agency, "Japan/U.S. Missile Defense Flight Test Successful," October 28, 2009 (09-News-0021). See also Christopher P. Cavas, "Japanese Destroyer Conducts Successful BMD Test," *NavyTimes.com*,

October 28, 2010, Test

This was the fourth Japanese flight test, and it involved a single ballistic missile target. MDA states that

> The Japan Maritime Self-Defense Force (JMSDF) and the United States Missile Defense Agency (MDA) announced the successful completion of an Aegis Ballistic Missile Defense (BMD) intercept flight test, in cooperation with the U.S. Navy, off the coast of Kauai in Hawaii.
>
> The event marked the fourth time that a JMSDF ship has engaged a ballistic missile target, including three successful intercepts, with the sea-based midcourse engagement capability provided by Aegis BMD.
>
> The JFTM-4 test event verified the newest engagement capability of the Japan Aegis BMD configuration of the recently upgraded Japanese destroyer, JS KIRISHIMA. At approximately 5:06 p.m. (HST), 12:06 p.m. Tokyo time on Oct. 29, 2010, a separating 1,000 km class ballistic missile target was launched from the Pacific Missile Range Facility at Barking Sands, Kauai, Hawaii.
>
> JS KIRISHIMA crew members detected and tracked the target. The Aegis Weapon System then developed a fire control solution and launched a Standard Missile -3 (SM-3) Block IA missile. Approximately three minutes later, the SM-3 successfully intercepted the target approximately 100 miles above the Pacific Ocean. JFTM-4 is a significant milestone in the growing cooperation between Japan and the U.S. in the area of missile defense.
>
> Also participating in the test was USS LAKE ERIE and USS RUSSELL, Aegis ships which cooperated to detect, track and conduct a simulated intercept engagement against the same target.[71]

April 15, 2011, Test

MDA states that this flight test "was the most challenging test to date, as it was the first Aegis BMD version 3.6.1 intercept against an intermediate-range target (range 1,864 to 3,418 [statute] miles) and the first Aegis BMD 3.6.1 engagement relying on remote tracking data." MDA states that

> The Missile Defense Agency (MDA), U.S. Navy sailors aboard the Aegis destroyer USS O'KANE (DDG 77), and Soldiers from the 94th Army Air and Missile Defense Command operating from the 613th Air and Space Operations Center at Hickam Air Force Base, Hawaii, successfully conducted a flight test of the Aegis Ballistic Missile Defense (BMD) element of the nation's Ballistic Missile Defense System, resulting in the intercept of a separating ballistic missile target over the Pacific Ocean. This successful

---

October 28, 2009; and Amy Butler and Michael Bruno, "SM-3 Scores Hit In Japanese Test," *Aerospace Daily & Defense Report*," October 29, 2009: 3.

[71] Missile Defense Agency, "Joint Japan-U.S. Missile Defense Flight Test Successful," October 29, 2010 (10-News-0016). See also Marina Malenic, "Japanese Aegis Destroyer Successfully Completes Missile-Intercept Test," *Defense Daily*, November 1, 2010: 6.

test demonstrated the capability of the first phase of the European Phased Adaptive Approach (EPAA) announced by the President in September, 2009.

At 2:52 a.m. EDT (6:52 p.m. April 15 Marshall Island Time), an intermediate-range ballistic missile target was launched from the Reagan Test Site, located on Kwajalein Atoll in the Republic of the Marshall Islands, approximately 2,300 miles southwest of Hawaii. The target flew in a northeasterly direction towards a broad ocean area in the Pacific Ocean. Following target launch, a forward-based AN/TPY-2 X-band transportable radar, located on Wake Island, detected and tracked the threat missile. The radar sent trajectory information to the Command, Control, Battle Management, and Communications (C2BMC) system, which processed and transmitted remote target data to the USS O'KANE. The destroyer, located to the west of Hawaii, used the data to develop a fire control solution and launch the SM-3 Block IA missile approximately 11 minutes after the target was launched.

As the IRBM target continued along its trajectory, the firing ship's AN/SPY-1 radar detected and acquired the ballistic missile target. The firing ship's Aegis BMD weapon system uplinked target track information to the SM-3 Block IA missile. The SM-3 maneuvered to a point in space as designated by the fire control solution and released its kinetic warhead. The kinetic warhead acquired the target, diverted into its path, and, using only force of a direct impact, destroyed the threat in a "hit-to-kill" intercept.

During the test the C2BMC system, operated by Soldiers from the 94th Army Air and Missile Defense Command, received data from all assets and provided situational awareness of the engagement to U.S. Pacific Command, U.S. Northern Command and U.S. Strategic Command.

The two demonstration Space Tracking and Surveillance Satellites (STSS), launched by MDA in 2009, successfully acquired the target missile, providing stereo "birth to death" tracking of the target.

Today's event, designated Flight Test Standard Missile-15 (FTM-15), was the most challenging test to date, as it was the first Aegis BMD version 3.6.1 intercept against an intermediate-range target (range 1,864 to 3,418 [statute] miles) and the first Aegis BMD 3.6.1 engagement relying on remote tracking data. The ability to use remote radar data to engage a threat ballistic missile greatly increases the battle space and defended area of the SM-3 missile.

Initial indications are that all components performed as designed. Program officials will spend the next several months conducting an extensive assessment and evaluation of system performance based upon telemetry and other data obtained during the test.[72]

September 1, 2011, Test

This flight test, which did not result in an intercept, was the first flight test of the SM-3 Block IB interceptor. MDA states that it

[72] Missile Defense Agency, "Sea-based Missile Defense Flight Test Results in Successful Intercept," April 15, 2011 (11-News-0007).

was unable to achieve the planned intercept of a ballistic missile target during a test over the Pacific Ocean exercising the sea-based element of the Ballistic Missile Defense System (BMDS).

At approximately 3:53 a.m. Hawaii Standard Time (9:53 a.m. EDT) a short-range ballistic missile target was launched from the U.S. Navy's Pacific Missile Range Facility on Kauai, Hawaii. Approximately 90 seconds later, a Standard Missile 3 (SM-3) Block 1B interceptor missile was launched from the cruiser USS LAKE ERIE (CG-70) but an intercept of the target was not achieved.

This was the first flight test of the advanced SM-3 Block 1B interceptor missile. Program officials will conduct an extensive investigation to determine the cause of the failure to intercept.[73]

May 9, 2012, Test

MDA states that this flight test "was the first successful live fire intercept test of the SM-3 Block IB interceptor and the second-generation Aegis BMD 4.0.1 weapon system." MDA states that

> The Missile Defense Agency (MDA) and U.S. Navy sailors aboard the USS LAKE ERIE (CG 70) successfully conducted a flight test of the Aegis Ballistic Missile Defense (BMD) system, resulting in the first intercept of a short-range ballistic missile target over the Pacific Ocean by the Navy's newest Missile Defense interceptor, the Standard Missile – 3 (SM-3) Block IB.
>
> At 8:18 p.m. Hawaiian Standard Time (2:18 a.m. EDT May 10) the target missile was launched from the Pacific Missile Range Facility, located on Kauai, Hawaii. The target flew on a northwesterly trajectory towards a broad ocean area of the Pacific Ocean. Following target launch, the USS LAKE ERIE detected and tracked the missile with its onboard AN/SPY-1 radar. The ship, equipped with the second-generation Aegis BMD 4.0.1 weapon system, developed a fire control solution and launched the Standard Missile-3 (SM-3) Block IB interceptor.
>
> The USS LAKE ERIE continued to track the target and sent trajectory information to the SM-3 Block IB interceptor in-flight. The SM-3 maneuvered to a point in space, as designated by the fire control solution, and released its kinetic warhead. The kinetic warhead acquired the target, diverted into its path, and, using only the force of a direct impact, engaged and destroyed the threat in a hit-to-kill intercept.
>
> Today's event, designated Flight Test Standard Missile-16 (FTM-16) Event 2a, was the first successful live fire intercept test of the SM-3 Block IB interceptor and the second-generation Aegis BMD 4.0.1 weapon system. Previous successful intercepts were conducted with the Aegis BMD 3.6.1 weapon system and the SM-3 Block IA interceptor, which are currently operational on U.S. Navy ships deployed across the globe....

---

[73] Missile Defense Agency, "Sea-Based Missile Defense Test Conducted," September 1, 2011 (11-News-0016). See also Amy Butler, "Upgraded Ballistic Missile Killer Fizzles In First Flight Test," *Aerospace Daily & Defense Report*, September 2, 2011: 3; and Mike McCarthy, "Sea-Based Missile Defense Test Fails," *Defense Daily*, September 2, 2011: 2-3.

> Initial indications are that all components performed as designed. Program officials will conduct an extensive assessment and evaluation of system performance based upon telemetry and other data obtained during the test.[74]

June 26, 2012, Test

MDA states that this flight test "was the second consecutive successful intercept test of the SM-3 Block IB missile and the second-generation Aegis BMD 4.0.1 weapon system." MDA states that

> The Missile Defense Agency (MDA) and U.S. Navy sailors in the USS LAKE ERIE (CG 70) successfully conducted a flight test of the Aegis Ballistic Missile Defense (BMD) system, resulting in the intercept of a separating ballistic missile target over the Pacific Ocean by the Navy's newest missile defense interceptor missile, the Standard Missile-3 (SM-3) Block IB.
>
> At 11:15 pm Hawaii Standard Time, June 26 (5:15 am EDT June 27), the target missile was launched from the Pacific Missile Range Facility, located on Kauai, Hawaii. The target flew on a northwesterly trajectory towards a broad ocean area of the Pacific Ocean. Following target launch, the USS LAKE ERIE detected and tracked the missile with its onboard AN/SPY-1 radar. The ship, equipped with the second-generation Aegis BMD 4.0.1 weapon system, developed a fire control solution and launched the SM-3 Block IB missile.
>
> The USS LAKE ERIE continued to track the target and sent trajectory information to the SM-3 Block IB missile in-flight. The SM-3 maneuvered to a point in space, as designated by the fire control solution, and released its kinetic warhead. The kinetic warhead acquired the target, diverted into its path, and, using only the force of a direct impact, engaged and destroyed the threat in a hit-to-kill intercept.
>
> Today's test event was the second consecutive successful intercept test of the SM-3 Block IB missile and the second-generation Aegis BMD 4.0.1 weapon system. The first successful SM-3 Block IB intercept occurred on May 9, 2012. Today's intercept is a critical accomplishment for the second phase of the President's European Phased Adaptive Approach consisting of the SM-3 Block IB interceptor employed in an Aegis Ashore system in Romania in 2015.
>
> Initial indications are that all components performed as designed resulting in a very accurate intercept.[75]

October 25, 2012, Test

MDA states that in this flight test,

---

[74] Missile Defense Agency, "Second-Generation Aegis Ballistic Missile Defense System Completes Successful Intercept Flight Test," May 9, 2012 (12-News-0007).

[75] Missile Defense Agency, "Second-Generation Aegis Ballistic Missile Defense System Completes Second Successful Intercept Flight Test," June 27, 2012 (12-News-0008).

> The Missile Defense Agency (MDA), U.S. Army soldiers from the 94th and 32nd Army Air and Missile Defense Command (AAMDC); U.S. Navy sailors aboard the USS FITZGERALD (DDG 62); and airmen from the 613th Air and Space Operations Center successfully conducted the largest, most complex missile defense flight test ever attempted resulting in the simultaneous engagement of five ballistic missile and cruise missile targets.
>
> An integrated air and ballistic missile defense architecture used multiple sensors and missile defense systems to engage multiple targets at the same time....
>
> The USS FITZGERALD successfully engaged a low flying cruise missile over water. The Aegis system also tracked and launched an SM-3 Block 1A interceptor against a Short-Range Ballistic Missile. However, despite indication of a nominal flight of the SM-3 Block 1A interceptor, there was no indication of an intercept of the SRBM.[76]

February 12, 2013, Test

MDA states that in this flight test,

> The Missile Defense Agency (MDA) and U.S. Navy sailors aboard the USS LAKE ERIE (CG 70) successfully conducted a flight test of the Aegis Ballistic Missile Defense (BMD) system, resulting in the intercept of a medium-range ballistic missile target over the Pacific Ocean by a Standard Missile-3 (SM-3) Block IA guided missile.
>
> At 11:10 p.m. HST (4:10 a.m. EST) a unitary medium-range ballistic missile target was launched from the Pacific Missile Range Facility, on Kauai, Hawaii. The target flew northwest towards a broad ocean area of the Pacific Ocean.
>
> The in-orbit Space Tracking and Surveillance System-Demonstrators (STSS-D) detected and tracked the target, and forwarded track data to the USS LAKE ERIE. The ship, equipped with the second-generation Aegis BMD weapon system, used Launch on Remote doctrine to engage the target.
>
> The ship developed a fire control solution from the STSS-D track and launched the SM-3 Block IA guided missile approximately five minutes after target launch. The SM-3 maneuvered to a point in space and released its kinetic warhead. The kinetic warhead acquired the target reentry vehicle, diverted into its path, and, using only the force of a direct impact, engaged and destroyed the target.
>
> Initial indications are that all components performed as designed. Program officials will assess and evaluate system performance based upon telemetry and other data obtained during the test.
>
> Today's event, designated Flight Test Standard Missile-20 (FTM-20), was a demonstration of the ability of space-based assets to provide mid-course fire control quality data to an Aegis BMD ship, extending the battlespace, providing the ability for longer range intercepts and defense of larger areas.[77]

---

[76] Missile Defense Agency, "Ballistic Missile Defense System Engages Five Targets Simultaneously During Largest Missile Defense Flight Test in History," October 25, 2012 (12-News-0011).

[77] Missile Defense Agency, "Aegis Ballistic Missile Defense Intercepts Target Using Space Tracking and Surveillance System-Demonstrators (STSS-D) Data," February 13, 2013 (13-News-0002). See also Troy

May 16, 2013, Test

MDA states that in this flight test,

> The Missile Defense Agency (MDA) and U.S. Navy sailors aboard the USS LAKE ERIE (CG-70) successfully conducted a flight test today of the Aegis Ballistic Missile Defense (BMD) system, resulting in the intercept of a separating ballistic missile target over the Pacific Ocean by the Aegis BMD 4.0 Weapon System and a Standard Missile-3 (SM-3) Block IB missile.
>
> At 5:25 p.m. (Hawaii Time, 11:25 p.m. EDT), May 15, a separating short-range ballistic missile target was launched from the Pacific Missile Range Facility, on Kauai, Hawaii. The target flew northwest towards a broad ocean area of the Pacific Ocean. Following target launch, the USS LAKE ERIE (CG-70) detected and tracked the missile with its onboard AN/SPY-1 radar. The ship, equipped with the second-generation Aegis BMD weapon system, developed a fire control solution and launched the SM-3 Block IB missile. The SM-3 maneuvered to a point in space based on guidance from Aegis BMD Weapons Systems and released its kinetic warhead. The kinetic warhead acquired the target reentry vehicle, diverted into its path, and, using only the force of a direct impact, engaged and destroyed the target.
>
> Initial indications are that all components performed as designed. Program officials will assess and evaluate system performance based upon telemetry and other data obtained during the test.
>
> This test exercised the latest version of the second-generation Aegis BMD Weapon System and Standard Missile, providing capability for engagement of longer-range and more sophisticated ballistic missiles.
>
> Last night's event, designated Flight Test Standard Missile-19 (FTM-19), was the third consecutive successful intercept test of the Aegis BMD 4.0 Weapon System and the SM-3 Block IB guided missile. Previous successful ABMD 4.0 SM-3 Block IB intercepts occurred on May 9, 2012 and June 26, 2012. Other Aegis BMD intercepts have employed the ABMD 3.6 and 4.0 with the SM-3 Block IA missile, which is currently operational on U.S. Navy ships deployed across the globe.[78]

September 10, 2013, Test

MDA states that in this flight test,

> The Missile Defense Agency (MDA), Ballistic Missile Defense System (BMDS) Operational Test Agency, Joint Functional Component Command for Integrated Missile Defense, and U.S. Pacific Command, in conjunction with U.S. Army soldiers from the Alpha Battery, 2nd Air Defense Artillery Regiment, U.S. Navy sailors aboard the guided

---

Clarke, "Space-Based Sensors Star in "Stellar Eyes" Missile Defense Test," *Navy News Service*, February 13, 2013.

[78] Missile Defense Agency, "Aegis Ballistic Missile Defense System Completes Successful Intercept Flight Test," May 16, 2013 (13-News-0005). See also Mike McCarthy, "Aegis Missile Intercept Successful," *Defense Daily*, May 17, 2013: 7-8; and Amy Butler, "MDA Conducts Two Successful Flight Tests," *Aerospace Daily & Defense Report*, May 17, 2013: 3.

missile destroyer USS Decatur (DDG-73), and U.S. Air Force airmen from the 613th Air and Operations Center successfully conducted a complex missile defense flight test, resulting in the intercept of two medium-range ballistic missile targets. The flight test was planned more than a year ago, and is not in any way connected to events in the Middle East.

The test was conducted in the vicinity of the U.S. Army Kwajalein Atoll/Reagan Test Site and surrounding areas in the western Pacific. The test stressed the ability of the Aegis Ballistic Missile Defense (BMD) and Terminal High Altitude Area Defense (THAAD) weapon systems to function in a layered defense architecture and defeat a raid of two near-simultaneous ballistic missile targets.

The two medium-range ballistic missile targets were launched on operationally realistic trajectories towards a defended area near Kwajalein. Along with overhead space assets providing launch alerts, an Army-Navy/Transportable Radar Surveillance and Control (AN/TPY-2) radar in Forward Based Mode detected the targets and relayed track information to the Command, Control, Battle Management, and Communications (C2BMC) system for further transmission to defending BMDS assets.

The USS Decatur with its Aegis Weapon System detected and tracked the first target with its onboard AN/SPY-1 radar. The Aegis BMD weapon system developed a fire control solution, launched a Standard Missile-3 (SM-3) Block IA missile, and successfully intercepted the target.

In a demonstration of BMDS layered defense capabilities, a second AN/TPY-2 radar in Terminal Mode, located with the THAAD weapon system, acquired and tracked the target missiles. THAAD developed a fire control solution, launched a THAAD interceptor missile, and successfully intercepted the second medium-range ballistic missile target. THAAD was operated by soldiers from the Alpha Battery, 2nd Air Defense Artillery Regiment. As a planned demonstration of THAAD's layered defense capabilities, a second THAAD interceptor was launched at the target destroyed by Aegis as a contingency in the event the SM-3 did not achieve an intercept.

Initial indications are that all components performed as designed. MDA officials will extensively assess and evaluate system performance based upon telemetry and other data obtained during the test.

The event, a designated Flight Test Operational-01 (FTO-01), demonstrated integrated, layered, regional missile defense capabilities to defeat a raid of two threat-representative medium-range ballistic missiles in a combined live-fire operational test. Soldiers, sailors, and airmen from multiple combatant commands operated the systems, and were provided a unique opportunity to refine operational doctrine and tactics while increasing confidence in the execution of integrated air and missile defense plans.[79]

September 18, 2013, Test

MDA states that in this flight test,

[79] Missile Defense Agency, "Successful Missile Defense Test Against Multiple Targets," September 10, 2013 (13-News-0007). See also Megan Eckstein, "Aegis BMDS, THAAD Successful In Complex MDA Flight Test," *Defense Daily*, September 11, 2013: 1; and Amy Butler, "MDA Goes Two For Two In Operational Test," *Aerospace Daily & Defense Report*, September 11, 2013: 4.

The Missile Defense Agency (MDA), U.S. Pacific Command, and U.S. Navy sailors aboard the USS Lake Erie (CG 70) successfully conducted a flight test today of the Aegis Ballistic Missile Defense (BMD) system, resulting in the intercept of a complex separating short-range ballistic missile target over the Pacific Ocean by the Aegis BMD 4.0 Weapon System and a Standard Missile-3 (SM-3) Block IB guided missile.

At approximately 2:30 p.m. Hawaii Standard Time (8:30 p.m. EDT), a complex separating short-range ballistic missile target was launched from the Pacific Missile Range Facility on Kauai, Hawaii. The target flew northwest towards a broad ocean area of the Pacific Ocean. Following target launch, the USS Lake Erie detected and tracked the missile with its onboard AN/SPY-1 radar. The ship, equipped with the second-generation Aegis BMD weapon system, developed a fire control solution and launched two SM-3 Block IB guided missiles to engage the target. The first SM-3 that was launched successfully intercepted the target warhead. This was the first salvo mission of two SM-3 Block IB guided missiles launched against a single separating target.

Program officials will assess and evaluate system performance based upon telemetry and other data obtained during the test.

This test exercised the latest version of the second-generation Aegis BMD Weapon System, capable of engaging longer range and more sophisticated ballistic missiles. This was an operationally realistic test, in which the target's launch time and bearing are not known in advance, and the target complex was the most difficult target engaged to date.[80]

October 3, 2013, Test

MDA states that in this flight test,

The Missile Defense Agency (MDA), U.S. Pacific Command, and U.S. Navy sailors aboard the USS Lake Erie (CG 70) successfully conducted an operational flight test of the Aegis Ballistic Missile Defense (BMD) system, resulting in the intercept of a medium-range ballistic missile target over the Pacific Ocean by the Aegis BMD 4.0 Weapon System and a Standard Missile-3 (SM-3) Block IB guided missile.

At approximately 7:33 p.m. Hawaii Standard Time, Oct. 3 (1:33 a.m. EDT, Oct.4), a medium-range ballistic missile target was launched from the Pacific Missile Range Facility on Kauai, Hawaii. The target flew northwest towards a broad ocean area of the Pacific Ocean. Following target launch, the USS Lake Erie detected and tracked the missile with its onboard AN/SPY-1 radar. The ship, equipped with the second-generation Aegis BMD weapon system, developed a fire control solution and launched the SM-3 Block IB guided missile to engage the target. The SM-3 maneuvered to a point in space and released its kinetic warhead. The kinetic warhead acquired the target reentry vehicle,

---

[80] Missile Defense Agency, "Aegis Ballistic Missile Defense System Completes Successful Intercept Flight Test," September 18, 2013 (13-News-0008). See also Mike McCarthy, "Pentagon Succeeds At Sea-Based Missile Defense Test," *Defense Daily*, September 20, 2013: 1; Amy Butler, "Aegis Intercepts In First-Ever Salvo Test," *Aerospace Daily & Defense Report*, September 20, 2013: 3; and Jason Sherman and John Liang, "Missile Defense Agency's SM-3 Block IB Intercepts Target In Salvo Fire," *Inside the Navy*, September 23, 2013.

diverted into its path, and, using only the force of a direct impact, engaged and destroyed the target.

Program officials will assess and evaluate system performance based upon telemetry and other data obtained during the test.

This test exercised the latest version of the second-generation Aegis BMD Weapon System, capable of engaging longer range and more sophisticated ballistic missiles.[81]

November 6, 2014, Test

MDA states that in this flight test,

> The Missile Defense Agency, U.S. Pacific Command, and U.S. Navy Sailors aboard the USS John Paul Jones (DDG 53) successfully conducted a flight test today of the Aegis Ballistic Missile Defense (BMD) system, resulting in three successful near-simultaneous target engagements over the Pacific Ocean by the Aegis Baseline (BL) 9.C1 (BMD 5.0 Capability Upgrade) Weapon System configured ship. One short-range ballistic missile target was intercepted by a Standard Missile-3 (SM-3) Block IB guided missile, while two low-flying cruise missile targets were engaged by Standard Missile-2 (SM-2) Block IIIA guided missiles near-simultaneously.
>
> At approximately 12:03 p.m. (Hawaii Standard Time, 5:03 p.m. Eastern Standard Time) one short-range ballistic missile target and two cruise missile targets were launched from the Pacific Missile Range Facility (PMRF) on Kauai, Hawaii. Following the target launches, the USS John Paul Jones, in Integrated Air and Missile Defense (IAMD) Radar Priority Mode, detected and tracked the missiles with its onboard AN/SPY-1 radar.
>
> The ship, equipped with the Aegis BMD weapon system, developed a fire control solution and launched one SM-3 Block IB guided missile to engage the ballistic missile target. The SM-3 missile maneuvered to a point in space and released its kinetic warhead. The kinetic warhead acquired the target's reentry vehicle, diverted into its path, and destroyed the target with the sheer energy and force of direct impact. The ship also launched two SM-2 Block IIIA guided missiles to successfully engage the cruise missile targets.
>
> Program officials will evaluate system performance based upon telemetry and other data obtained during the test.
>
> This test, designated Flight Test Standard Missile-25 (FTM-25), was the first live-fire event of the Aegis Weapon System in IAMD Radar Priority Mode, engaging a ballistic missile target and a raid of cruise missile targets.[82]

---

[81] Missile Defense Agency, "Aegis Ballistic Missile Defense System Completes Successful Intercept Flight Test," October 4, 2013 (13-News-0009). See also Michael Fabey, "Aegis Completes Another Intercept Test," *Aerospace Daily & Defense Report*, October 7, 2013: 2; Jason Sherman, "SM-3 Block IB Completes IOT&E With A Bang, Full-Rate Production Review," *Inside the Navy*, October 7, 2013; Mike McCarthy, "Aegis Missile Defense Test Scores Hit," *Defense Daily*, October 7, 2013: 4.

[82] Missile Defense Agency, "Aegis Ballistic Missile Defense System Completes Successful Intercept Flight Test," November 6, 2014 (14-News-0012). See also Andrea Shalal, "U.S. Aegis System Zaps Cruise, Ballistic Missile Targets in Test," *Reuters (www.reuters.com)*, November 6, 2014; Mike McCarthy, "Aegis BMD Hits Three Targets In Simultaneous Test," *Defense Daily*, November 10, 2014.

June 25, 2015, Test

MDA's summary table of Aegis BMD flight tests[83] shows this as a test that did not result in the launch of an SM-3. MDA as of August 3, 2015, had not issued a news release discussing this event. MDA's count of 31 successful intercepts in 37 launches through July 29, 2015, does not appear to include this test, suggesting that this was considered a "no test" event—a test in which there was a failure that was not related to the Aegis BMD system or the SM-3 interceptor. A June 26, 2015, news report states the following:

> The U.S. Missile Defense Agency on Friday said a target malfunction caused it to abort a key intercept test of the Aegis Ashore missile defense system, built by Lockheed Martin Corp, that is due to be installed in Romania this year.
>
> "Due to a target malfunction, the test wasn't conducted and an interceptor wasn't launched," said Rick Lehner, a spokesman for the U.S. Defense Department agency....
>
> It was not immediately clear what caused the target to malfunction, or when the test would be rescheduled.[84]

October 4, 2015, Test

MDA as of November 10, 2015, had not issued a news release discussing this event. MDA's count of 32 successful intercepts in 39 launches through November 1, 2015, does not appear to include this test, suggesting that this was considered a "no test" event—a test in which there was a failure that was not related to the Aegis BMD system or the SM-3 interceptor.

October 20, 2015, Test

Regarding this test, the Navy states the following:

> USS Ross (DDG 71) successfully intercepted a ballistic missile in the North Atlantic Ocean during the Maritime Theater Missile Defense (MTMD) Forum's At Sea Demonstration (ASD) Oct. 20, 2015.
>
> This is first time a Standard Missile-3 (SM-3) Block IA guided interceptor was fired on a non-U.S. range and the first intercept of a ballistic missile threat in the European theater.
>
> For the scenario, a short-range Terrier Orion ballistic missile target was launched from Hebrides Range and was inflight simultaneously with two anti-ship cruise missiles fired at the coalition task group. Ross fired a SM-3 and successfully engaged the ballistic

---

[83] "*Aegis Ballistic Missile Defense Test Firing Record*," accessed August 3, 2015, at http://www.mda.mil/global/documents/pdf/aegis_tests.pdf. [The URL now begins with https:/i].

[84] Andrea Shalal, "U.S. Skips Aegis Aegis Ashore Missile Test After Target Malfunction," *Reuters*, June 26, 2015. See also "First Aegis Ashore Intercept Test Aborted. Does this Raise Issues for Planned 2015 Deployment Date for the Romanian Aegis Ashore Site?" *Mostly Missile Defense*, June 27, 2015.

missile target in space. In its air defense role, USS The Sullivans (DDG 68) fired a SM-2, which is the first time a SM-2 was fired on the Hebrides Range....

"ASD-15 shows that with communication, collaboration and commitment nations can come together and flawlessly defend against a complex threat scenario." [said] Vice Adm. James Foggo, Commander, U.S. 6th Fleet....

ASD-15 is a U.K.-hosted, U.S.-facilitated, multi-national demonstration of coalition Integrated Air and Missile Defense capability....

There are a number of firsts associated with this event including:

- First intercept of a ballistic missile target in the European theater
- First SM-3 fired on a non-U.S. range
- The first firing of an SM-2 and SM-3 on the Hebrides Range, United Kingdom
- First use of multi-national beyond line of sight link architecture for IAMD purposes in the European theater
- First international ship (Netherlands and Spain) transmissions of BMD cues to a U.S. BMD guided missile destroyer
- First time coalition IAMD used in a scenario with simultaneous attack from anti-ship cruise and ballistic missiles.

This test demonstrates the commitment of the United States to the defense of Europe through our four Aegis ships forward deployed to Rota, Spain, and shore station in Romania.

The 10 MTMD Forum member nations are: Australia, Canada, France, Germany, Italy, The Netherlands, Norway, Spain, United Kingdom, and the United States.

Eight nations provided ships and aircraft for ASD-15 including Canada, France, Italy, The Netherlands, Norway, Spain, United Kingdom, and the United States with Germany providing personnel to augment the Forum's multi-national Combined Task Group staff.

The tactical data link used in ASD-15 covers over 5.7 million square miles.

USS Mount Whitney (LCC-20), flag ship for U.S. 6th Fleet, served as the viewing platform for officials representing participating coalition nations during ASD-15; delegates from seven MTMD Forum nations, Denmark, and Japan watched the missile intercept on a live video feed aboard the ship.

The Maritime Theater Missile Defense forum was established in 1999 as a co-operative body for participating navies to develop improved cooperation and promote interoperability in sea-based missile defense.[85]

November 1, 2015, Test

Regarding this test, MDA states the following:

---

[85] U.S. Naval Forces Europe-Africa/U.S. 6th Fleet Public Affairs, "USS Ross Successfully Intercepts Ballistic Missile Target During Coalition Test," *Navy News*, October 20, 2015.

The U.S. Missile Defense Agency (MDA), Ballistic Missile Defense System (BMDS) Operational Test Agency, Joint Functional Component Command for Integrated Missile Defense, U.S. European Command, and U.S. Pacific Command conducted a complex operational flight test of the BMDS demonstrating a layered defense architecture.

The test, designated Flight Test Operational-02 Event 2a, was conducted in the vicinity of Wake Island and surrounding areas of the western Pacific Ocean. The test stressed the ability of Aegis Ballistic Missile Defense (BMD) and Terminal High Altitude Area Defense (THAAD) weapon systems to negate two ballistic missile threats while Aegis BMD simultaneously conducted an anti-air warfare operation.

This was a highly complex operational test of the BMDS which required all elements to work together in an integrated layered defense design to detect, track, discriminate, engage, and negate the ballistic missile threats.

BMDS assets included: a THAAD battery consisting of a THAAD Fire Control and Communications (TFCC) unit, THAAD launcher, and an Army Navy/Transportable Radar Surveillance and Control Model 2 (AN/TPY-2) radar in terminal mode; a second AN/TPY2 radar in forward-based mode; Command, Control, Battle Management and Communications (C2BMC); and the USS JOHN PAUL JONES (DDG-53) Aegis BMD-configured ship with its onboard AN/SPY-1 radar.

At approximately 11:05 pm EDT (October 31), a Short Range Air Launch Target (SRALT) was launched by a U.S. Air Force C-17 aircraft southeast of Wake Island. The THAAD AN/TPY-2 radar in terminal mode detected the target and relayed track information to the TFCC to develop a fire control solution and provide track information for use by other defending BMDS assets. The THAAD weapon system developed a fire control solution, launched a THAAD interceptor missile, and successfully intercepted the SRALT target.

While THAAD was engaging the SRALT, an extended Medium Range Ballistic Missile (eMRBM) was air-launched by another Air Force C-17. The eMRBM target was detected and tracked by multiple BMDS assets including the AN/TPY-2 in forward-based mode, and the USS JOHN PAUL JONES with its AN/SPY-1 radar. Shortly after eMRBM launch, a BQM-74E air-breathing target was also launched and tracked by the USS JOHN PAUL JONES.

As a demonstration of layered defense capabilities, both Aegis BMD and THAAD launched interceptors to engage the eMRBM. The USS JOHN PAUL JONES successfully launched a Standard Missile-3 (SM-3) Block IB Threat Upgrade guided missile, but an anomaly early in its flight prevented a midcourse intercept. However, the THAAD interceptor, in its terminal defense role, acquired and successfully intercepted the target. Concurrently, Aegis BMD successfully engaged the BQM-74E air-breathing target with a Standard Missile-2 Block IIIA guided missile. A failure review is currently underway to investigate the SM-3 anomaly.

Several other missile defense assets observed the launches and gathered data for future analysis. Participants included the Command, Control, Battle Management, and Communications (C2BMC) Experimental Lab (X-Lab), C2BMC Enterprise Sensors

Laboratory (ESL), and the Space Tracking and Surveillance System-Demonstrators (STSS-D).

The MDA will use test results to improve and enhance the BMDS.[86]

December 10, 2015, Test

Regarding this test, MDA states the following:

> The Missile Defense Agency (MDA) and the Ballistic Missile Defense System (BMDS) Operational Test Agency, in conjunction with U.S. Pacific Command, U.S. European Command, and Joint Functional Component Command for Integrated Missile Defense, successfully conducted the first intercept flight test today (December 9, Hawaii Standard Time) of a land-based Aegis Ballistic Missile Defense (BMD) weapon system and Standard Missile (SM)-3 Block IB Threat Upgrade guided missile, launched from the Aegis Ashore Missile Defense Test Complex at the Pacific Missile Range Facility (PMRF), Kauai, Hawaii.
>
> During the test, a target representing a medium-range ballistic missile was air-launched from a U.S. Air Force C-17 aircraft over the broad ocean area southwest of Hawaii. An AN/TPY-2 radar in Forward Based Mode, located at PMRF, detected the target and relayed target track information to the Command, Control, Battle Management, and Communication (C2BMC) system. The Aegis Weapon System at the Aegis Ashore site received track data from C2BMC and used its component AN/SPY-1 radar to acquire, track, and develop a fire control solution to engage the target. The Aegis Weapon System then launched the SM-3 Block IB Threat Upgrade guided missile from its Vertical Launch System. The SM-3's kinetic warhead acquired the target reentry vehicle, diverted into its path, and destroyed the target using the kinetic force of a direct impact.
>
> The primary purpose of the test, designated Flight Test Operational-02 Event 1a, was to assess the operational effectiveness of the Aegis Ashore capability as part of a larger BMDS architecture. Aegis Ashore uses a nearly identical configuration of the Vertical Launch System, fire control system, and SPY-1 radar currently in use aboard Aegis BMD cruisers and destroyers deployed at sea around the world.
>
> Vice Admiral James D. Syring, MDA Director, said, "Today's test demonstrated that the same Aegis Ballistic Missile Defense capability that has been fielded at sea and operational for years, will soon be operational ashore as part of the European Phased Adaptive Approach (EPAA) Phase 2 capability in Romania. I am very proud of the tremendous effort by the entire government/industry team in executing this vitally important mission for our Nation and our allies."[87]

---

[86] Missile Defense Agency, "Ballistic Missile Defense System Demonstrates Layered Defense While Conducting Multiple Engagements in Operational Test," November 1, 2015 (15-NEWS-0008).

[87] Missile Defense Agency, "Standard Missile Completes First Intercept Test from Aegis Ashore Test Site," December 10, 2015 (15-NEWS-0011).

February 3, 2017, Test

Regarding the intercept of February 3, 2017, MDA states the following:

> The U.S. Missile Defense Agency (MDA), the Japan Ministry of Defense (MoD), and U.S. Navy sailors aboard USS John Paul Jones (DDG 53) successfully conducted a flight test Feb. 3 (Hawaii Standard Time), resulting in the first intercept of a ballistic missile target using the Standard Missile-3 (SM-3) Block IIA off the west coast of Hawaii....
>
> At approximately 10:30 p.m., Hawaii Standard Time, Feb. 3 (3:30 a.m. Eastern Daylight Time, Feb. 4) a medium-range ballistic missile target was launched from the Pacific Missile Range Facility at Kauai, Hawaii. John Paul Jones detected and tracked the target missile with its onboard AN/SPY-1D(V) radar using the Aegis Baseline 9.C2 weapon system. Upon acquiring and tracking the target, the ship launched an SM-3 Block IIA guided missile which intercepted the target.
>
> "Today's test demonstrates a critical milestone in the cooperative development of the SM3 Block IIA missile," said MDA Director Vice Adm. Jim Syring. "The missile, developed jointly by a Japanese and U.S. government and industry team, is vitally important to both our nations and will ultimately improve our ability to defend against increasing ballistic missile threats around the world."
>
> Based on preliminary data the test met its primary objective. Program officials will continue to evaluate system performance based upon telemetry and other data obtained during the test.
>
> The flight test, designated SM-3 Block IIA Cooperative Development (SCD) Project Flight Test, Standard Missile (SFTM)-01, was the third flight test of the SM-3 Block IIA guided missile, and the first intercept test. This test also marks the first time an SM-3IIA was launched from an Aegis ship and the first intercept engagement using the Aegis Baseline 9.C2 (BMD 5.1) weapon system.[88]

June 21, 2017, Test

Regarding the intercept test of June 21, 2017, MDA states the following:

> The U.S. Missile Defense Agency and the Japan Ministry of Defense conducted a development flight test today of a new Standard Missile-3 (SM-3) BLock IIA missile off the coast of Hawaii.
>
> A planned intercept was not achieved....
>
> At approximately 7:20 p.m., Hawaii Standard Time, June 21 (1:20 am Eastern Daylight Time, June 22), a medium-range ballistic target missile was launched from the Pacific Missile Range Facility at Kauai, Hawaii. The USS John Paul Jones (DDG 53) detected and tracked the target missile with its onboard AN/SPY-1 radar using the Aegis Baseline 9.C2 weapon system. Upon acquiring and tracking the target, the ship launched an SM-3 Block IIA guided missile, but the missile did not intercept the target.

---

[88] Missile Defense Agency, "U.S., Japan Successfully Conduct First SM-3 Block IIA Intercept Test," February 3, 2017 (17-NEWS-0002).

Program officials will conduct an extensive analysis of the test data. Until that review is complete, no additional details will be available.[89]

A July 24, 2017, press report stated the following:

A U.S. Missile Defense Agency review of a failed ballistic missile intercept test showed that a mistaken input into the combat system by a sailor on the destroyer John Paul Jones caused the missile to self-destruct before reaching the target.

A tactical datalink controller, in charge of maintaining encrypted data exchanges between ships and aircraft, accidentally identified the incoming ballistic missile target as a friendly in the system, causing the SM-3 missile to self-destruct in flight, according to a source familiar with the test.

The head of MDA did not comment on the sailor error, but said in a statement that the ongoing review confirmed it wasn't an issue with the SM-3 Block IIA missile or the Navy's Aegis combat system.

"Though the review is still in process, the SM-3 IIA interceptor and Aegis Combat System have been eliminated as the potential root cause," of the failure, said Air Force Lt. Gen. Sam Greaves, the director of MDA.

"We are conducting an extensive review as part of our standard engineering and test processes, and it would be inappropriate to comment further until we complete the investigation."[90]

October 15, 2017, Test

Regarding the intercept test of October 15, 2017, MDA states the following:

Ships from Canada, France, Germany, Italy, the Netherlands, Spain, the United Kingdom, and the United States participated in a live-fire integrated air and missile defense (IAMD) scenario, defending against a ballistic missile target as well as three anti-ship cruise missiles Oct. 15 as part of exercise Formidable Shield 2017 (FS17). Naval Striking and Support Forces NATO (STRIKFORNATO) is conducting Formidable Shield on behalf of the U.S. 6th Fleet. The U.S. Missile Defense Agency is also a major participant in this exercise.

During the collective self-defense scenario, the Arleigh Burke-class guided-missile destroyer USS Donald Cook (DDG 75) successfully detected, tracked and intercepted a medium-range ballistic missile target with a Standard Missile-3 Block IB guided missile. Simultaneously, the Spanish frigate SPS Alvaro de Bazan (F101) fired an Evolved SeaSparrow Missile (ESSM) against an incoming anti-ship cruise missile while the Netherlands frigate HNLMS Tromp (F803) fired ESSMs against a pair of incoming anti-ship cruise missiles. This was the first time NATO's smart defense concept was

[89] Missile Defense Agency, "Aegis Missile Defense Test Conducted," June 21, 2017 (17-NEWS-0006).

[90] David B. Larter, "Sailor Error Led to Failed US Navy Ballistic Missile Intercept Test," *Defense News*, July 24, 2017.

demonstrated with ships serving as air defense units protecting naval ballistic missile defense units.

Following that event, the U.S. Missile Defense Agency and U.S. Navy sailors aboard USS McFaul (DDG 74) successfully test fired a Standard Missile-6 (SM-6). That flight test, designated Standard Missile Controlled Test Vehicle (SM CTV)-03, demonstrated the successful performance of an SM-6 launched from an Aegis Ballistic Missile Defense capable DDG and was conducted as part of the system's flight certification process. The SM-6 test was not part of the Formidable Shield exercise, but was conducted in coordination with that event to leverage the available range assets....

Formidable Shield is designed to improve allied interoperability in an IAMD environment, using NATO command-and-control reporting structures and datalink architecture. FS17 is the inaugural iteration of this exercise....

More than 14 ships, 10 aircraft, and approximately 3,300 personnel from Belgium, Canada, Denmark, France, Germany, Italy, the Netherlands, Spain, the U.K., and the U.S., are participating in FS17 on the U.K. Ministry of Defense's Hebrides Range located on the Western Isles of Scotland....

U.S. ships participating in Formidable Shield include the Arleigh Burke-class guided-missile destroyers Donald Cook, USS Mitscher (DDG 57), USS Winston S. Churchill (DDG 81), and the Louis and Clark-class dry cargo ship USNS Medger Evers (T-AKE 13).

Formidable Shield 2017 began Sept. 24, and is scheduled to conclude Oct. 18, 2017. This exercise is planned to be a recurring, biennial event, and is designed to assure allies, deter adversaries, and demonstrate our commitment to collective defense of the NATO alliance. Formidable Shield and exercise Joint Warrior 17-2, a U.K.-led, multinational exercise in a maritime training environment for allies to improve interoperability and prepare forces for combined operations, are occurring concurrently.[91]

January 31, 2018, Test

Regarding the intercept test of January 31, 2018, MDA states the following:

The Missile Defense Agency and U.S. Navy sailors manning the Aegis Ashore Missile Defense Test Complex (AAMDTC) conducted a live-fire missile flight test Jan. 31 using a Standard-Missile (SM)-3 Block IIA missile launched from the Pacific Missile Range Facility, Kauai, Hawaii. This was a developmental and operational test of a new capability and utilized a missile variant not yet in production. The primary objective of the test, to intercept an air-launched intermediate-range ballistic missile target with an SM-3 Block IIA missile, was not achieved. However, much was still learned that demonstrated an increase in the effective range of the overall ballistic missile defense system.

Several firsts were accomplished as a result of this mission, which included using both ground and space-based sensors to remotely cue the launch of the interceptor by the

[91] Missile Defense Agency, "Formidable Shield 2017: Ship Engages BMD Target During NATO Exercise, MDA and Navy Conduct SM-6 Test Launch," October 15, 2017 (17-NEWS-0010).

Aegis weapon system. This was also the first time an SM-3 Block IIA missile was launched from land using the Aegis Ashore test complex. The test also demonstrated a highly complex multi-domain command, control, battle management and communications system, which was used by operational crews to execute the mission.

"We always make progress every time we conduct a test," said MDA Director Lt. Gen. Sam Greaves. "While we are disappointed that we did not demonstrate a successful intercept, we did demonstrate significant advances in capability and collected valuable test data that will allow us to further improve our capability and capacity of the ballistic missile defense system. We are committed to protecting and defending our nation, its warfighters, friends and allies against all ranges of ballistic missiles in all phases of flight."

MDA will conduct an extensive investigation to determine the cause or causes of any anomalies that may have prevented a successful intercept.[92]

An October 29, 2018, press report states:

The January flight test failure of a Standard Missile-3 Block IIA guided missile interceptor does not have "fleetwide" implications affecting other SM-3 variants or other missile systems in the U.S. inventory....

In April, Michael Griffin, under secretary of defense for research and engineering, said the then-ongoing investigation into the cause of the Jan. 31 Flight Test Standard Missile-29 (FTM-29) failure indicated a "highly standardized" component used in "other fleets," raising the prospect that remedial action could be required across multiple weapon systems....

"The investigators determined that the problem was isolated to the SM-3 Block IIA," Michelle Baldanza, a spokeswoman for Griffin's office, told Inside Defense on Oct. 3. "We are confident that this is not a fleetwide problem."[93]

September 11, 2018, Test

Regarding the intercept test of September 11, 2018, MDA states the following:

---

[92] Missile Defense Agency, "Test Conducted From Aegis Ashore Missile Defense Test Complex," January 31, 2018 (18-NEWS-0001). See also David B. Larter, "Another US Navy Ballistic Missile Intercept Reportedly Fails in hawaii," *Defense News*, January 31, 2018; Richard Abott, "SM-3 IIA Missile Defense Test In Hawaii Reportedly Fails, Defense Tester Report Finds Lower Confidence," *Defense Daily*, February 1, 2018: 4-6; David B. Larter, "Reality Check: Failures Happen, Even in Missile Defense Testing," *Defense News*, February 1, 2018; Ben Werner, "Pentagon Confirms SM-3 Block IIA Missile Missed its Target in Test This Week," *USNI News*, February 1, 2018; Jason Sherman, "MDA Acknowledges Aegis BMD Flight Test Failure," *Inside the Navy*, February 5, 2018 (which identifies the target as an "air-launched intermediate-range ballistic missile target"); "Failed Missile Test Off of Kauai Costs the US $130 Million," *Associated Press*, February 20, 2018; Daniel Cebul, "Missile Defense Failure Might Prove a Good Thing for SM-3 IIA in the Long Run, Says MDA Head," *Defense News*, March 6, 2018; Anthony Capaccio, "Blame a Spark Plug for U.S.-Japan Missile Failure, Pentagon Says," *Bloomberg*, October 1, 2018; Rich Abott, "MDA Says SM3 IIA Test Failure Caused By Third Stage Ignition Part," *Defense Daily*, October 3, 2018: 7; Jason Sherman, "MDA Pins SM-3 Block IIA Flight Test Failure on Device Needed to Ignite Rocket," *Inside the Navy*, October 5, 2018.

[93] Jason Sherman, "Last January's SM-3 Block IIA Failure Carries No 'Fleetwide' Implications," *Inside the Navy*, October 29, 2018.

The Japan Maritime Self-Defense Force (JMSDF) and the United States Missile Defense Agency (MDA) announce the successful completion of an Aegis Ballistic Missile Defense (BMD) intercept flight test, in cooperation with the U.S. Navy, off the coast of Kauai in Hawaii. The event, designated Japan Flight Test Mission-05 (JFTM-05), was conducted in order to demonstrate a successful engagement of a target missile from the JS ATAGO using the sea-based midcourse engagement capability provided by Aegis BMD.

The JFTM-05 test event verified the newest BMD engagement capability of the Japan (J6) Aegis BMD configuration of the recently upgraded Japan destroyer, JS ATAGO (DDG177). At approximately 10:37pm HST on September 11, 2018 a simple separating, ballistic missile target was launched from the Pacific Missile Range Facility at Barking Sands, Kauai, Hawaii. JS ATAGO crew members detected and tracked the target. The Aegis Weapon System then developed a fire control solution and a Standard Missile -3 Block IB Threat Upgrade (SM-3 Blk IB TU) missile was launched. The SM-3 successfully intercepted the target above the Pacific Ocean. JFTM-05 is a significant milestone in the growing cooperation between Japan and the U.S. in the area of missile defense.

"This successful test is a major milestone verifying the capabilities of an upgraded Aegis BMD configuration for Japan's destroyers," said MDA Director Lt. Gen. Sam Greaves. "This success provides confidence in the future capability for Japan to defeat the developing threats in the region. My congratulations to the Japan Maritime Self-Defense Force, our MDA team, and our industry partners. We are committed to assisting the Government of Japan in upgrading its national missile defense capability against emerging threats."[94]

October 26, 2018, Test

Regarding the intercept test of October 26, 2018, MDA states the following:

The U.S. Missile Defense Agency (MDA), and U.S. Navy sailors aboard USS John Finn (DDG-113) successfully conducted an intercept of a medium-range ballistic missile target with a Standard Missile-3 (SM-3) Block IIA missile during a flight test off the west coast of Hawaii....

On October 26, 2018, the target missile was launched from the Pacific Missile Range Facility at Kauai, Hawaii. The USS John Finn (DDG-113) detected and tracked the target missile with its onboard AN/SPY-1 radar using the Aegis Baseline 9.C2 weapon system. Upon acquiring and tracking the target, the ship launched an SM-3 Block IIA guided missile which intercepted the target.

"This was a superb accomplishment and key milestone for the SM-3 Block IIA return to flight," said MDA Director Lt. Gen. Sam Greaves. "My congratulations to the

---

[94] Missile Defense Agency, "Japan Missile Defense Flight Test Successful Through Intercept," September 12, 2018 (18-NEWS-0005). See also Jason Sherman, "Japan's Newest Destroyer Intercepts Ballistic Missile Target in Test with United States," *Inside the Navy*, September 17, 2018; Patrick Tucker, "Japan's New Ship-Based Interceptor Shoots Down a Ballistic Missile in Test," *Defense One*, September 12, 2018.

entire team, including our sailors, industry partners, and allies who helped achieve this milestone."

Based on observations and initial data review, the test met its objectives. Program officials will continue to evaluate system performance.[95]

December 10, 2018, Test

Regarding the intercept test of October 26, 2018, MDA states the following:

The Missile Defense Agency (MDA) and U.S. Navy sailors manning the Aegis Ashore Missile Defense Test Complex (AAMDTC) at the Pacific Missile Range Facility (PMRF) at Kauai, Hawaii, successfully conducted Flight Test Integrated-03 (FTI-03). This was an operational live fire test demonstrating the Aegis Weapon System Engage On Remote capability to track and intercept an Intermediate Range Ballistic Missile (IRBM) target with an Aegis Ashore-launched Standard Missile-3 (SM-3) Block IIA interceptor.

FTI-03 consisted of an IRBM target, air-launched by a U.S. Air Force C-17 from the broad ocean area thousands of miles southwest of the Aegis Ashore Test site that launched the SM-3 Block IIA Interceptor. The engagement leveraged a ground, air and space-based sensor/command and control architecture linked by the Ballistic Missile Defense System's Command and Control, Battle Management, and Communications (C2BMC) suite.

"Today's successful flight test demonstrated the effectiveness of the European Phased Adaptive Approach Phase 3 architecture. It also was of great significance to the future of multi-domain missile defense operations and supports a critical initial production acquisition milestone for the SM-3 Block IIA missile program," said MDA Director Lt. Gen. Sam Greaves. "This system is designed to defend the United States, its deployed forces, allies, and friends from a real and growing ballistic missile threat. I offer my congratulations to all members of the team, military, civilian, contractors and allies who helped make this possible."

Based on preliminary data, the test met its objective, and program officials will continue to evaluate system performance based upon telemetry and other data obtained during the test.[96]

---

[95] Missile Defense Agency, "U.S. Successfully Conducts SM-3 Block IIA Intercept Test," October 26, 2018 (18-NEWS-0006). See also Aaron Mehta, "After Consecutive Failures, Watch US Navy Intercept Test Missile with SM-3 Weapon," *Defense News*, October 26, 2018; Jason Sherman, "SM-3 Block IIA, A $1 Billion Development Project, Hits Target in Do-Over Test," *Inside Defense (Daily News)*, October 26, 2018, which states that the test was designed FTM45; Rich Abott, "MDA Succeeds in Fourth SM-3 IIA Intercept After Two Failures," *Defense Daily*, October 29, 2018, which also states that the test was designated FTM-45.

[96] Missile Defenes Agency, "SM-3 Block IIA Launched From Aegis Ashore Successfully Intercepts Intermediate Range Ballistic Missile Target During Operational Test," December 11, 2018 (18_NEWS-0007). See also Rich Abott, "MDA Scores SM-3 IIA Hit Using Japan Aegis Model," *Defense Daily*, December 12, 2018; David B. Larter, "US Navy, Missile Defense Agency Shoot Down an Intermediate-Range Ballistic Missile in Space," *Defense News*, December 11, 2018; Marcus Weisgerber, "Test Validates New US Interceptor for European, Japanese Missile Shields," *Defense One*, December 11, 2018.

Details on Selected Endo-Atmospheric (SM-2 Block IV and SM-6) Flight Tests Since July 2015

May 24, 2006, Test

Regarding the intercept of May 24, 2006, MDA states the following:

The U.S. Navy, in cooperation with the Missile Defense Agency, today successfully conducted a ballistic missile defense demonstration involving the intercept of a target missile in the terminal phase (the last few seconds) of flight. The test involved an Aegis cruiser modified to detect, control and engage a ballistic missile target with a modified Standard Missile - 2 (SM-2) Block IV. The Pearl Harbor-based Aegis cruiser USS Lake Erie (CG 70) conducted the mission against a short-range target missile launched from the Pacific Missile Range Facility, Barking Sands, Kauai, Hawaii. It was the first sea-based intercept of a ballistic missile in its terminal phase.

The modified Aegis Weapon System and the modified SM-2 Block IV provided the firing ship the capability to guide the missile to achieve either; 1) a direct body to body hit between the interceptor and the threat or, 2) a near-direct hit where the high pressure, heat and fragments are placed on the threat by a blast fragmentation warhead. This warhead is similar in concept to that used in the deployed Israeli Arrow system. In today's test, the threat missile was completely destroyed by the combined effects of these two mechanisms.

"This is another example of the ongoing cooperative spirit between the Navy and the Missile Defense Agency," said Rear Admiral Barry McCullough, Director, Surface Warfare, on the staff of the Chief of Naval Operations.

"We believe it is an important step towards the desired end-state of a robust sea-based terminal ballistic missile defense capability," McCullough added, "and it begins to meet an immediate near-term concern of our Combatant Commanders." "The only terminal phase program we currently have that is operational is the Patriot Advanced Capability 3 (PAC-3)," he added, "and considerations to put those aboard ships are still under review."

There is currently no sea-based terminal ballistic missile defense capability. The Navy Area Theater Ballistic Missile Defense (TBMD) Program, had been under development, but was terminated in December 2001. In ballistic missile defense, the modified Aegis Weapon System, with a modified SM-2 Block IV missile provides a near term, limited emergency capability against a very specific segment of the ballistic missile threat. The Navy and MDA consider it vital to develop a more robust capability for terminal ballistic missile defense of the joint sea base and friendly force embarkation points ashore.

"There is a significant number of SM-2 Block IV missiles available, which may be modified and deployed on Navy ships modified to perform a BMD mission," said Air Force Lieutenant General Henry "Trey" Obering, Missile Defense Agency director. "While talking with the Navy and the Combatant Commanders, on how and when, we might be able to make that happen," Lieutenant General Obering added, "MDA will continue to improve its development of the midcourse Aegis sea-based ballistic missile defense capability, which utilizes the Standard Missile – 3 (SM-3), and has successfully achieved 6 intercepts in 7 flight tests."[97]

June 5, 2008, Test

Regarding the intercept of June 5, 2008, MDA states the following:

Air Force Lieutenant General Henry "Trey" Obering III, Missile Defense Agency (MDA) director, announced the successful completion of the latest flight test of the sea-based Aegis Ballistic Missile Defense (BMD) element, conducted jointly with the U.S. Navy off the coast of Kauai, Hawaii. The event, designated as Flight Test Maritime-14 (FTM-14), marked the fourteenth overall successful intercept, in sixteen attempts, for the Aegis BMD program and the second successful intercept of a terminal phase (last few seconds of flight) target by a modified Standard Missile - 2 Block IV (SM-2 Blk IV) interceptor. The mission was completed by the cruiser USS Lake Erie (CG 70), using the tactically -certified Aegis BMD shipboard weapon system, modified for a terminal capability, and the modified SM2 Blk IV. This is the 35th successful terminal and midcourse defense intercept in 43 tests since 2001.

Aegis BMD is the sea-based mid-course component of the MDA's Ballistic Missile Defense System (BMDS) and is designed to intercept and destroy short to intermediate-range ballistic missile threats. In 2006, the program's role was expanded to include a sea-based terminal defense effort, using a modified version of the SM-2 Blk IV. Unlike other missile defense technologies now deployed or in development, the SM-2 Blk IV does not use "hit to kill" technology (directly colliding with the target) to destroy the target missile. Rather, it uses a blast fragmentation device that explodes in direct proximity to the target to complete the intercept and destroy the target.

At 8:13 a.m. Hawaii Standard Time (2:13 p.m. Eastern Daylight Time) a short range target was launched from a mobile launch platform 300 miles west of the Pacific Missile Range Facility (PMRF), Barking Sands, Kauai, Hawaii. Moments later, the USS Lake Erie's Aegis BMD Weapon System detected and tracked the target and developed fire control solutions.

Approximately four minutes later, the USS Lake Erie's crew fired two SM-2 Blk IV missiles, and two minutes later they successfully intercepted the target inside the earth's atmosphere, about 12 miles above the Pacific Ocean and about 100 miles west-northwest of Kauai.

FTM-14 test objectives included evaluation of: the BMDS ability to intercept and kill a short range ballistic missile target with the Aegis BMD, modified with the terminal

---

[97] Missile Defense Agency, "First at-Sea Demonstration of Sea-Based Terminal Capability Successfully Completed," May 24, 2006 (06-FYI-0079). See also Gregg K. Kakesako, "Missile Defense System Makes History," *Honolulu Star-Bulletin*, May 25, 2006; Audrey McAvoy, "Ship Shoots Down Test Missile For The First Time," *NavyTimes.com*, May 25, 2006; "Navy, MDA Announce First Terminal Sea-Based Intercept," *Aerospace Daily & Defense Report*, May 26, 2006; Zachary M. Peterson, "Navy Conducts First Sea-Based Terminal Phase Missile Defense Test," *Inside the Navy*, May 29, 2006; and Jeremy Singer, "Sea-Based Terminal May Boost U.S. Missile Defense Capability," *Space News*, June 12, 2006.

mission capability; the modified SM-2 Blk IV missile using SPY-1 cue; and system-level integration of the BMDS.[98]

March 26, 2009, Test

Regarding the intercept of March 26, 2009, the Navy states the following:

Commander, U.S. 3rd Fleet, Vice Adm. Samuel J. Locklear announced the completion of the fleet operational exercise, Stellar Daggers, March 26.

The scheduled event took place March 24 and 26. Command and control of the participants in Stellar Daggers resided with U.S. 3rd Fleet based in San Diego.

San Diego-based Aegis destroyer, USS Benfold (DDG 65) engaged multiple targets during this multi-event exercise with Standard Missile-2 (SM-2) Block IIIA and modified SM-2 BLK IV missiles. The overall objective of Stellar Daggers was to test the Aegis system's sea-based ability to simultaneously detect, track, engage and destroy multiple incoming air and ballistic missile threats during terminal or final phase of flight.

During the event, Benfold's Aegis Weapons System successfully detected and intercepted a cruise missile target with a SM-2 BLK IIIA, while simultaneously detecting and intercepting an incoming short range ballistic missile (SRBM) target with a modified SM2 BLK IV. This is the first time the fleet has successfully tested the Aegis system's ability to intercept both an SRBM in terminal phase and a low-altitude cruise missile target at the same time.[99]

July 28-29, 2015, Test

Regarding the intercepts of July 28 and 29, 2015, MDA states the following:

The Missile Defense Agency (MDA), U.S. Pacific Command, and U.S. Navy Sailors aboard the USS John Paul Jones (DDG 53) successfully conducted a series of four flight test events exercising the Aegis Ballistic Missile Defense (BMD) element of the nation's Ballistic Missile Defense System (BMDS). The flight test, designated Multi-Mission Warfare (MMW) Events 1 through 4, demonstrated successful intercepts of short-range ballistic missile and cruise missile targets by the USS John Paul Jones, configured with Aegis Baseline 9.C1 (BMD 5.0 Capability Upgrade) and using Standard Missile (SM)-6 Dual I and SM-2 Block IV missiles. All flight test events were conducted at the Pacific Missile Range Facility (PMRF), Kauai, Hawaii.

MDA Director Vice Adm. James D. Syring said, "This important test campaign not only demonstrated an additional terminal defense layer of the BMDS, it also proved the robustness of the multi-use SM-6 missile on-board a Navy destroyer, further reinforcing the dynamic capability of the Aegis Baseline 9 weapon system."

Event 1

---

[98] Missile Defense Agency, "Successful Sea-Based Missile Defense Intercept," June 5, 2008 (08-NEWS-0068). See also Dave Ahearn, "Aegis, SM-2 Interceptors Kill Target Missile In Terminal-Phase Success," *Defense Daily*, June 6, 2008.

[99] "Navy Completes Air and Ballistic Missile Exercise," *Navy News Service*, March 26, 2009.

On July 28, at approximately 10:30 p.m. Hawaii Standard Time (July 29, 4:30 a.m. Eastern Daylight Time), a short-range ballistic missile (SRBM) target was launched from PMRF in a northwesterly trajectory. The USS John Paul Jones, positioned west of Hawaii, detected, tracked, and launched a SM-6 Dual I missile, resulting in a successful target intercept.

Event 2

On July 29, at approximately 8:15 p.m. Hawaii Standard Time (July 30, 2:15 a.m. Eastern Daylight Time), a short-range ballistic missile (SRBM) target was launched from PMRF in a northwesterly trajectory. The USS John Paul Jones detected, tracked, and launched a SM-2 Block IV missile, resulting in a successful target intercept.

Event 3

On July 31, at approximately 2:30 p.m. Hawaii Standard Time, (8:30 p.m. Eastern Daylight Time) an AQM-37C cruise missile target was air-launched to replicate an air-warfare threat. The USS John Paul Jones detected, tracked, and successfully engaged the target using an SM-6 Dual I missile.

Event 4

On August 1, at approximately 3:45 p.m. Hawaii Standard Time, (9:45 p.m. Eastern Standard Time), a BQM-74E cruise missile target was launched from PMRF. The USS John Paul Jones detected, tracked, and successfully engaged the target using an SM-6 Dual I missile. The SM-6's proximity-fuze warhead was programmed not to detonate after reaching the lethal distance from the target, thus providing the ability to recover and reuse the BQM-74E target....

MMW Event 1 was the first live fire event of the SM-6 Dual I missile.

MMW Events 1 and 2 were the 30th and 31st successful ballistic missile defense intercepts in 37 flight test attempts for the Aegis BMD program since flight testing began in 2002.[100]

December 14, 2016, Test

Regarding the intercept of December 14, 2016, MDA states the following:

The Missile Defense Agency and sailors aboard USS John Paul Jones (DDG 53), an Aegis baseline 9.C1 equipped destroyer, today successfully fired a salvo of two SM-6 Dual I missiles against a complex medium-range ballistic missile target, demonstrating the Sea Based Terminal endo-atmospheric defensive capability and meeting the test's primary objective.

The test was conducted off the coast of Hawaii just after midnight on Dec. 14.

"This test demonstrated the capabilities MDA and the Navy are delivering to our fleet commanders," said MDA Director Vice Adm. Jim Syring. "The SM-6 missile and the Aegis Weapon System continue to prove that they are critical components of our nation's multilayered, robust ballistic missile defense system."...

---

[100] Missile Defense Agency, "Aegis Ballistic Missile Defense System Completes Successful Series of Intercept Flight Test Events," August 3, 2015 (15-NEWS-0007).

Program officials will continue evaluating system performance based upon telemetry and other data obtained during the test.[101]

A December 16, 2016, press report states the following:

The Missile Defense Agency (MDA) said its new Sea Based Terminal (SBT) system achieved its second ballistic missile intercept during a Dec. 14 test over the Pacific Ocean.

During the test, the USS John Paul Jones (DDG-53)... fired a salvo of two Raytheon [RTN] Standard Misisle-6 (SM-6) interceptors in immediate succession against a medium-range ballistic missile target launched from the Pacific Missile Range Facility on Kauai, Hawaii. The first interceptor was not armed and was designed to collect test data, MDA said. The second interceptor, which carried an explosive warhead, intercepted the Lockheed Martin-built target....

MDA called the target "complex" but declined to elaborate. However, according to the Missile Defense Advocacy Alliance, the target emulated China's Dong-Feng 21 (DF-21), a ballistic missile equipped with a maneuverable re-entry vehicle and designed to destroy U.S., aircraft carriers.

The event, designated Flight Test Standard Missile-27 (FTM-27), was SBT's first salvo test and its second intercept in as many tries.[102]

A March 15, 2017, press report quoted Mike Campisi, Raytheon's SM-6 senior director, as stating the following: "We had two missiles in the air and we wanted to make sure that we were in fact pulling in on the target and looking at target versus looking at the other missile that's in the air.

Simulations all said the missile would never look at the other missile in the air however, but it's nice to prove that."[103]

August 29, 2017, Test

Regarding the intercept of August 29, 2017, MDA states the following:

The Missile Defense Agency and U.S. Navy sailors aboard the USS John Paul Jones (DDG 53) successfully conducted a complex missile defense flight test, resulting in the intercept of a medium-range ballistic missile (MRBM) target using Standard Missile-6 (SM-6) guided missiles during a test off the coast of Hawaii today [August 29].

John Paul Jones detected and tracked a target missile launched from the Pacific Missile Range Facility on Kauai, Hawaii with its onboard AN/SPY-1 radar, and onboard SM-6 missiles executed the intercept.

---

[101] Missile Defense Agency, "MDA Conducts SM-6 MRBM Intercept Test," December 14, 2016 (16-NEWS-0012).

[102] Marc Selinger, "Missile Defense Agency Scores Second Intercept With Sea Based Terminal System," *Defense Daily*, December 16, 2016: 3-4.

[103] Kris Osborn, "Breakthrough - Missile Defense Agency Fires 2 SM-6 Interceptors at Once - Testing New Seeker Technology," *Scout Military*, March 15, 2017.

"We are working closely with the fleet to develop this important new capability, and this was a key milestone in giving our Aegis BMD ships an enhanced capability to defeat ballistic missiles in their terminal phase," said MDA Director Lt. Gen. Sam Greaves. "We will continue developing ballistic missile defense technologies to stay ahead of the threat as it evolves."

This test, designated Flight Test Standard Missile-27 Event 2 (FTM-27 E2), marks the second time that an SM-6 missile has successfully intercepted a medium-range ballistic missile target.[104]

APPENDIX B. HOMEPORTING OF U.S. NAVY AEGIS BMD SHIPS AT ROTA, SPAIN

This appendix presents additional background information on the homeporting of four BMD capable Aegis destroyers at Rota, Spain.

As part of the October 5, 2011, U.S.-Spain joint announcement of the plan, the Prime Minister of Spain, Jose Luis Rodriguez Zapatero, stated the following, in part:

> This meeting marks a step forward on the path that we set for ourselves less than a year ago at the Lisbon Summit, aiming to make NATO an Alliance that is "more effective, engaged and efficient than ever before," in the words of [NATO] Secretary-General Rasmussen.
>
> At that historic Summit, decisions of enormous importance for the future of the Alliance were taken, such as the New Strategic Concept to face the new challenges of the 21st century, and the establishment of a new command structure that is leaner and more flexible, and improved.
>
> Besides these two important innovations, and as a consequence of them, the allies decided to develop an Anti-Missile Defence System....
>
> As you will recall, as a consequence of this new structure launched in Lisbon, Spain obtained an installation of great importance within NATO's Command and Control Structure: the Combined Air Operations Centre (CAOC) in Torrejón de Ardoz, Spain.
>
> This Centre, together with the Centre in Uedem, Germany, will form part of the air command and control system which is to include the anti-missile defence that the Alliance is going to implement.
>
> Together with this land-based component of the new air defence system, I can inform you that Spain is also going to support, starting in 2013, an important part of the system's naval element.
>
> In recent months, the different options have been studied, and finally, it was decided that Spain should be the site for this component of the system, due to its geostrategic location and its position as gateway to the Mediterranean.

---

[104] Missile Defense Agency, "Aegis BMD System Intercepts Target Missile," August 29, 2017 (17-NEWS-0009).

> Specifically, the United States is going to deploy, as its contribution to NATO's Anti-Missile Defence System, a total of four vessels equipped with the AEGIS system, to be based in Rota.
>
> This means that Rota is going to become a support centre for vessel deployment, enabling them to join multinational forces or carry out NATO missions in international waters, particularly in the Mediterranean....
>
> Moreover, this initiative will have a positive impact, in socio-economic terms, on our country, and most especially on the Bay of Cadiz.
>
> Permanently basing four vessels in Rota will require investing in the Base's infrastructure, and contracts with service providers, thus generating approximately a thousand new jobs, both directly and indirectly.
>
> For the shipyards, and for Spain's defence industry, the foreseeable impact will also be highly positive, as the USA is considering conducting the vessels' maintenance and upkeep at the nearby San Fernando shipyards, in the province of Cadiz. In addition, there will be significant transfer of state-of-the-art technology, from which Spain can benefit.[105]

As part of the same joint announcement, Secretary of Defense Leon Panetta stated the following, in part:

> With four Aegis ships at Rota, the alliance is significantly boosting combined naval capabilities in the Mediterranean, and enhancing our ability to ensure the security of this vital region. This relocation of assets takes place as part of the United States' ongoing effort to better position forces and defensive capabilities in coordination with our European allies and partners.
>
> This announcement should send a very strong signal that the United States is continuing to invest in this alliance, and that we are committed to our defense relationship with Europe even as we face growing budget constraints at home....
>
> Alongside important agreements that were recently concluded with Romania, Poland, and Turkey, Spain's decision represents a critical step in implementing the European Phased Adaptive Approach, as our leaders agreed to in Lisbon....
>
> Beyond missile defense, the Aegis destroyers will perform a variety of other important missions, including participating in the Standing NATO Maritime Groups, as well as joining in naval exercises, port visits, and maritime security cooperation activities....
>
> The agreement also enables the United States to provide rapid and responsive support to the U.S. Africa and U.S. Central Commands, as needed.[106]

---

105 "*Announcement on Missile Defence Cooperation by NATO Secretary General Anders Fogh Rasmussen, the Prime Minister of Spain, Jose Luis Rodriguez Zapatero and US Defense Secretary Leon Panetta*," October 5, 2011, accessed October 6, 2011, at http://www.nato.int/cps/en/SID-107ADE55-FF83A6B8/natolive/opinions_78838.htm.

106 "*Announcement on Missile Defence Cooperation by NATO Secretary General Anders Fogh Rasmussen, the Prime Minister of Spain, Jose Luis Rodriguez Zapatero and US Defense Secretary Leon Panetta*," October 5, 2011, accessed October 6, 2011, at http://www.nato.int/cps/en/SID-107ADE55-FF83A6B8/natolive/

An October 5, 2011, press report stated the following:

> A senior U.S. defense official said making the [ships'] base at Rota, on Spain's southwestern Atlantic coast near Cadiz, would reduce the numbers of [BMD-capable Aegis] ships needed for the [EPAA] system.
>
> "You [would] probably need 10 of these ships if they were based in the eastern U.S. to be able to ... transit across the ocean back and forth to [keep the same number on] patrol in the Med," he said.
>
> The U.S. official said the United States was committed to having at least one ship on station at all times in the eastern Mediterranean, where their anti-missile missiles would be most effective. Having them based in Rota would enable more than one to be in the eastern Mediterranean as needed.
>
> The ships also would be part of the pool of vessels available to participate in standing NATO maritime groups, which are used to counter piracy and for other missions, he said.[107]

An October 10, 2011, press report stated the following:

> "Our plan is to have the first couple [of ships] there in 2014 and the next two in about 2015," said Cmdr. Marc Boyd, spokesman for [U.S. Navy] 6th Fleet. Boyd added: "It's really early in the process and we haven't selected any of the ships yet." Boyd said the shift will bring an estimated 1,300 sailors and Navy civilians and 2,100 dependents to Naval Station Rota, which would double the base's ranks. Naval Station Rota spokesman Lt. j.g. Jason Fischer said the base now has 1,067 sailors....
>
> The three piers at the base primarily support Navy ships passing through on port calls.
>
> Boyd said 6th Fleet is considering plans to add base infrastructure and maintenance facilities to support the ships, as well as additional housing for crews, "but the base is pretty suited as it is now."[108]

APPENDIX C. ALLIED PARTICIPATION AND INTEREST IN AEGIS BMD PROGRAM

This appendix presents additional background information on allied participation and interest in the Aegis BMD program for countries other than Japan.

An October 12, 2018, press report states:

---

opinions_78838.htm. See also "*SECDEF Announces Stationing of Aegis Ships at Rota, Spain*," accessed October 6, 2011, at http://www.navy.mil/ search/display.asp?story_id=63109.

[107] David Brunnstrom and David Alexander, "Spain to Host U.S. Missile Defense Ships," *Reuters*, October 5, 2011. Ellipsis as in original.

[108] Sam Fellman, "U.S. To Base Anti-Missile Ships in Spain," *Defense News*, October 10, 2011: 76.

The South Korean military has decided to buy ship-based SM-3 interceptors to thwart potential ballistic missile attacks from North Korea, a top commander of the Joint Chiefs of Staff revealed Oct. 12.

"The decision was made actually during a top JCS meeting in September last year," said Maj. Gen. Kim Sun-ho, the head of Joint Chiefs of Staff's force buildup planning bureau, in response to a lawmaker's question about the SM-3 missile procurement.

"The type of the ship-based anti-ballistic missile to be procured is an SM-3 class," Kim said during a parliamentary audit of the JCS. "The interceptor will be responsible for shooting down an incoming ballistic missile in the upper tier of the KAMD system."

KAMD refers to the Korea Air and Missile Defense network designed to take down low-flying missiles in the terminal phase. For lower-altitude interceptions, American-built Patriot missiles and locally developed medium-range surface-to-air missiles, dubbed M-SAM, have been deployed in the field....

The South Korean military believes SM-3 interceptors will be effective against an electromagnetic pulse attack originating from a high altitude.

The timetable for adopting the SM-3 has not been laid out, according to the JCS. In the meantime, a preliminary study on the procurement of SM-class interceptors is underway.[109]

An October 3, 2016, press report states that MDA is examining how allied countries in Europe could be brought into the European Aegis missile defense architecture. The report states that

MDA is studying how the Netherlands' new SMART-L long-range naval radar could be integrated into U.S. ballistic missile defense architectures, namely the Aegis Ashore system in Europe, according to Rear Adm. Johnny Wolfe, the program executive officer for Aegis BMD at MDA. He said the agency is also looking at how to loop the United Kingdom's Type 45 destroyers and Spain's Aegis destroyers—which do not have BMD capabilities of their own—into the U.S. network.[110]

A September 6, 2016, press report states:

A trio of planned South Korean guided missile destroyers will be built with the capability to intercept ballistic missile threats, USNI News has learned.

The addition of the capability will give the Republic of Korea (RoK) Navy a powerful organic BMD capability in addition to U.S. Army ground-based interceptors peppered throughout South Korea.

Under the plan, the three remaining ships in the Sejong the Great-class will be able to simultaneously intercept traditional air warfare threats while adding a ballistic missile

[109] Jeff Jeong, "South Korea to Buy Ship-Based Interceptors to Counter Ballistic Missile Threats," *Defense News*, October 12, 2018.

[110] Justin Doubleday, "MDA Looking to Integrate European Allies into Missile Defense Architecture," *Inside the Navy*, October 3, 2016. See also Ellen Mitchell, "Missile Defense Agency Seeks More Aegis Integration in Europe," *Politico Pro Defense Report*, September 30, 2016.

defense capability through a series of hardware and software upgrades over the current class of ship, several sources confirmed to USNI News.

The destroyers will be fitted with the U.S. Navy's Baseline 9 version of the Aegis Combat System that combines modern computing architecture to allow the ship's AN/SPY-1D(v) radar to detect and track aircraft, cruise missiles and ballistic missiles at the same time.

The capability will likely be paired with Raytheon Standard Missile 3 BMD interceptors the ships can pair with the combat system to detect and destroy medium-range ballistic missile threats. Several Korean press outlets have reported the military is seeking to install SM-3s on the three new ships.

Officials with Aegis combat system developer Lockheed Martin told USNI News the new Korean ships would have an "integrated air and missile defense" (IAMD) capability installed aboard but would not elaborate on any other details of the combat system.[111]

An October 26, 2015, press report states:

The U.S. Navy and its NATO counterparts are discussing how to make maritime ballistic missile defense (BMD) training a routine event in Europe, in the hopes that countries will grow more comfortable working with one another in this warfare area and even invest in greater capabilities, the head of American ballistic missile defense in Europe told USNI News.

Last week's Maritime Theater Missile Defense (MTMD) Forum Integrated Air and Missile Defense (IAMD) At Sea Demonstration [i.e., the October 20, 2015 Aegis BMD flight test] was the first of its kind but will not be the last—the U.S. Navy is both planning a 2016 follow-up to coincide with the annual Rim of the Pacific (RIMPAC) exercise, and working with NATO to develop an ongoing maritime ballistic missile defense exercise program, Capt. Jeffrey Wolstenholme, commodore of Task Force 64, told USNI News in an interview from aboard USS Ross (DDG-71) in the U.S. 6th Fleet area of operations.

Wolstenholme said BMD had for a long time been considered a land-based mission set. The U.S. Army and Air Force, as well as their counterparts in Europe, have a variety of assets across the continent to track and engage incoming missiles – including the Raytheon Patriot surface-to-air missile system and the Lockheed Martin Terminal High-Altitude Area Defense (THAAD) system.

"The (MTMD) forum was started because of the emphasis that was starting to be placed on maritime ballistic missile defense," he said.

"We have Patriot missile defense capabilities, THAAD missile defense capabilities that are primarily in the Army and Air Force realm. Maritime has always kind of played second fiddle to that, but with the advent of the Aegis ship and what we have brought

[111] Sam LaGrone, "New South Korean Destroyers to Have Ballistic Missile Defense Capability," *USNI News*, September 6, 2016. See also Sam LaGrone, "Report: South Kore Wants BMD Capability for Guided Missile Destroyers," *USNI News*, August 15, 2016; Bradley Perrett and Kim Minseok, "Adding Interceptors, SM-3 May Join Missile Defenses in South Korea," *Aviation Week & Space Technology*, August 15-28, 2016: 49.

forward with the ballistic missile defense capability within in the U.S. Navy, now maritime is really coming to the forefront.

"And the other nations are starting to get involved in this warfare area as well," he continued.

"We're seeing a lot of development in the Netherlands. The Spanish are showing a lot of interest, as well as the United Kingdom and the Italians. And to some degree the French, who have been watching this."

Though NATO is not affiliated with the MTMD Forum, most of the 10 forum members are in NATO—Australia, Canada, France, Germany, Italy, The Netherlands, Norway, Spain, United Kingdom and the United States. Australia did not participate in the demo and Germany sent personnel to support the exercise but not any military platforms.

NATO is in the midst of discussions about how to improve theater missile defense, Wolstenholme said, and was watching the nine-country live fire demonstration closely.

"There's a lot of discussion going on throughout the NATO community. In fact, just earlier this month there was a conference in Spain ... and there was a lot of discussion about where do we go next after this At-Sea Demo in developing an exercise program," he said.

"And there's several proposals being discussed right now to figure out how we get this stood up and make it more mature."....

The exercise included the first launch of a Standard Missile-3 in Europe, and securing the region for the ballistic missile target launch and the SM-3 intercept was no easy undertaking—commercial air traffic in and out of Europe typically flies right over the Hebrides Range in Scotland and had to be diverted to the south, and U.S. Navy P-3s and P-8s and U.K. E-3Ds scanned the water to ensure the seas were clear of all boat traffic....

Mary Keifer, Lockheed Martin's Aegis in-service and fleet readiness program director, said after the at-Sea demonstration that the company was working with NATO and MTMD Forum members to improve their ships on a budget. After working with the Spanish Navy in 2007 to demonstrate a carry-on/carry-off temporary solution to help Spain's Aegis-equipped ships track ballistic missiles, Keifer said the company again worked with Spain ahead of the demonstration to do a partial upgrade to some Aegis BMD tracking capabilities.[112]

A July 28, 2014, press report states:

The Italian navy is working to develop the ballistic missile defense (BMD) capability of its Orizzonte-class air-defense ships and pave the way for BMD systems to be installed on a new class of ship to be launched in the early 2020s.

Software engineers at the Italian navy's programming center—known as Maricenprog— near the navy's main dockyard at Taranto, have been developing tactical

[112] Megan Eckstein, "NATO Hopes To Boost Collective Maritime BMD Capability Through Exercises, Investments," *USNI News*, October 26, 2015.

BMD capabilities for the ship as part of the country's participation in the wider NATO tactical BMD program. The Italian defense ministry supports the effort with the land-based TPS-77 radar system and the SAMP-T ground-based air defense system, but wants to back up these efforts at sea with the Orizzonte or Horizon-class ships.

According to Gianpaolo Blasi, director of Maricenprog, the program has already completed two of what NATO describe as Ensemble Tests (ET), which pave the way for entry into the NATO BMD program. The navy is preparing for a trial due to take place in 2015 that will see the Orizzonte-class vessel ITN Doria supporting and defending another—as yet unconfirmed—BMD-capable ship that will track and potentially engage a ballistic missile target. During the trials the Doria will act as shotgun, defending the missile-tracking vessel from conventional air threats that the other ship cannot deal with as it tracks the ballistic missile.

The Doria will be able to transmit details of the engagement around the fleet through a tactical data link modified to carry BMD data.[113]

A June 13, 2014, press report states:

Talks between the U.S. and Australia have given fresh momentum to Washington's plans to create a larger ballistic-missile defense shield for its allies in Asia.

According to a U.S. statement overnight, discussions between President Barack Obama and visiting Australian Prime Minister Tony Abbott resulted in a commitment from Canberra for help in pushing forward with expanded missile-defense plans as a counter to North Korea....

Washington's statement on Thursday [June 12] said the U.S. was now examining ways for Australia to participate in a bigger regional system using the country's coming fleet of missile destroyers equipped with advanced Aegis radar capability.

"We are...working to explore opportunities to expand cooperation on ballistic missile defense, including working together to identify potential Australian contributions to ballistic-missile defense in the Asia-Pacific region," the U.S. statement said.

Australia is building a new fleet of warships that could be equipped to shoot down hostile missiles, as part of an ambitious military buildup that includes investments in new stealth-fighter aircraft, cruise missiles, amphibious carriers and submarines. The revamp will cost close to 90 billion Australian dollars (US$85 billion) over a decade.

"This might mean the Australian Defence Force could end up mounting advanced missiles on its Aegis-equipped air-warfare destroyers," said security analyst James Brown of Australia's Lowy Institute.[114]

---

[113] Tony Osborne, "Italian Navy Paves Way For Ship-Based BMD Capability," *Aerospace Daily & Defense Report*, July 28, 2014: 1-2.

[114] Rob Taylor, "U.S. and Australia to Cooperate on Asian Missile-Defense Plans," *Wall Street Journal* (http://online.wsj.com), June 13, 2014.

A September 16, 2013, press report states:

One of the UK Royal Navy's new Type 45 destroyers is conducting tests to establish whether the warships could provide British forces with theater ballistic-missile defense (TBMD) capabilities for the first time, according to the head of the Royal Navy.

First Sea Lord Adm. Sir George Zambellas said during a speech to industry executives and military personnel on the opening day of the DSEi defense exhibition that the "type is on trials in the Pacific to explore the ballistic-missile defense capabilities that are ready to be exploited, bringing strategic opportunities to the vessel."

The Type 45 destroyer Daring, one of six Type 45s built by BAE Systems for the Royal Navy, has been in the Pacific for several weeks, having departed its Portsmouth base this summer for a wide-ranging nine-month deploy-ment, which the Royal Navy said in May would include science and technology trials. The work is being done as part of a US Missile Defense Agency (MDA) research and development test....

In May, the UK Defence Ministry confirmed it was talking to Aster 30 partners France and Italy about developing an extended-range version of a missile already used by the French and Italian armies to intercept incoming missiles While there is no program to adapt the Type 45 to include TBMD capability, the trials support the possibility of such a move once a decision whether to go down that route is made by the British government.[115]

A March 18, 2013, press report states:

Raytheon has discussed a possible pooling arrangement with three navies in northern Europe to make its SM-3 ballistic missile inter-ceptor more affordable, according to a senior company executive.

Speaking after a successful test of a new data link enabling the SM-3 to communicate with X-band radars operated by Dutch, Danish and Ger-man warships, George Mavko, director of European missile defense at Raytheon Missile Systems, said the idea of a pooling arrangement had been raised by the company, even though none of the countries are pursuing procurement at this point....

While all three European navies have expressed an interest in the capability of the SM-3 to engage ballistic missiles at ranges outside the atmosphere, none appear close to actually procuring the missiles....

Instead, led by the Dutch, the initial moves appear focused on updating naval X-band radars and other systems so they can provide target data to SM-3 missiles even if they can't prosecute their own attack....

Aside from the pooling idea, Raytheon also recently opened discussions with the U.S. Missile Defense Agency over co-production of SM-3 systems in Europe to sweeten any future deal, Mavko said....

---

[115] Andrew Chuter, "UK Royal Navy Examines BMD Capabilities," *Defense News*, September 16, 2013: 38.

Small bits of the missile are already produced in Europe, although it was "too early to imply the U.S. is willing to release any major subsystems to other countries for co-production," Mavko said....

Raytheon has been cooperating with the Dutch Navy for several years, exploring the potential of the SM-3 to talk to X-band radars. The Dutch have co-funded a study with the U.S. government on the feasibil-ity of a dual-band data link; the study is due to be extended into a second phase. The German government has agreed to participate this time.[116]

A March 11, 2013, press report states:

The Eurosam SAMP/T surface-to-air missile system has destroyed a representative theater ballistic missile during a test in France.

The March 6 test saw a joint Italian and French team engage an aircraft-launched target using an Aster 30 missile fired from the Biscarosse missile test center on the Bay of Biscay coast.

According to French government defense procurement agency the DGA, the operational evaluation firing was jointly carried out by the Italian 4th Artillery Regiment of Mantova with the French military airborne test center (CEAM) of Mont-de-Marsan. In a change from previous interceptions, the SAMP/T used Link 16 data links to provide target information. The test also was the first to use what Eurosam calls a NATO environment in terms of command and control of the weapon, rather than simply using French sensors.

The company says the firing was as "close to what would be an operational use for an anti-theater ballistic missile mission under the aegis of the alliance Active Layered Theater Ballistic Missile Defense program."

The company adds, "The NATO Ballistic Missile Defense Operations Cell, located in Ramstein, Germany, was in the loop via Link 16 network."[117]

Another March 11, 2013, press report states:

Joint US and European testing of command, control, communications and radar systems are underway to demonstrate the feasibility of integration of European radars and command and control systems into a future missile defense systems based on the planned European Phased Adaptive Approach (EPAA) utilizing the several AEGIS destroyers or cruisers to be based in Spain, land-based SM-3 interceptors to be stationed in Romania and Poland, along with SPY-2 radars sites. These assets are to be complemented by a number of European deployed radar sites.

In recent weeks tests were carried out to evaluate such integration. Last week Raytheon reported about a recent trial that showed that a radar used by Dutch, German

[116] Andrew Chuter, "Raytheon Pushes European SM-3 Missile Pool," *Defense News*, March 18, 2013: 4.

[117] Tony Osborne, "European SAMP/T Destroys Ballistic Missile In Test," *Aerospace Daily & Defense Report*, March 11, 2013: 3.

and Danish navies could provide target information to the interceptor. The current radar installed on the Dutch frigates is incompatible with the A EGIS/SM -3 link operating over S-band. The demonstration which took place at the Den Helder military test range validated a datalink that al lows the missile to receive information from the Thales sensor while retaining the ability to communicate with Aegis combat ships used by the U.S. Navy. Generally, The Dutch, German and Danish navies datalinks are operating on X bands, while Norway, Spain and the U.S. operate A EG IS frigates communicating with their interceptors over the S band. To avoid unique configurations of missiles, Raytheon has developed a dual -band datalink which enables the same missile to communicate in both bands. This dual -band datalink was first tested in 2011.[118]

A March 8, 2013, press report states:

The British Royal Navy is exploring the possibility of outfitting its newest class of destroyers with a ballistic missile defense capability.

The Defence Ministry said this week it wants to examine the potential for the Type 45 destroyers to play a role in defending the United Kingdom and al lies from the threat of ballistic missiles. The ministry said it will build on its relationship with the Pentagon's Missile Defense Agency to look at the option....

The joint Defence Ministry and industry-run U.K. Missile Defence Center (M DC) plans to take part in a trial that for the first time will use a Type 45 in a research and development program with their American counterparts.

That will involve testing the Sampson radar, which is part of the Sea V i per missile system, in detecting and tracking ballistic missiles, the ministry said.

The is no program to deploy ballistic missile defense on Type 45s but the M DC has in recent years been exploring the option for the destroyers.

"It will be a step change to be able to work so closely with such a ship in an emerging area of defense," MDC head Simon Pavitt said in a statement. "Working with an operational platform will make a significant difference to our level of understanding and could contribute both financially and technically towards any future program."[119]

An October 2012 article stated:

The Royal Netherlands Navy's (RNLN's) four De Zeven Provincien-class LCF air defence and command frigates are to receive a substantially upgraded and rearchitectured SMART-L D-band volume search radar that will give the ships a ballistic missile defence (BMD) early warning capability.

---

[118] Tamir Eshel, "Integrating European Radars with AEGIS/SM-3 Missile Defenses," *Defense Update* (http://defenseupdate.com), March 11, 2013, accessed March 20, 2013, at http://defense-update.com/20130311_integrating-europeanradars-with-aegissm-3-missile-defenses.html.

[119] Mike McCarthy, "U.K. Examining Sea-Based Missile Defense," *Defense Daily*, March 8, 2013: 10. See also "British Destroyer to Participate in U.S. Missile Defense Trials," *Defense Update* (http://defense-update.com), March 7, 2013, accessed March 27, 2013, at http://defense-update.com/20130307_british-destroyer-to-participate-in-u-smissile-defense-trials.html.

Thales Nederland received a EUR116 million (USD145 million) contract from the Netherlands' Defence Materiel Organisation (DMO) in June 2012 for the new extended-range sensor known as 'SMART-L EWC'. This new variant of SMART-L, which builds on the results of a previous Extended Long Range (ELR) capability demonstration, will push instrumented range out to 2,000 km; improve elevation coverage; introduce new wave forms and processing optimised for the detection and tracking of very-high-velocity ballistic missile targets at altitude; and enable estimation of trajectories, launch sites and points of impact. At the same time, all SMART-L volume air search functionality will be retained.[120]

A journal article published in the summer of 2012 states:

Today the steady growth of Aegis-capable ships in the U.S. Navy—as well as an increasing number of world navies fielding such ships—presents new opportunities and challenges....

... the Aegis BMD capabilities present in the navies of U.S. allies and friends can now provide the Global Maritime Partnership with a means to address the "high end" of the kill chain with combined, coordinated, ballistic-missile defense: the Aegis BMD Global Enterprise.

This potential is already manifest in the Asia-Pacific region in the close working relationship between the United States and Japan. Korea and Australia could well join this Aegis network soon, giving the four governments the means to address not only territorial BMD but also coordinated BMD of fleet units operating together. In Europe, plans are well along to provide robust territorial defense of European nations with ALTBMD [active layered theater BMD] and the EPAA. Together, these systems provide a nascent BMD capability today and promise an even more robust capability as the EPAA evolves over the next decade and a half.

But as demonstrated in Iraq, Afghanistan, and now Libya, NATO and the nations of Europe have equities often well beyond the territorial boundaries of the European continent. Also, a European military deployed beyond Europe's borders will always have a naval component. This is therefore a propitious time to begin to link European allies more completely into an Aegis BMD Global Enterprise in much the same way the U.S. Navy is linked to its Asia-Pacific partners—Japan today, Korea soon, and thereafter Australia in the near future—in a high-end Aegis BMD Global Maritime Partnership....

The diffusion of Aegis BMD capability abroad is occurring quietly. Governments that have made naval force-structure investment decisions based primarily on inwardly focused national interests have discovered that their investments also enable them to combine their resources in collective defense....

This effort to create a broad BMD enterprise builds on the current participation of allied navies in the Aegis program. This global effort started with a foreign military sales relationship with Japan, subsequently expanded to relationships with Australia and

[120] Kate Tringham, "Warning Signs: Netherlands Evolves SMART-L Radar For Ballistic Missile Defence Mission," *Jane's International Defence Review*, October 2012: 28-29.

Korea, and now includes a commercial connection with Spain as well as an enterprise between Norway and Spain.22 Several other states have expressed interest in acquiring the Aegis weapon system and Aegis BMD. Importantly, Australia and other countries that are acquiring the Aegis system are stipulating that the systems they buy must have the capability of adding BMD in the future....

In Europe, the decision as to whether and how to connect the European NATO allies' short- and medium-range theater missile-defense systems to the U.S. long-range missile defense system will be critical to the coherence of alliance-wide BMD. A high level of commitment to international partnership on the parts of both the United States and its allies—already evinced by ALTBMD and C2BMC shared situational-awareness tests—will encourage interoperability initiatives. This interoperability will, in turn, help ensure the success of the U.S. Phased Adaptive Approach....

Close cooperation in the area of Aegis BMD between the United States and Japan, possibly Korea, and potentially Australia does not in itself qualify as an "Aegis BMD Global Enterprise." But to include European nations in an Aegis-afloat enterprise of capabilities approaching those planned for the ALTBMD/EPAA system would....

European navies are now deployed worldwide fulfilling the vision of a Global Maritime Partnership: supporting operations in Iraq and Afghanistan, fighting in Libya, conducting antipiracy patrols in the Horn of Africa and elsewhere, and supporting humanitarian assistance operations around the world. There could be no more propitious time to begin to link more completely European allies in an Aegis BMD Global Enterprise, in much the same way the U.S. Navy is now linked to its Asia-Pacific partners in a high-end Aegis BMD Global Maritime Partnership....

But it is unlikely that such a venture would succeed without ongoing U.S. leadership, the same sort of leadership that is supporting sea-based Aegis BMD for territorial and fleet ballistic-missile defense today in the northeast Pacific as well as sea-based and land-based ballistic territorial missile defense in Europe. Clearly, U.S. leadership could be what accelerates the morphing of a now-nascent Aegis BMD Global Enterprise in Europe into a global Aegis BMD afloat capability....

There is a growing worldwide commitment to Aegis ballistic-missile defense, a commitment with broad potential to field an international global enterprise capable of defending against the most imminent, and growing, threat to nations and navies, on land and at sea alike—the threat of ballistic missiles, particularly those armed with weapons of mass destruction.[121]

A May 7, 2012, press report states:

The German Navy's fleet of frigates could be upgraded to deploy Raytheon's [RTN] Standard Missile-3 to participate in NATO's ballistic missile defense program if the modifications were approved by the government, Germany's top naval officer recently said.

[121] Brad Hicks, George Galdorisi, and Scott C. Truver, "The Aegis BMD Global Enterprise," *Naval War College Review*, Summer 2012: 65-80.

Vice Admiral Axel Schimpf, the counterpart to the U.S. Navy's chief of naval operations, said in a recently published article that the F124 frigates are capable of being upgraded to play a vital role in ballistic missile defense (BMD).

"The German Navy, with the F124 Frigates in their current configuration, has a weapon system at their disposal which forms the basis for capability enhancements for (German) armed forces' participation in various roles," according to a translation of an article he penned in Marine Forum, a publication of the German Maritime Institute.

One option, Schimpf said, would be to upgrade the F124s' SMART-L and Active Phased Array Radar (APAR) combat management system, along with the Mk-41 vertical launch system to accommodate the SM-3....

The enhancements would be one way for Germany to participate in the Obama administration's European Phased Adaptive Approach (EPAA) embraced by NATO, and could be done in cooperation with Denmark or the Netherlands, Schimpf said....

The German government has not made on decisions on whether to adapt its frigates for ballistic missile defense, and Germany's role in EPAA is the source of ongoing political discussions in Berlin ahead of NATO's May 20-21 summit in Chicago....

Only a handful of NATO allies deploy the Aegis combat system on ships, and Germany is not one of them. Germany's combat system does not operate on an S-band frequency used on Aegis. Raytheon, however, says it has developed a duel band data link that would allow the combat system on allied ships to talk to the SM-3 and guide it to targets.[122]

An October 3, 2011, press report stated that:

The Netherlands, which has had a longtime interest in a missile shield, is pressing ahead to build up its own capabilities. The Dutch defense ministry plans to expand the capabilities of the Thales Smart-L radar on Dutch frigates to take on BMD roles. The program's value is estimated at €100-250 million, including logistics support and spares.

Other European navies using the sensor may follow the Dutch lead.

Dutch Defense Minister Hans Hillen notes that the Smart-L effort would help address the BMD sensor shortage within the NATO alliance. Citing NATO's decision last year to take a more expansive approach to BMD, Hillen says Smart-L could give the ALTBMD [Active Layered Theater BMD] command-and control backbone the required long-range target-detection analysis to help identify where a threat originates.

The Netherlands has already carried out a sensor trial for the expanded role in cooperation with the U.S. Navy. The move does not include the purchase of Raytheon Standard Missile SM-3 interceptors.

Both hardware and software modifications to the combat management system are needed. All four [of the Dutch navy's] De Zeven Provincien-class frigates would be modified to ensure that two can be deployed, even as one is in maintenance and the fourth is being readied for operations.

---

[122] Mike McCarthy, "Raytheon's SM-3 An Option For German Role In Missile Defense, Admiral Says," *Defense Daily*, May 7, 2012: 9.

Thales is due to complete a series of studies to prepare for the acquisition of the upgrade in the third quarter of 2012. The goal is to have the first frigates ready for operations by 2017. All four should be upgraded by the end of that year.

Although the Netherlands is leading the program, other Smart-L users, including the German navy and Denmark, have been monitoring the effort. France also has shown interest in the system, Hillen said in a letter to legislators.

France also wants to upgrade its Aster 30 interceptor to give it a basic BMD capability, although a formal contract has not been awarded....

Raytheon, meanwhile, is still fighting to win a foothold for its Standard Missile 3 (SM-3) in Europe. The company continues its push to persuade continental navies to embrace the SM-3 Block 1B for missile defense roles, and says it has largely validated the dual-mode data link that would be key to the concept.

The data link would feature both S- and X-band capability—the former to support the Aegis radar system used by the U.S. and others, and the latter for the Smart-L/APAR (active phased array radar) combination used, for instance, by the Dutch navy.[123]

A September 2011 press report states:

The gulf in sea-based ballistic missile defence (BMD) capability between the navies of NATO's European member states and the US Navy (USN) was brought into stark relief by the recent deployment of the Ticonderoga-class cruiser USS Monterey to the Mediterranean and Black Sea region, as the first element of the United States' European Phased Adaptive Approach (EPAA) for missile defence....

However, this situation is about to change as European NATO nations are committing their naval assets to BMD in response to evolving alliance policy towards developing a BMD architecture to protect the continent from perceived threats emanating from the Middle East.

NATO embarked on an Active Layered Theatre Ballistic Missile Defence System (ALTBMDS) programme in September 2005, following a two-year feasibility study. Its initial focus was the protection of deployed alliance forces and high-value assets against short- and medium-range threats. At the November 2010 Lisbon Summit, political leaders from NATO states committed to expanding that remit to include the defence of the alliance's European territory.

ALTBMD is providing a C2 framework on which to build a scalable and adaptable BMD 'system of systems' architecture, integrating new national systems as they are committed to the alliance and enabling a complete lower- and upper-layer capability covering Europe to be fielded. The first of these, Capability 1, with initial operational capability planned for the 2012 timeframe, integrates C2 infrastructure, sensors and

---

[123] Robert Wall, Amy Svitak, and Amy Butler, "Supporting Role," *Aviation Week & Space Technology*, October 3, 2011: 28-29. A shorter version of the story was published as Robert Wall, "Dutch Press Forward On Ship-Based Missile Defense Effort," *Aerospace Daily & Defense Report*, September 27, 2011: 4. See also Menno Steketee, "Dutch Frigates to Gain BMD Capability," *Jane's Navy International* (*Janes.com*), September 28, 2011. (The print version of the report appeared under the same article title in the November 2011 issue of *Jane's Navy International*, page 8.

ground-based Patriot interceptors. The expansion to provide upper-layer defence is due to achieve full operational capability between 2015 and 2016.

The US contribution to this architecture is the EPAA set out by the Obama administration in September 2009....

There is evidence that the EPAA has acted as a spur for some European nations to make a more coherent contribution to the NATO BMD construct, particularly in the maritime domain, as they seek to maintain sovereignty in the development and integration of indigenous BMD systems and defence of their territories.

A number of classes of the latest generation of anti-air warfare (AAW) combatants with the potential to acquire a BMD capability are either operational or entering service in the navies of Denmark, France, Germany, Italy, the Netherlands, Norway, Spain and the UK. These offer the attributes of flexibility in deployment, mobility and sustainability inherent in naval platforms and could operate as effective sensor nodes even without an organic intercept capability.

They would be able to forward deploy close to the origin of the threat and act as force multipliers in this role by providing early warning of launches and cueing of off-board interceptor systems with the provision of timely and accurate impact point prediction and missile tracks, together with launch point prediction for counter-targeting.[124]

APPENDIX D. TARGET FOR SIMULATING ENDO-ATMOSPHERIC FLIGHT OF DF-21 ASBM

A past oversight issue for Congress concerns the lack of a target for simulating the endo-atmospheric (i.e., final) phase of flight of China's DF-21 anti-ship ballistic missile. DOD's Director, Operational Test and Evaluation (DOT&E), in a December 2011 report (DOT&E's annual report for FY2011), stated the following:

Anti-Ship Ballistic Missile Target

A threat representative Anti-Ship Ballistic Missile (ASBM) target for operational open-air testing has become an immediate test resource need. China is fielding the DF-21D ASBM, which threatens U.S. and al lied surface warships in the Western Pacific. While the Missile Defense Agency has exo-atmospheric targets in development, no program currently exists for an endo-atmospheric target. The endo-atmospheric ASBM target is the Navy's responsibility, but it is not currently budgeted. The Missile Defense Agency estimates the non-recurring expense to develop the exo-atmospheric target was $30 million with each target costing an additional $30 million; the endo-atmospheric target will be more expensive to produce according to missile defense analysts. Numerous Navy acquisition programs will require an ASBM surrogate in the coming

[124] Charles Hollosi, "European Fleets Respond to Ballistic Missile Threats," *Jane's Navy International*, September 2011: 23-24, 26-30.

years, although a limited number of targets (3-5) may be sufficient to validate analytical models.[125]

A February 28, 2012, press report stated the following:

"Numerous programs will require" a test missile to stand in for the Chinese DF-21D, "including self-defense systems used on our carriers and larger amphibious ships to counter anti-ship ballistic missiles," [Michael Gilmore, the Pentagon's director of operational test and evaluation] said in an e-mailed statement....

"No Navy target program exists that adequately represents an anti-ship ballistic missile's trajectory," Gilmore said in the e-mail. The Navy "has not budgeted for any study, development, acquisition or production" of a DF-21D target, he said.

Lieutenant Alana Garas, a Navy spokeswoman, said in an e-mail that the service "acknowledges this is a valid concern and is assessing options to address it. We are unable to provide additional details."...

Gilmore, the testing chief, said his office first warned the Navy and Pentagon officials in 2008 about the lack of an adequate target. The warnings continued through this year, when the testing office for the first time singled out the D F-21D in its annual public report....

The Navy "can test some, but not necessarily all, potential means of negating anti - ship ballistic missiles," without a test target, Gilmore said.[126]

The December 2012 report from DOT&E (i.e., DOT&E's annual report for FY2012) did not further discuss this issue; a January 21, 2013, press report stated that this is because the details of the issue are classified.[127]

A December 16, 2016, press report states the following (emphasis added):

The Missile Defense Agency (MDA) said its new Sea Based Terminal (SBT) system achieved its second ballistic missile intercept during a Dec. 14 test over the Pacific Ocean.

During the test, the USS John Paul Jones (DDG-53)... fired a salvo of two Raytheon [RTN] Standard Misisle-6 (SM-6) interceptors in immediate succession against a medium-range ballistic missile target launched from the Pacific Missile Range Facility on Kauai, Hawaii. The first interceptor was not armed and was designed to collect test data, MDA said. The second interceptor, which carried an explosive warhead, intercepted the Lockheed Martin-built target....

MDA called the target "complex" but declined to elaborate. However, according to the Missile Defense Advocacy Alliance, the target emulated China's Dong-Feng 21 (DF-

---

[125] Department of Defense, Director, Operational Test and Evaluation, *FY 2011 Annual Report*, December 2011, p. 294.

[126] Tony Capaccio, "Navy Lacks Targets To Test U.S. Defenses Against China Missile," *Bloomberg Government* (*bgov.com*), February 28, 2012.

[127] Christopher J. Castelli, "DOD Testing Chief Drops Public Discussion Of ASBM Target Shortfall," *Inside the Navy*, January 21, 2013.

21), a ballistic missile equipped with a maneuverable re-entry vehicle and designed to destroy U.S., aircraft carriers.

The event, designated Flight Test Standard Missile-27 (FTM-27), was SBT's first salvo test and its second intercept in as many tries.[128]

[128] Marc Selinger, "Missile Defense Agency Scores Second Intercept With Sea Based Terminal System," *Defense Daily*, December 16, 2016: 3-4.

In: Conventional Weapons Systems
Editor: Stewart Keller
ISBN: 978-1-53615-414-6

Chapter 3

COAST GUARD POLAR SECURITY CUTTER (POLAR ICEBREAKER) PROGRAM: BACKGROUND AND ISSUES FOR CONGRESS*

Ronald O'Rourke

SUMMARY

The Coast Guard Polar Security Cutter (PSC) program, previously known as the polar icebreaker (PIB) program, is a program to acquire three new heavy polar icebreakers, to be followed years from now by the acquisition of up to three new medium polar icebreakers. The Coast Guard wants to begin construction of the first new heavy polar icebreaker in FY2019 and have it enter service in 2023. The PSC program has received about $359.6 million in acquisition funding through FY2018, including $300 million provided through the Navy's shipbuilding account and $59.6 million provided through the Coast Guard's acquisition account. The Coast Guard's proposed FY2019 budget requests $750 million in Coast Guard acquisition funding for the program.

The acquisition cost of a new heavy polar icebreaker had earlier been estimated informally at roughly $1 billion, but the Coast Guard and Navy now believe that three heavy polar icebreakers could be acquired for a total cost of about $2.1 billion, or an average of about $700 million per ship. The first ship will cost more than the other two because it will incorporate design costs for the class and be at the start of the production learning curve for the class. When combined with the program's $359.6 million in prior-year funding, the $750 million requested for FY2019 would fully fund the procurement of the first new heavy polar icebreaker and partially fund the procurement of the second.

The operational U.S. polar icebreaking fleet currently consists of one heavy polar icebreaker, *Polar Star*, and one medium polar icebreaker, *Healy*. In addition to *Polar Star*, the Coast Guard has a second heavy polar icebreaker, *Polar Sea*. *Polar Sea*,

* This is an edited, reformatted and augmented version of Congressional Research Service Publication No. RL34391, dated Updated December 10, 2018.

however, suffered an engine casualty in June 2010 and has been nonoperational since then. *Polar Star* and *Polar Sea* entered service in 1976 and 1978, respectively, and are now well beyond their originally intended 30-year service lives. The Coast Guard has used *Polar Sea* as a source of spare parts for keeping *Polar Star* operational.

A Department of Homeland Security (DHS) Mission Need Statement (MNS) approved in June 2013 states that "current requirements and future projections ... indicate the Coast Guard will need to expand its icebreaking capacity, potentially requiring a fleet of up to six icebreakers (3 heavy and 3 medium) to adequately meet mission demands in the high latitudes...."

The current condition of the U.S. polar icebreaker fleet, the DHS MNS, and concerns among some observers about whether the United States is adequately investing in capabilities to carry out its responsibilities and defend its interests in the Arctic, have focused policymaker attention on the question of whether and when to acquire one or more new heavy polar icebreakers as replacements for *Polar Star* and *Polar Sea*.

On March 2, 2018, the U.S. Navy, in collaboration with the U.S. Coast Guard under the polar icebreaker integrated program office, released a request for proposal (RFP) for the advance procurement and detail design for the Coast Guard's heavy polar icebreaker, with options for detail design and construction for up to three heavy polar icebreakers.

Issues for Congress for FY2019 for the PSC program include, inter alia, whether to approve, reject, or modify the Coast Guard's FY2019 acquisition funding request; whether to use a contract with options or a block buy contract to acquire the ships; whether to continue providing at least some of the acquisition funding for the PSC program through the Navy's shipbuilding account; technical, schedule, and cost risk in the PSC program; and whether to procure heavy and medium polar icebreakers to a common basic design.

INTRODUCTION

This report provides background information and issues for Congress on the Coast Guard's program for acquiring new polar icebreakers. The program was previously known as the polar icebreaker (PIB) program. It is now called the Polar Security Cutter (PSC) program, though many observers may continue, as a matter of convenience, to refer to it as the polar icebreaker program. The Coast Guard's proposed FY2019 budget requests $750 million in acquisition funding for the program.

The issue for Congress is whether to approve, reject, or modify the Administration's FY2019 acquisition funding request for the PSC program, and, more generally, whether to approve, reject, or modify the Coast Guard's overall plan for procuring new polar icebreakers. Congress's decisions on this issue could affect Coast Guard funding requirements, the Coast Guard's ability to perform its polar missions, and the U.S. shipbuilding industrial base.

For a brief discussion of the Coast Guard's Great Lakes icebreakers, see *Appendix F*. A separate CRS report covers acquisition of general-purpose cutters for the Coast Guard.[1] Another CRS report provides an overview of various issues relating to the Arctic.[2]

BACKGROUND

Missions of U.S. Polar Icebreakers

The permanent statute that sets forth the Coast Guard's primary duties—14 U.S.C. 2—states that, among other primary duties, the Coast Guard shall "develop, establish, maintain, and operate, with due regard to the requirements of national defense, aids to maritime navigation, icebreaking facilities, and rescue facilities for the promotion of safety on, under, and over the high seas and waters subject to the jurisdiction of the United States," and "pursuant to international agreements, develop, establish, maintain, and operate icebreaking facilities on, under, and over waters other than the high seas and waters subject to the jurisdiction of the United States...."

U.S. polar ice operations support 9 of the Coast Guard's 11 statutory missions.[3] The roles of U.S. polar icebreakers can be summarized as follows:

- conducting and supporting scientific research in the Arctic and Antarctic;
- defending U.S. sovereignty in the Arctic by helping to maintain a U.S. presence in U.S. territorial waters in the region;
- defending other U.S. interests in polar regions, including economic interests in waters that are within the U.S. exclusive economic zone (EEZ) north of Alaska;
- monitoring sea traffic in the Arctic, including ships bound for the United States; and
- conducting other typical Coast Guard missions (such as search and rescue, law
- enforcement, and protection of marine resources) in Arctic waters, including U.S. territorial waters north of Alaska.

---

[1] CRS Report R42567, *Coast Guard Cutter Procurement: Background and Issues for Congress*, by Ronald O'Rourke.

[2] CRS Report R41153, *Changes in the Arctic: Background and Issues for Congress*, coordinated by Ronald O'Rourke.

[3] The nine missions supported by polar ice operations are search and rescue; maritime safety; aids to navigation; ice operations; marine environmental protection; living marine resources; other law enforcement (protect the exclusive economic zone [EEZ]); ports, waterways and costal security; and defense readiness. The two missions not supported by polar ice operations are illegal drug interdiction and undocumented migrant interdiction. (Department of Homeland Security, *Polar Icebreaking Recapitalization Project Mission Need Statement, Version 1.0*, approved by DHS June 28, 2013, p. 10.)

Operations to support National Science Foundation (NSF) research activities in the Arctic and Antarctic have accounted in the past for a significant portion of U.S. polar icebreaker operations.[4] Supporting NSF research in the Antarctic has included performing an annual mission, called Operation Deep Freeze, to break through the Antarctic ice so as to resupply McMurdo Station, the large U.S. Antarctic research station located on the shore of McMurdo Sound, near the Ross Ice Shelf. In terms of the maximum thickness of the ice to be broken, the annual McMurdo resupply mission generally poses the greatest icebreaking challenge for U.S. polar icebreakers, though Arctic ice can frequently pose its own significant icebreaking challenges for U.S. polar icebreakers.

Although polar ice is diminishing due to climate change, observers generally expect that this development will not eliminate the need for U.S. polar icebreakers, and in some respects might increase mission demands for them. Even with the diminishment of polar ice, there are still significant ice-covered areas in the polar regions, and diminishment of polar ice could lead in coming years to increased commercial ship, cruise ship, and naval surface ship operations, as well as increased exploration for oil and other resources, in the Arctic—activities that could require increased levels of support from polar icebreakers, particularly since waters described as "ice free" can actually still have some amount of ice.[5] Changing ice conditions in Antarctic waters have made the McMurdo resupply mission more challenging since 2000.[6]

The Coast Guard's strategy document for the Arctic region, released on May 21, 2013, states that "The United States must have adequate icebreaking capability to support research that advances fundamental understanding of the region and its evolution," and that "The Nation must also make a strategic investment in icebreaking capability to enable access to the high latitudes over the long-term."[7]

Current U.S. Polar Icebreakers

The U.S. polar icebreaker fleet currently includes four ships—three Coast Guard ships and one ship operated by the NSF. The ships are described briefly below.

---

[4] This passage, beginning with "The roles of...", originated in an earlier iteration of this CRS report and was later transferred by the Government Accountability Office (GAO) with minor changes to Government Accountability Office, *Coast Guard[:]Efforts to Identify Arctic Requirements Are Ongoing, but More Communication about Agency Planning Efforts Would Be Beneficial*, GAO-10-870, September 2010, p. 53.

[5] For more on changes in the Arctic due to diminishment of Arctic ice, see CRS Report R41153, *Changes in the Arctic: Background and Issues for Congress*, coordinated by Ronald O'Rourke.

[6] National Research Council, *Polar Icebreakers in a Changing World: An Assessment of U.S. Needs*, Washington, 2007, pp. 6-7, 14, 63.

[7] *United States Coast Guard Arctic Strategy*, Washington, May 2013, p. 35; accessed May 24, 2013, at http://www.uscg.mil/seniorleadership/DOCS/CG_Arctic_Strategy.pdf.

Coast Guard Polar Icebreakers

The Coast Guard's three polar icebreakers—*Polar Star*, *Polar Sea*, and *Healy*—are multimission ships that can break through ice, support scientific research operations, and perform other missions typically performed by Coast Guard ships. Although usually referred to as icebreakers, they are, more generally speaking, Coast Guard cutters.[8]

Heavy Polar Icebreakers Polar Star and Polar Sea

Polar Star (WAGB-10) and *Polar Sea* (WAGB-11),[9] sister ships built to the same general design (*Figure 1* and *Figure 2*), were acquired in the early 1970s as replacements for earlier U.S. icebreakers. They were designed for 30-year service lives, and were built by Lockheed Shipbuilding of Seattle, WA, a division of Lockheed that also built ships for the U.S. Navy, but which exited the shipbuilding business in the late 1980s.

Source: Coast Guard photo accessed at http://www.uscg.mil/pacarea/cgcpolarsea/ history.asp on April 21, 2011.

Figure 1. Polar Star and Polar Sea. (Side by side in McMurdo Sound, Antarctica)

The ships are 399 feet long and displace about 13,200 tons.[10] They are among the world's most powerful nonnuclear-powered icebreakers, with a capability to break through ice up to 6 feet thick at a speed of 3 knots. Because of their icebreaking capability, they are considered heavy polar icebreakers. In addition to a crew of 134, each ship can embark a scientific research staff of 32 people.

---

[8] Cutters are commissioned Coast Guard vessels greater than 65 feet in length.

[9] The designation WAGB means Coast Guard icebreaker. More specifically, W means Coast Guard ship, A means auxiliary, G means miscellaneous purpose, and B means icebreaker.

[10] By comparison, the Coast Guard's new National Security Cutters—its new high-endurance cutters—are about 418 feet long and displace roughly 4,000 tons.

Polar Star was commissioned into service on January 19, 1976, and consequently is now several years beyond its intended 30-year service life. Due to worn-out electric motors and other problems, the Coast Guard placed the ship in caretaker status on July 1, 2006.[11] Congress in FY2009 and FY2010 provided funding to repair *Polar Star* and return it to service for 7 to 10 years; the repair work, which reportedly cost about $57 million, was completed, and the ship was reactivated on December 14, 2012.[12]

Polar Sea was commissioned into service on February 23, 1978, and consequently is also several years beyond its originally intended 30-year service life. In 2006, the Coast Guard completed a rehabilitation project that extended the ship's expected service life to 2014. On June 25, 2010, however, the Coast Guard announced that *Polar Sea* had suffered an engine casualty, and the ship was unavailable for operation after that.[13] The Coast Guard placed *Polar Sea* in commissioned, inactive status on October 14, 2011. The Coast Guard transferred certain major equipment from *Polar Sea* to *Polar Star* to facilitate *Polar Star*'s return to service,[14] and continues to use *Polar Sea* as a source of spare parts for *Polar Star*.

Medium Polar Icebreaker Healy

Healy (WAGB-20) (*Figure 3*) was funded in the early 1990s as a complement to *Polar Star* and *Polar Sea*, and was commissioned into service on August 21, 2000. The ship was built by Avondale Industries, a shipyard located near New Orleans, LA, that built numerous Coast Guard and Navy ships, and which eventually became part of Huntington Ingalls Industries (HII).[15] (HII subsequently wound down shipbuilding activities at Avondale, and the facility is no longer building ships.[16])

Although it is referred to as a medium polar icebreaker, *Healy* is actually a bit larger than *Polar Star* and *Polar Sea*—it is 420 feet long and displaces about 16,000 tons. Compared to *Polar Star* and *Polar Sea*, *Healy* has less icebreaking capability (which is

---

[11] Source for July 1, 2006, date: U.S. Coast Guard email to CRS on February 22, 2008. The Coast Guard's official term for caretaker status is "In Commission, Special."

[12] See, for example, Kyung M. Song, "Icebreaker Polar Star Gets $57 Million Overhaul," *Seattle Times*, December 14, 2012.

[13] "Icebreaker POLAR SEA Sidelined By Engine Troubles," *Coast Guard Compass (Official Blog of the U.S. Coast Guard)*, June 25, 2010. See also "USCG Cancels Polar Icebreaker's Fall Deployment," *DefenseNews.com*, June 25, 2010; Andrew C. Revkin, "America's Heavy Icebreakers Are Both Broken Down," *Dot Earth (New York Times blog)*, June 25, 2010.

[14] Source: October 17, 2011, email to CRS from Coast Guard Congressional Affairs office. Section 222 of the Coast Guard and Maritime Transportation Act of 2012 (H.R. 2838/P.L. 112-213 of December 20, 2012) prohibited the Coast Guard from removing any part of Polar Sea and from transferring, relinquishing ownership of, dismantling, or recycling the ship until it submitted a business case analysis of the options for and costs of reactivating the ship and extending its service life to at least September 30, 2022, so as to maintain U.S. polar icebreaking capabilities and fulfill the Coast Guard's high latitude mission needs, as identified in the Coast Guard's July 2010 High Latitude Study. The business case analysis was submitted to Congress with a cover date of November 7, 2013. For more on the High Latitude Study, see *Appendix A*.

[15] HII was previously owned by Northrop Grumman, during which time it was known as Northrop Grumman Shipbuilding.

[16] See, for example, Marc Selinger, "Avondale Shipyard's Fate Remains Unclear," *Defense Daily*, April 21, 2015: 5.

why it is referred to as a medium polar icebreaker), but more capability for supporting scientific research. The ship can break through ice up to 41/2 feet thick at a speed of 3 knots, and embark a scientific research staff of 35 (with room for another 15 surge personnel and 2 visitors). The ship is used primarily for supporting scientific research in the Arctic.

Source: Coast Guard photo accessed at http://www.uscg.mil/history/webcutters/Healy_CGC_1_300.jpg on April 21, 2011.

Figure 3. Healy.

National Science Foundation (NSF) Polar Research Ships

Nathaniel B. Palmer

Nathaniel B. Palmer was built for the NSF in 1992 by North American Shipbuilding, of Larose, LA. Called *Palmer* for short, it is operated for NSF by Edison Chouest Offshore (ECO) of Galliano, LA, a firm that owns and operates research ships and offshore deepwater service ships.[17] *Palmer* is 308 feet long and has a displacement of about 6,500 tons. It has a crew of 22 and can embark a scientific staff of 27 to 37.[18] It was purpose-built as a single-mission ship for conducting and supporting scientific research in the Antarctic. It is capable of breaking ice up to 3 feet thick at speeds of 3 knots, which is sufficient for breaking through the ice conditions found in the vicinity of the Antarctic

[17] For more on ECO, see the firm's website at http://www.chouest.com/.

[18] Sources vary on the exact number of scientific staff that can be embarked on the ship. For some basic information on the ship, see http://www.nsf.gov/od/opp/support/nathpalm.jsp; http://www.usap.gov/vesselScienceAndOperations/documents/prvnews_june03.pdfprvnews_june03.pdf; http://nsf.gov/od/opp/antarct/treaty/pdf/plans0607/15plan07.pdf; http://www.nsf.gov/pubs/1996/nsf9693/fls.htm; and http://www.hazegray.org/worldnav/usa/nsf.htm.

Peninsula, so as to resupply Palmer Station, a U.S. research station on the peninsula. The ship might be considered less an icebreaker than an oceanographic research ship with enough icebreaking capability for the Antarctic Peninsula. *Palmer's* icebreaking capability is not considered sufficient to perform the McMurdo resupply mission.

Laurence M. Gould

Like *Palmer*, the polar research and supply ship *Laurence M. Gould* was built for NSF by North American Shipping. It was completed in 1997 and is operated for NSF on a long-term charter from ECO. It is 230 feet long and has a displacement of about 3,800 tons. It has a crew of 16 and can embark a scientific staff of 26 to 28 (with a capacity for 9 more in a berthing van). It can break ice up to 1 foot thick with continuous forward motion. Like *Palmer*, it was built to support NSF operations in the Antarctic, particularly operations at Palmer Station on the Antarctic Peninsula.

Sikuliaq

Sikuliaq (see-KOO-lee-auk), which is used for scientific research in polar areas, was built by Marinette Marine of Marinette, WI, and entered service in 2015. It is operated for NSF by the College of Fisheries and Ocean Sciences at the University of Alaska Fairbanks as part of the U.S. academic research fleet through the University National Oceanographic Laboratory System (UNOLS). *Sikuliaq* is 261 feet long and has a displacement of about 3,600 tons. It has a crew of 22 and can embark an additional 26 scientists and students. The ship can break ice 2.5 or 3 feet thick at speeds of 2 knots. The ship is considered less an icebreaker than an ice-capable research ship.

Summary

Table 1 summarizes the above ships. In addition to the ships shown in *Table 1*, another U.S.- registered polar ship with icebreaking capability— the Arctic oil-exploration support ship *Aiviq*— was used by Royal Dutch Shell oil company to support an oil exploration and drilling effort (now ended) in Arctic waters off Alaska. The ship, which completed construction in 2012, is owned by ECO and chartered by Royal Dutch Shell. It was used primarily for towing and laying anchors for drilling rigs, but is also equipped for responding to oil spills.

Required Numbers of U.S. Polar Icebreakers

June 2013 DHS Polar Icebreaker Mission Need Statement

The Department of Homeland Security (DHS) in June 2013 approved a Mission Need Statement (MNS) for the polar icebreaker recapitalization project. The MNS states (emphasis added):

This Mission Need Statement (MNS) establishes the need for polar icebreaker capabilities provided by the Coast Guard, to ensure that it can meet current and future mission requirements in the polar regions....

Table 1. Coast Guard and NSF Polar Ships

	Coast Guard				**NSF**	
	Polar Star	*Polar Sea*	*Healy*	*Palmer*	*Laurence M. Gould*	*Sikuliaq*
Currently operational?	Yes	No	Yes	Yes	Yes	Yes
Entered service	1976	1978	2000	1992	1997	2015
Length (feet)	399	399	420	308	230	261
Displacement (tons)	13,200	13,200	16,000	6,500	3,780	3,665
Icebreaking capability (ice thickness in feet) at 3 knots or other speed	6 feet	6 feet	4.5 feet	3 feet	1 foot at continuous forward motion	2.5 or 3 feet at 2 knots
Icebreaking capability using back and ram (ice thickness in feet)	21 feet	21 feet	8 feet	n/a	n/a	n/a
Operating temperature	-60° Fahren-heit	-60°	-50°	n/a	n/a	n/a
		Fahren-heit	Fahren-heit			
Crew (when operational)	155[a]	155[a]	85[b]	22	16	22
Additional scientific staff	32	32	35[c]	27-37	26 to 28[d]	26

Sources: Prepared by CRS using data from U.S. Coast Guard, National Research Council, National Science Foundation, DHS Office of Inspector General, and (for Palmer) additional online reference sources. n/a is not available.

[a] Includes 24 officers, 20 chief petty officers, 102 enlisted, and 9 in the aviation detachment.

[b] Includes 19 officers, 12 chief petty officers, and 54 enlisted.

[c] In addition to 85 crew members 85 and 35 scientists, the ship can accommodate another 15 surge personnel and 2 visitors.

[d] Plus 9 more in a berthing van.

Current requirements and future projections based upon cutter demand modeling, as detailed in the HLMAR [High Latitude Mission Analysis Report], indicate the Coast Guard will need to expand its icebreaking capacity, potentially requiring a fleet of up to six icebreakers (3 heavy and 3 medium) to adequately meet mission demands in the high latitudes.... The analysis took into account both the Coast Guard statutory mission requirements and additional requirements for year-round presence in both polar regions detailed in the Naval Operations Concept (NOC) 2010.... The analysis also evaluated employing single and multi-crewing concepts.... Strategic home porting analysis based upon existing infrastructure and distance to operational areas provided the final input to determine icebreaker capacity demand.[19]

---

[19] Department of Homeland Security, *Polar Icebreaking Recapitalization Project Mission Need Statement, Version 1.0*, approved by DHS June 28, 2013, pp. 1, 2, 9, 10, 11, 12.

While the MNS can be viewed as an authoritative U.S. government statement regarding required numbers of U.S. polar icebreakers, it can be noted that the key sentence in the above-quoted passage from the MNS (i.e., the sentence in bold) includes the terms "potentially" and "up to." These terms, which are often overlooked in discussions of required numbers of U.S. polar icebreakers, make the key sentence less ironclad as a requirements statement than it would have been if the terms had not been included, and could be interpreted as an acknowledgment that the requirement might amount to something less than three heavy and three medium polar icebreakers.

It can also be noted, as stated in the above-quoted passage from the MNS, that the MNS was informed by the High Latitude Mission Analysis Report (HILMAR), and that the HLMAR took into account not only Coast Guard statutory mission requirements, but additional Department of Defense (DOD) requirements for year-round presence in both polar regions as detailed in the 2010 Naval Operations Concept (NOC). This is potentially significant, because DOD appears to have subsequently dropped its 2010 requirement for year-round presence in the polar regions.[20]

The use in the MNS of the terms "potentially" and "up to," combined with DOD's decision to drop its requirement for year-round presence in the polar regions, together raise a question, other things held equal, as to whether required numbers of U.S. polar icebreakers might be something less than three heavy and three medium polar icebreakers. It is also possible, however, that there have been other changes since the MNS was issued in 2013 that would have the effect, other things held equal, of *increasing* U.S. requirements for polar icebreakers. The net result of this situation appears uncertain.

In addition to HILMAR, a number of other studies have been conducted in recent years to assess U.S. requirements for polar icebreakers and options for sustaining and modernizing the Coast Guard's polar icebreaker fleet. The findings of some of these studies are presented in *Appendix A*.

---

[20] A September 25, 2017, GAO report on polar icebreakers states (emphasis added):

> In December 2016, DOD reported to Congress that it had no specific defense requirement for icebreaking capability because Navy Arctic requirements are met by undersea and air assets which can provide year-round presence.
>
> -- DOD reported in April 2017 that its only potential defense requirement—for the Thule Air Force Base resupply [mission] in Greenland—is met by the Canadian Coast Guard through a Memorandum of Understanding with USCG.
>
> -- USCG's 2013 Polar Icebreaker Mission Needs Statement identified polar icebreaker capacity needs as partly based on the 2010 Naval Operations Concept—[a document that provides] joint maritime security strategy implementation guidance for the Navy, Marine Corps, and USCG— which stated that U.S. naval forces had a demand for year-round polar icebreaking presence in the Arctic and Antarctic.
>
> -- In April 2017, DOD joint staff officials confirmed that DOD and Naval defense strategy had been updated and does not include icebreaking requirements. DOD officials in charge of operations in the Pacific said that although they do not have a requirement for a heavy icebreaker, icebreakers play a key role in aiding the icebreaking mission to McMurdo.
>
> (Government Accountability Office, *Coast Guard: Status of Polar Icebreaking Fleet Capability and Recapitalization Plan*, GAO-17-698R, September 25, 2017, p. 20 (briefing slide 11.))

Table 2. Major Icebreakers of the World as of May 1, 2017 (Includes some icebreakers designed for Baltic use)

	Total all types, in inventory (+ under construction + planned)	In inventory, government owned or operated			In inventory, privately owned and operated		
		45,000 or more BHP	20,000 to 44,999 BHP	10,000 to 19,999 BHP	45,000 or more BHP	20,000 to 44,999 BHP	10,000 to 19,999 BHP
Russia	46 (+11 + 4)	6 (all nuclear powered; 2 not operational)	16 (1 nuclear powered; 5 designed for Baltic use)	7		9	8
Finland	10		7 (4 designed for Baltic use)	1			2
Canada	7 (+2 +5)		2	5			
Sweden	7 (+0 +3)		4 (3 designed for Baltic use)				3
United States	5 (+0 +3)	2 (*Polar Star* and *Polar Sea*; *Polar Sea* not operational)	1 (*Healy*)			1 (*Aiviq*)	1 (*Palmer*)
Denmark	4						4 (all 4 designed for Baltic use)
China	3 (+1 +0)			3			
Estonia	2			2 (both designed for Baltic use)			
Norway	1 (+1 +0)			1			
Germany	1 (+0 +1)						1
Chile	1 (+0 +1)			1			
Australia	1 (+0 +1)			1			
Latvia	1			1 (designed for Baltic use)			
Japan	1		1				
South Korea	1			1			
South Africa	1			1			
Argentina	1			1 (not operational)			
United Kingdom	0 (+1 +0)						

Source: Table prepared by CRS based on U.S. Coast Guard chart showing data compiled by the Coast Guard as of May 1, 2017, accessed September 14, 2017, at http://www.dco.uscg.mil/Portals/9/DCO%20Documents/Office% 20of%20 Waterways%20and%20Ocean%20 Policy/20170501%20major%20icebreaker%20chart.pdf?ver=2017-06-08-091723-907.

Notes: BHP = the brake horsepower of the ship's power plant. A ship with 45,000 or more BHP might be considered a heavy polar icebreaker, a ship with 20,000 to 44,999 BHP might be considered a medium polar icebreaker, and a ship with 10,000 to 19,999 BHP might be considered a light polar icebreaker or an ice-capable polar ship.

Coast Guard Testimony

At a November 17, 2015, hearing before the Europe, Eurasia, and Emerging Threats subcommittee and the Western Hemisphere subcommittee of the House Foreign Affairs Committee, then-Vice Admiral Charles Michel, the Vice Commandant of the Coast Guard, stated in his prepared statement that "Polar icebreakers are critical to supporting key national priorities laid out in the National Security Presidential Directive on Arctic Region policy and the National Strategy for the Arctic Region."[21] During the discussion portion of the hearing, Michel testified that the "Coast Guard needs at least two heavy icebreakers to provide year-round assured access and self-rescueability in the polar regions."[22]

At a June 14, 2016, hearing before the Coast Guard and Maritime Transportation subcommittee of the House Transportation and Infrastructure Committee, Admiral Michel testified that "our commandant also testified that we need self-rescue capability for our heavy icebreaker and that includes the existing *Polar Star* that we have out there now. So that means at least two, the High Latitude study says three heavy polar icebreakers is what the Coast Guard's requirement is. So that's kind of where we're talking about for heavy icebreakers."[23] Coast Guard officials have reiterated this point from time to time in subsequent testimony.

October 2016 Coast Guard Request for Information (RFI)

On October 26, 2016, the Coast Guard released a request for information (RFI) to receive industry feedback on its notional polar icebreaker acquisition approach and schedule. The summary of the RFI, dated October 25, 2016, states that "The United States Coast Guard has a need for three Heavy Polar Icebreakers and three Medium Polar Icebreakers with the priority being Heavy Polar Icebreakers."[24]

September 2017 GAO Report

A September 25, 2017, Government Accountability Office (GAO) report on polar icebreakers states that

> the Coast Guard has been unable to address all polar icebreaking requests since 2010. For example, the Coast Guard reported fulfilling 78 percent (25 of 32) of U.S. government agency requests for polar icebreaking services during fiscal year 2010 through 2016.

---

[21] Testimony of Vice Admiral Charles D. Michel, Vice Commandant, U.S. Coast Guard, on "Arctic Operations" Before the House Foreign Affairs Committee—Western Hemisphere & Europe, Eurasia, and Emerging Threats Subcommittees, November 17, 2015, p. 3.

[22] Transcript of hearing.

[23] Transcript of hearing.

[24] *Summary of RFI*, October 25, 2016, page 2, accessed November 10, 2016, at https://www.uscg.mil/acquisition/icebreaker/pdf/Acquisition-Strategy-RFI.pdf.

Coast Guard officials cited various factors affecting the Coast Guard's ability to meet all requests, particularly the unavailability of its heavy polar icebreakers.[25]

July 2018 GAO Report

A July 2018 GAO report stated that

> the Coast Guard operates one medium icebreaker, the Healy, which has an expected end of service life in 2029. Despite the requirement for three medium icebreakers, Coast Guard officials said they are not currently assessing acquisition of the medium polar icebreakers because they are focusing on the heavy icebreaker acquisition and plan to assess the costs and benefits of acquiring medium polar icebreakers at a later time.[26]

Polar Icebreakers Operated by Other Countries

In discussions of U.S. polar icebreakers, some observers note the size of the polar icebreaking fleets operated by other countries. Countries with interests in the polar regions have differing requirements for polar icebreakers, depending on the nature and extent of their polar interests and activities. *Table 2* shows a Coast Guard summary of major icebreakers around the world; the figures in the table include some icebreakers designed for use in the Baltic Sea.

Coast Guard Polar Security Cutter (PSC) Program

Overview

The PSC program was initiated in the Coast Guard's FY2013 budget submission, and envisages the acquisition of three new heavy polar icebreakers, to be followed years from now by the acquisition of up to three new medium polar icebreakers. The Coast Guard wants to begin construction of the first new heavy polar icebreaker in FY2019 and have it enter service in 2023. The Coast Guard's proposed FY2019 budget requests $750 million in Coast Guard acquisition funding for the program.

Program Name

The program was previously known as the polar icebreaker (PIB) program. It is now called the Polar Security Cutter (PSC) program, though many observers may continue, as a matter of convenience, to refer to it as the polar icebreaker program.[27]

---

[25] Government Accountability Office, *Coast Guard: Status of Polar Icebreaking Fleet Capability and Recapitalization Plan*, GAO-17-698R, September 25, 2017, pp. 2-3. A similar statement appears on page 4.

[26] Government Accountability Office, *Coast Guard Acquisitions[:] Actions Needed to Address Longstanding Portfolio Management Challenges*, GAO-18-454, July 2018, p. 13.

[27] A September 27, 2018, press report states:

> The U.S. Coast Guard changed the name of its heavy icebreaker program to highlight its importance to national security, as funding for the first-in-class ship may be in jeopardy.

Desired Capabilities for New Polar Icebreaker

The Coast Guard's key performance parameters (KPPs) for new heavy polar icebreakers include the following:

- an ability to break through 6 feet of ice at 3 knots (threshold) or 8 feet of ice at three knots (objective);[28]
- an ability to break through ridged ice of 21 feet;
- an ability to operate without replenishment (i.e., resupply) for 80 days (threshold) or 90 days (objective); and
- an ability to exchange voice and data with DHS, Coast Guard, Defense Department units, and other stakeholders.[29]

The Coast Guard states that the desired capabilities for a new polar icebreaker are similar to the capabilities of *Polar Star* and *Polar Sea* in the following general ways:

Now dubbed the Polar Security Cutter, Coast Guard leadership and backers on Capitol Hill are determined to secure funding for the planned new class of heavy icebreaker—the first for the Coast Guard in more than four decades—by marketing its vital role protecting the nation's sovereignty.

"When we talk about icebreaking capability, that doesn't sell very well to all audiences," Rear Adm. Melvin Bouboulis, the Coast Guard's Assistant Commandant for Engineering and Logistics, said during the recent American Society of Naval Engineers (ASNE) Fleet Maintenance and Modernization Symposium.

"We understand that some folks think just it goes and breaks ice, but we've purposely changed the name of that program to Polar Security Cutter because it is really the U.S. presence in the Arctic regions and preserving our national interest and security in those areas."

The Coast Guard's heavy icebreaker mission has traditionally supported scientific research. But the new icebreaker's request for proposals released earlier this year hinted at a possible future national security mission. After much speculation about whether the ship would be armed, Coast Guard officials said in the RFP [Request for Proposals] they wanted the ability to add deck-mounted weapons to the icebreaker in the future.

The name-change was talked about by Adm. Karl Schultz almost as soon as he became Coast Guard commandant in June....

In August, Schultz suggested a program name change when appearing at an event co-hosted by the U.S. Naval Institute and the Center for Strategic and International Studies.

"You know right now it's the heavy icebreaker, the polar icebreaker. I think the paradigm has shifted a little bit," Schultz said. "I'm having a conversation with my team about, maybe that's the polar security cutter. I mean, that's really what we're talking about: we're talking about national sovereign interests up there, we're talking about competition. ... The Coast Guard is the face of that competition, and we've really got to press into that."...

The Coast Guard's proposed icebreaker is now officially called the Polar Security Cutter, Lt. Amy Midgett, a Coast Guard spokeswoman, confirmed to USNI News. The hull designation will be WMSP. W is the standard prefix for Coast Guard vessels, and MSP stands for Maritime Security-Polar, Brian Olexy, a Coast Guard spokesman, told USNI News. The intended missions of the icebreaker will remain the same—supporting scientific research—and designating the ship a security cutter does not alter how it is funded.

(Ben Werner and Sam LaGrone, "Coast Guard Renames New Icebreaker Program 'Polar Security Cutter,'" USNI News, September 27, 2018.

See also Sydney J. Freedberg Jr., "With Funding In Peril, Coast Guard Pushes Icebreaker As 'Polar Security Cutter,'" *Breaking Defense*, October 29, 2018.

[28] The terms *threshold* and *objective* are acquisition terms. Threshold can be translated roughly as minimum required capability. Objective can be translated roughly as maximum or preferred capability (if feasible and affordable).

[29] Coast Guard PSC program industry day briefing entitled "*Polar Icebreaker (PIB) Acquisition Program Industry Engagement*," slide 23, accessed April 4, 2016, at http://www. uscg.mil/ACQUISITION/icebreaker/pdf/Industry%20Day%2018%20March%202016.pdf.

- the ability to conduct long-range, high-endurance, independent operations with heavy icebreaking capability;
- flexibility in personnel support spaces and systems;
- interoperability to support interagency and interservice mission execution.[30]

The Coast Guard states that the desired capabilities for a new polar icebreaker differ from the capabilities of *Polar Star* and *Polar Sea* in the following general ways:

- features for improved reliability, maintainability, supportability, operational availability, and system redundancy;
- features for meeting modern environmental standards;
- features for improved ship control;
- features for modern human habitability and human systems integration; and
- space, weight, and power margins (i.e., growth margin) for accepting specialized capabilities.[31]

Notional Program Schedule

On October 26, 2016, the Coast Guard released a request for information (RFI) to receive industry feedback on its notional polar icebreaker acquisition approach and schedule. The summary of the RFI, dated October 25, 2016, presents a notional schedule for acquiring three heavy polar icebreakers under which procurement of long leadtime materials (LLTM) for the three ships would start in the fourth quarter of FY2019, the second quarter of FY2021, and the second quarter of FY2022, respectively, and the ships would be delivered in the fourth quarter of FY2023, the second quarter of FY2025, and the second quarter of FY2026, respectively. Each ship would be commissioned into service a few weeks or months after it is delivered.[32]

Acquisition Cost

Steps Taken by Coast Guard and Navy to Reduce Acquisition Cost

Coast Guard and Navy have taken certain steps to reduce the time and cost for designing and building new Coast Guard heavy polar icebreakers, including the following:

---

[30] Coast Guard PSC program industry day briefing entitled "*Polar Icebreaker (PIB) Acquisition Program Industry Engagement*," slide 24, accessed April 4, 2016, at http://www. uscg.mil/ACQUISITION/icebreaker/ pdf/Industry%20Day%2018%20March%202016.pdf.

[31] Coast Guard PSC program industry day briefing "*Polar Icebreaker (PIB) Acquisition Program Industry Engagement*," slide 26, accessed April 4, 2016, at http://www. uscg.mil/ACQUISITION/icebreaker/ pdf/Industry%20Day%2018%20March%202016.pdf.

[32] *Summary of RFI*, October 25, 2016, 3 pp., accessed November 10, 2016, at https://www.uscg.mil/ acquisition/icebreaker/pdf/Acquisition-Strategy-RFI.pdf.

- The Coast Guard and Navy in August 2016 established an integrated program office for the PSC program, facilitating Navy efforts to share with the Coast Guard ship-design and ship-acquisition best practices that can reduce the time and cost of designing and building ships.[33]
- All five shipbuilders that have been awarded contracts for design studies and analysis for a new polar icebreaker (see "Recent Acquisition Actions" below) are employing a so-called "parent design" approach, meaning that they intend to modify existing polar icebreaker designs (parent designs) rather than develop entirely new designs from scratch (i.e., "clean-sheet" designs).[34] This can reduce the time and cost needed to design a new polar icebreaker, and the technical risk and cost risk involved in building the ship.
- The Coast Guard and Navy have carefully reviewed the specific operational requirements for new heavy polar icebreakers, and have adjusted some of those requirements to help reduce their acquisition cost.
- The design for the heavy polar icebreakers will rely less on military specifications (MilSpecs) and more on civilian commercial shipbuilding specifications than it might have under a more traditional military ship-acquisition approach.

Estimated Acquisition Cost has Declined Substantially

As a result of the above steps by the Coast Guard and Navy—and perhaps also model test tank results that have substantially reduced the required power of the new icebreakers' propulsion plant[35]—the estimated acquisition cost for new heavy polar

---

[33] A September 25, 2017, GAO report states that "the Coast Guard and Navy formalized this partnership through a January 2017 Memorandum of Understanding." (Government Accountability Office, *Coast Guard: Status of Polar Icebreaking Fleet Capability and Recapitalization Plan*, GAO-17-698R, September 25, 2017, p. 4.) A May 2018 GAO report states that "in 2017, DHS, the USCG, and Navy entered into several agreements that outline oversight roles, among other things. For example, these agreements state that the program will follow DHS acquisition policies with DHS leadership serving as the acquisition decision authority for program milestones. However, the Navy will review and approve acquisition documents before the program seeks DHS approval. These agreements also state that the program's contracting actions could be funded by either USCG or Navy appropriations, and the source of the appropriations will award the contract." (Government Accountability Office, *Homeland Security Acquisitions[:] Leveraging Programs' Results Could Further DHS's Progress to Improve Portfolio Management*, GAO-18-339SP, May 2018, p. 86.)

[34] Source: March 16, 2018, Coast Guard-Navy briefing to CRS and the Congressional Budget Office (CBO) on the PSC program.

[35] A June 22, 2018 press report states:

> The U.S. Coast Guard is collaborating with the National Research Council of Canada (NRC) to access its renowned ice tank facilities in St. John's, Newfoundland, to model and evaluate the specifications needed to design the new heavy icebreakers....
>
> The testing at the NRC ice tank in St. John's has already led to some critical changes to the ship capability criteria.
>
> "We confirmed that with modern icebreaker hull form and propulsion configuration design we could reduce the estimated required power," [Neil Meister, technical director of the U.S. Coast Guard Polar Icebreaker Acquisition Program] said. "In comparison to polar-class icebreakers they have 60,000 installed horse power and we see that you can meet the same icebreaking capability requirement with about 40 per cent less power."

(Levon Sevunts, "U.S. Coast Guard Turns to Canada for Help with Designing Its New Heavy Icebreaker," Radio Canada International, June 22, 2018.)

icebreakers has been substantially reduced. The procurement cost of a new heavy polar icebreaker had earlier been estimated informally at roughly $1 billion, but the Coast Guard and Navy informed CRS and CBO in March 2018 that they now believe that three polar icebreakers could be acquired for a total cost of about $2.1 billion, or an average of about $700 million per ship.[36] (The first ship will cost more than the other two because it will incorporate design costs for the class and be at the start of the production learning curve for the class.) The March 2, 2018, Request for Proposals (RFP) for the PSC program states that "For informational purposes only, the government has established an estimate for the HPIB [heavy polar icebreaker] shipbuilder costs in the amount of $746M [million] for the lead ship... with an average ship price of $615M across three HPIBs...."[37] Other information reported by GAO identifies a smaller reduction in procurement cost, to something more than $900 million per ship.[38]

For additional background information on earlier estimates of polar icebreaker acquisition costs, see *Appendix B*.

Program Funding

The PSC program has received about $359.6 million in acquisition funding through FY2018, including $300 million provided through the Navy's shipbuilding account—$150 million in FY2017, and another $150 million in FY2018—and $59.6 million in funding provided through the Coast Guard's acquisition account. The Coast Guard's proposed FY2019 budget requests $750 million in Coast Guard acquisition funding for the program.

The request for $750 million reflected a change that the Trump Administration made to its proposed FY2019 budget not long before that budget was submitted to Congress. Prior to this change, the Administration's plan was to request $30 million in Coast Guard

---

[36] Source: March 16, 2018, Coast Guard-Navy briefing to CRS and CBO on the polar icebreaker program. For further discussion, see the section entitled "Estimated Acquisition Cost Has Declined Substantially" in CRS Report RL34391, *Coast Guard Polar Icebreaker Program: Background and Issues for Congress*, by Ronald O'Rourke.

[37] Naval Sea Systems Command (HQ), Solicitation N0002418R2210, March 2, 2018, page 257 of 294. See also Government Accountability Office, *Coast Guard Acquisitions[:] Polar Icebreaker Program Needs to Address Risks before Committing Resources*, GAO-18-600, September 2018, pp. 42-43. The GAO report also states that DHS and the Coast Guard estimate the total lifecycle cost of a three-ship PSC program at $8,545 million (objective) and $9,827 million (threshold) (page 13), and that the ships' detail design and construction costs account for about 25% of these figures, with the remaining 75% or so accounted for by remaining acquisition costs, 30 years of annual operating and support (O&S) costs, and end-of-life ship-disposal costs (page 31). Twenty-five percent of $8,545 million and $9,827 million would equate to design and construction costs of $2,136.3 million (an average of about $712 million per ship) and $2,456.8 million (an average of about $819 million per ship).

[38] A May 2018 GAO report stated that the acquisition program baseline (APB) approved for the polar icebreaker program in January 2018 estimated the program's acquisition cost at $3,207 million, and that the "current estimate" of the program's acquisition as of January 2018 was $2,789 million, or an average of about $930 million per ship. (Government Accountability Office, *Homeland Security Acquisitions[:] Leveraging Programs' Results Could Further DHS's Progress to Improve Portfolio Management*, GAO-18-339SP, May 2018, p. 85.) See also Government Accountability Office, *Coast Guard Acquisitions[:] Actions Needed to Address Longstanding Portfolio Management Challenges*, GAO-18-454, July 2018, which states on page 18 that "The polar icebreaker program has an estimated total acquisition cost of more than $3 billion...."

acquisition funding for the program. Some documents in the Coast Guard's FY2019 budget submission that were prepared before this change occurred reflect this $30 million figure rather than the $750 million figure.

For additional background information on funding for the PSC program, see *Appendix C.*

Contract Type

The baseline plan for the PSC program calls for acquiring new polar icebreakers using a contract with options. Coast Guard and Navy officials, however, are open to the idea of instead using a block buy contract to acquire the ships, and have requested information on this possibility as part of the request for proposals (RFP) for the PSC program that was released on March 2, 2018 (see next section).

Recent Acquisition Actions

Recent acquisition actions in the PSC program include the following:

- On February 22, 2017, the Coast Guard awarded five firm fixed-price contracts for heavy polar icebreaker design studies and analysis. The objective of the studies was to identify design and systems approaches to reduce acquisition cost and production timelines.
- On April 4, 2017, the Coast Guard released its draft heavy polar icebreaker system specifications in a request for information (RFI). The request sought questions, comments and feedback related to heavy polar icebreaker technology risks, sustainability, producibility and affordability.
- In May 2017, the Coast Guard began model testing of potential hull designs and propulsion configurations for the polar icebreaker. The testing was anticipated to be completed by March 2018, with the results to be used to inform the final specifications for the icebreakers.[39]
- On October 19, 2017, the U.S. Navy, in collaboration with the Coast Guard under the polar icebreaker integrated program office, released a draft request for proposal (RFP) for detail design and construction of a heavy polar icebreaker.
- In January 2018, DHS approved the initial acquisition program baseline (APB) for the PSC program. The APB establishes cost, schedule, and performance goals for the program.[40]
- On March 2, 2018, the U.S. Navy, in collaboration with the U.S. Coast Guard under the polar icebreaker integrated program office, released an RFP for the

[39] Source: Government Accountability Office, *Homeland Security Acquisitions[:] Leveraging Programs' Results Could Further DHS's Progress to Improve Portfolio Management*, GAO-18-339SP, May 2018, p. 86.

[40] Source: Government Accountability Office, *Homeland Security Acquisitions[:] Leveraging Programs' Results Could Further DHS's Progress to Improve Portfolio Management*, GAO-18-339SP, May 2018, p. 85.

advance procurement and detail design for the Coast Guard's heavy polar icebreaker, with options for detail design and construction for up to three heavy polar icebreakers.

Regarding the February 22, 2017, award of the five contracts for polar icebreaker design studies and analysis, the Coast Guard stated:

The Coast Guard today awarded five firm fixed-price contracts for heavy polar icebreaker design studies and analysis. The contracts were awarded to Bollinger Shipyards LLC of Lockport, Louisiana; Fincantieri Marine Group LLC of Washington, D.C.; General Dynamics/National Steel and Shipbuilding Company of San Diego; Huntington Ingalls Inc. of Pascagoula, Mississippi; and VT Halter Marine Inc. of Pascagoula. The total value of the award is approximately $20 million.

The objective of the studies is to identify design and systems approaches to reduce acquisition cost and production timelines. In addition to a requirement to develop heavy polar icebreaker designs with expected cost and schedule figures, the contracts require the awardees to examine major design cost drivers; approaches to address potential acquisition, technology and production risks; and benefits associated with different types of production contract types.

The heavy polar icebreaker integrated program office, staffed by Coast Guard and Navy personnel, will use the results of the studies to refine and validate the draft heavy polar icebreaker system specifications. The use of design studies is an acquisition best practice influenced by the Navy's acquisition experience with the landing craft, utility (LCU) amphibious transport ship and T-AO(X) [aka TAO-205] fleet oiler,[41] which are being acquired under accelerated acquisition schedules.

"These contracts will provide invaluable data and insight as we seek to meet schedule and affordability objectives," said Rear Adm. Michael Haycock, the Coast Guard's director of acquisition programs and program executive officer. "Our nation has an urgent need for heavy polar icebreaking capability. We formed an integrated program office with the Navy to take advantage of their shipbuilding experience. This puts us in the best possible position to succeed in this important endeavor."

"The Navy is committed to the success of the heavy icebreaker program and is working collaboratively with our Coast Guard counterparts to develop a robust acquisition strategy that drives affordability and competition, while strengthening the industrial base," said Jay Stefany, executive director, Amphibious, Auxiliary and Sealift Office, Program Executive Office, Ships. "Our ability to engage early with our industry partners will be critical to delivering this capability to our nation."

The studies are expected to take 12 months to complete, with study results provided incrementally during that time. The Coast Guard plans to release a draft request for proposal (RFP) for detail design and construction by the end of fiscal year 2017, followed by release of the final RFP in fiscal year 2018. The integrated program office plans to

[41] For more on the TAO-205 program, see CRS Report R43546, *Navy John Lewis (TAO-205) Class Oiler Shipbuilding Program: Background and Issues for Congress*, by Ronald O'Rourke.

award a single contract for design and construction of the lead heavy polar icebreaker in fiscal year 2019, subject to appropriations.[42]

Regarding the March 2, 2018, release of the RFP for the PSC program, the Navy and Coast Guard state:

> The RFP is for Advance Planning and Engineering Efforts, with options for the Detail Design and Construction (DD&C) of up to three (3) Heavy Polar Icebreaker (HPIB) cutters....
>
> To enable ongoing program planning and responses to Congressional inquiries, the Coast Guard and Navy HPIB IPO desire input from prime offerors related to the benefits of Congressional authorization of Block Buy and/or Economic Order Quantity.[43] Submission of this information is voluntary and will not be used to evaluate any proposal submitted by the offeror in response to this RFP. Email submissions providing dollarized estimated savings per ship for authorization provided for 1) all three cutters and 2) only the second and third cutters should be emailed to the Bidders Question contacts identified below with the email title "HPIB Block Buy/EOQ Input—Contractor Name." Submissions within 60 days of RFP release are preferred.[44]

Foreign Cooperation and Participation

U.S. Coast Guard Cooperation with Canadian Coast Guard

A February 9, 2017, U.S. Coast Guard news release states the following:

> The U.S. and Canadian governments on Feb. 7 established a partnership that will enable the U.S. Coast Guard heavy polar icebreaker acquisition program to test and validate potential heavy polar icebreaker design models at Canada's National Research Council (NRC) in St John's, Newfoundland.
>
> The testing, which includes analyses of maneuverability in ice and icebreaking resistance and powering, will be used to further inform the baseline requirements for new heavy polar icebreakers, expand current icebreaker design and operational knowledge, and support the urgent need to recapitalize U.S. heavy icebreaking capability. The partnership is being facilitated by the Department of Homeland Security Science and Technology Directorate and was developed under the Agreement Between the U.S. and Canada for Cooperation in Science and Technology for Critical Infrastructure Protection

---

[42] "*Acquisition Update: Coast Guard Awards Multiple Contracts for Heavy Polar Icebreaker Industry Studies*," February 22, 2017, accessed March 20, 2017, at https:// www.uscg.mil/ acquisition/ newsroom/ updates/ icebreaker022217.asp.

[43] Economic order quantity (EOQ) purchases, which can take place as part of a block buy contract, are up-front batch orders of selected components of the end items (in this case, ships) that are to be procured under the contract.

[44] Federal Business Opportunities (FedBizOpps.gov), Heavy Polar Icebreaker (HPIB) Detail Design and Construction (DD&C), Solicitation Number: N00024-18-R-2210, March 2, 2018.

and Border Security, enacted in 2004. Model and test activities at the NRC are scheduled to formally begin in April 2017.

The NRC is home to one of the world's largest ice tank facilities, which is used to measure the performance and evaluate the safety of ice-going ships and structures in controlled model-scale conditions. The NRC ice tank is capable of modeling a wide range of marine ice conditions, including first-year and multiyear ice, pack ice, ridged ice and glacial ice.

In addition to the modeling work that will be conducted at the NRC, the Coast Guard and Navy will conduct additional model test work to evaluate the performance of the icebreaker in open water at the Naval Surface Warfare Center, Carderock Division, in Bethesda, Maryland.[45]

Parent Design Approach Could Involve One or More Foreign Designs

As mentioned above, all five shipbuilders that have been awarded contracts for design studies and analysis for a new polar icebreaker are employing a so-called "parent design" approach, meaning that they intend to modify existing polar icebreaker designs (parent designs) rather than develop entirely new designs from scratch (i.e., "clean-sheet" designs). It is possible that one or more of the parent designs are foreign designs. If a parent design being used by a shipbuilder is a foreign design, that shipbuilder might do so in cooperation with the foreign shipbuilder that developed the original parent design.

General Dynamics Teaming with VARD

A January 17, 2017, press report states the following:

> General Dynamics is teaming with Norwegian ship designer and manufacturer VARD for the Coast Guard's heavy Polar Icebreaker Program, the company says. GD's exhibit booth at last week's Sea Air Space Symposium displayed a graphic with the two companies' names and an artist's concept of a Coast Guard Icebreaker. VARD is owned by Italy's FINCANTIERI. GD says VARD is one of the premiere global designers of polar icebreakers. GD's NASSCO shipbuilding division is leading the company's effort on the icebreaker program, for which the Coast Guard early this year plans to award initial design study contracts.[46]

---

[45] "*Acquisition Update: U.S., Canada Governments Partner on Model and Test Activities to Support Heavy Polar Icebreaker Acquisition Efforts*," February 9, 2017, accessed March 20, 2017, at https://www.uscg.mil/acquisition/newsroom/updates/icebreaker020917.asp.

[46] "Icebreaker Team," *Defense Daily*, January 17, 2017: 2. See also Levon Sevunts, "U.S. Coast Guard Turns to Canada for Help with Designing Its New Heavy Icebreaker," *Radio Canada International*, June 22, 2018.

Issues for Congress

FY2019 Funding Request

One issue for Congress for FY2019 is whether to approve, reject, or modify the Coast Guard's FY2019 acquisition funding request for the PSC program. In considering this issue, Congress may consider, among other things, whether the Coast Guard has accurately priced the work it is proposing to do in FY2019.

One potential option for Congress would be to reduce the Coast Guard's FY2019 acquisition funding request so that the total amount of funding provided through FY2019 would be enough to fully fund the procurement of the first new heavy polar icebreaker, but not partially fund the second ship in the program. This might involve reducing the FY2019 acquisition funding request by roughly $150 million.[47] Under this option, funding for the second ship in the program could be provided starting in FY2020 or a subsequent year.

Another potential option for Congress would be to increase the FY2019 acquisition funding request so that the total amount of funding provided through FY2019, in addition to fully funding the procurement of the first new heavy polar icebreaker, would additionally fully fund (rather than partially fund) the procurement of the second ship in the program. Based on a total acquisition cost of $2.1 billion for three heavy polar icebreakers, fully funding the second ship in the program in FY2019 might involve increasing the FY2019 funding request by $400 million to $500 million. Under this option, funding for the third ship in the program could be provided starting in FY2020 or a subsequent year.

Contract with Options vs. Block Buy Contract

Another potential issue for Congress is whether to use a contract with options or a block buy contract to acquire the ships. As noted earlier, the baseline plan for the PSC program calls for acquiring ships using a contract with options, but Coast Guard and Navy officials are open to the idea of instead using a block buy contract to acquire the ships, and have requested information on this possibility as part of the request for proposals (RFP) for the PSC program that was released on March 2, 2018.

---

[47] The Coast Guard's proposed FY2019 budget was submitted before Congress finalized action on the Coast Guard's FY2018 budget. In its action on the FY2018 budget, Congress approved the Coast Guard's request for $19 million in Coast Guard acquisition funding for the program, and provided $150 million in unrequested acquisition funding for the program in the Navy's shipbuilding account. If the FY2019 request for $750 million is intended solely to complete the funding for the first ship, and if this figure does not assume that more than $19 million would be provided for the program in FY2018, then approving the $750 million request would provide $150 million more than needed to fully fund the first ship.

Although a contract with options covers multiple years, it operates more like a form of annual contracting, and it does not generate the kinds of savings that are possible with a block buy contract. Compared to a contract with options, a block buy contract would reduce the government's flexibility regarding whether and when to acquire the second and third ships, and what design to build them to,[48] and in return reduce the combined acquisition cost of the ships covered by the contract. The Navy has used block buy contracts to reduce procurement costs of Virginia-class attack submarines and (in more recent years) Littoral Combat Ships (LCSs) and John Lewis (TAO-205) class oilers.[49] CRS estimates that compared to costs using a contract with options, using a block buy contract that included economic order quantity (EOQ) purchases (i.e., up-front batch purchases) of materials and components for three heavy polar icebreakers would reduce the combined acquisition cost of the three ships by upwards of 7%, which could equate to a savings of upwards of $150 million.

A congressionally mandated July 2017 National Academies of Sciences, Engineering, and Medicine (NASEM) report on acquisition and operation of polar icebreakers states (emphasis as in original):

> 3. Recommendation: USCG should follow an acquisition strategy that includes block buy contracting with a fixed price incentive fee contract and take other measures to ensure best value for investment of public funds.
>
> Icebreaker design and construction costs can be clearly defined, and a fixed price incentive fee construction contract is the most reliable mechanism for controlling costs for a program of this complexity. This technique is widely used by the U.S. Navy. To help ensure best long-term value, the criteria for evaluating shipyard proposals should incorporate explicitly defined lifecycle cost metrics....
>
> A block buy authority for this program will need to contain specific language for economic order quantity purchases for materials, advanced design, and construction activities. A block buy contracting program with economic order quantity purchases enables series construction, motivates competitive bidding, and allows for volume

---

[48] Stated more fully, from a congressional perspective, trade-offs in using block buy contracting include the following:
- reduced congressional control over year-to-year spending, and tying the hands of future Congresses;
- reduced flexibility for making changes in Coast Guard acquisition programs in response to unforeseen changes in strategic or budgetary circumstances (which can cause any needed funding reductions to fall more heavily on acquisition programs not covered by multiyear contracts);
- a potential need to shift funding from later fiscal years to earlier fiscal years to fund economic order quantity (EOQ) purchases (i.e., up-front batch purchases) of components;
- the risk of having to make penalty payments to shipbuilders if multiyear contracts need to be terminated due to unavailability of funds needed to the continue the contracts; and
- the risk that materials and components purchased for ships to be acquired in future years might go to waste if those ships are not eventually acquired.

[49] See CRS Report R41909, Multiyear Procurement (MYP) and Block Buy Contracting in Defense Acquisition: Background and Issues for Congress, by Ronald O'Rourke and Moshe Schwartz; CRS Report RL33741, *Navy Littoral Combat Ship (LCS) Program: Background and Issues for Congress*, by Ronald O'Rourke; and CRS Report R43546, *Navy John Lewis (TAO-205) Class Oiler Shipbuilding Program: Background and Issues for Congress*, by Ronald O'Rourke.

purchase and for the timely acquisition of material with long lead times. It would enable continuous production, give the program the maximum benefit from the learning curve, and thus reduce labor hours on subsequent vessels....

If advantage is taken of learning and quantity discounts available through the recommended block buy contracting acquisition strategy, the average cost per heavy icebreaker is approximately $791 million, on the basis of the acquisition of four ships.[50]

Funding Coast Guard Polar Icebreakers through Navy's Shipbuilding Account

Another potential issue for Congress is whether to continue providing at least some of the acquisition funding for the PSC program through the Navy's shipbuilding account, known formally as the Shipbuilding and Conversion Navy (SCN) appropriation account. A May 2018 GAO report states that agreements between DHS, the Coast Guard, and the Navy that were made following the establishment of the Coast Guard-Navy integrated program office for the PSC program "state that the program's contracting actions could be funded by either USCG or Navy appropriations, and the source of the appropriations will award the contract."[51] As noted earlier, of the $359.6 million in acquisition funding that has been provided for the PSC program through FY2018, $300 million (about 83%) has been provided through the SCN account—$150 million in FY2017, and another $150 million in FY2018.

Although providing funding for Coast Guard ships through the SCN account creates some complexity in tracking and executing funding for Coast Guard ship acquisition, and can raise a question as to whether that funding would otherwise go toward the acquisition of Navy ships, it has been used in the past for funding Coast Guard ships other than heavy polar icebreakers:

- Healy was funded largely (about 89%) through the SCN account.[52]

---

[50] National Academies of Sciences, Engineering, and Medicine, Division on Earth and Life Studies and Transportation Research Board, *Acquisition and Operation of Polar Icebreakers: Fulfilling the Nation's Needs*, Letter Report, with cover letter dated July 11, 2017, pp. 14, 15.

[51] Government Accountability Office, *Homeland Security Acquisitions[:] Leveraging Programs' Results Could Further DHS's Progress to Improve Portfolio Management*, GAO-18-339SP, May 2018, p. 86.

[52] The somewhat complicated funding history for the ship is as follows: The Coast Guard's proposed FY1990 budget requested $244 million for the acquisition of an icebreaker. The FY1990 DOD appropriations act (H.R. 3072/P.L. 101- 165 of November 21, 1989) provided $329 million for the ship in the SCN account. (See pages 77 and 78 of H.Rept. 101-345 of November 13, 1989.) This figure was then reduced by $4.2 million by a sequester carried out under the Balanced Budget And Emergency Deficit Control Act of 1985, also known as the Gramm-Rudman-Hollings Act (H.J.Res. 372/P.L. 99-177 of December 12, 1985). Another $50 million was rescinded by the Dire Emergency Supplemental Appropriations for Disaster Assistance, Food Stamps, Unemployment Compensation Administration, and Other Urgent Needs, and Transfers, and Reducing Funds Budgeted for Military Spending Act of 1990 (H.R. 4404/P.L. 101-302 of May 25, 1990). An additional $59 million for the ship was then appropriated in the FY1992 DOD Appropriations Act (H.R. 2521/P.L. 102-172 of November 26, 1991). Also, an additional $40.4 million in acquisition funding for the ship was provided

- Thirty-three of the Coast Guard's 49 Island-class 110-foot patrol boats (i.e., about 67% of the boats) were procured under a Navy contract. The contract was for the construction of 21 of the boats, and included FY1990 SCN funds and prior year DOD nonexpiring funding. During the construction phase of the contract, the Navy exercised options under the contract for the construction 12 additional boats using FY1990 SCN funding.[53]

Subsections (a), (b), and (c) of Section 122 of the FY2018 National Defense Authorization Act (H.R. 2810/P.L. 115-91 of December 12, 2017) states:

> SEC. 122. Icebreaker vessel.
>
> (a) Authority to procure one polar-class heavy icebreaker.—
>
> (1) IN GENERAL.—There is authorized to be procured for the Coast Guard one polar-class heavy icebreaker vessel.
>
> (2) CONDITION FOR OUT-YEAR CONTRACT PAYMENTS.—A contract entered into under paragraph (1) shall provide that any obligation of the United States to make a payment under the contract for a fiscal year after fiscal year 2018 is subject to the availability of appropriations or funds for that purpose for such later fiscal year.
>
> (a) Limitation on availability of funds for procurement of icebreaker vessels.—None of the funds authorized to be appropriated by this Act or otherwise made available for the Department of Defense for any fiscal year that are unobligated as of the date of the enactment of this Act may be obligated or expended for the procurement of an icebreaker vessel other than the one polar-class heavy icebreaker vessel authorized to be procured under subsection (a)(1).
>
> (b) Contracting authority.—
>
> (1) COAST GUARD.—If funds are appropriated to the department in which the Coast Guard is operating to carry out subsection (a)(1), the head of contracting activity for the Coast Guard shall be responsible for contracting actions carried out using such funds.
>
> (2) NAVY.—If funds are appropriated to the Department of Defense to carry out subsection (a)(1), the head of contracting activity for the Navy, Naval Sea Systems Command shall be responsible for contracting actions carried out using such funds.
>
> (3) INTERAGENCY ACQUISITION.—Notwithstanding paragraphs (1) and (2), the head of contracting activity for the Coast Guard or head of contracting activity for the

---

through a series of annual appropriations in the Coast Guard's Acquisition, Construction, and Improvements (AC&I) account (as it was known prior to FY2019) from FY1988 through FY2001. The resulting net funding for the ship was thus $374.2 million, of which $333.8 million, or 89.2%, was DOD funding, and $40.4 million, or 10.8%, was Coast Guard acquisition funding. (Source: Undated Coast Guard information paper provided to CRS by Coast Guard legislative liaison office, March 3, 2016.)

[53] Source: Navy information paper dated August 15, 2017, provided to CRS by Navy Office of Legislative Affairs on August 23, 2017.

Navy, Naval Sea Systems Command (as the case may be) may authorize interagency acquisitions that are within the authority of such head of contracting activity.[54]

Regarding Section 122, the conference report (H.Rept. 115-404 of November 9, 2017) on H.R. 2810/P.L. 115-91 states:

Icebreaker vessel (sec. 122)

The House bill contained provisions (sec. 122, 123, and 1012) that would authorize the Secretary of the Navy to act as a general agent for the Secretary of the Department in which the Coast Guard is operating and enter into a contract for icebreaker vessels; prohibit funds for the Department of Defense from being used for the procurement of an icebreaker vessel; and amend section 2218 of title 10, United States Code, to authorize funds associated with the National Defense Sealift Fund for the construction of icebreaker vessels.

The Senate amendment contained a similar provision (sec. 1048).

The Senate recedes with an amendment that would authorize one polar-class heavy icebreaker vessel, prohibit funds for the Department of Defense from being used for the procurement of an icebreaker vessel other than this one polar-class heavy icebreaker vessel, clarify contracting authorities, and require a Comptroller General report.

The conferees recognize the national importance of recapitalizing the U.S. icebreaker fleet and the extraordinary circumstances that necessitated use of Department of Defense funding to procure the first polar-class heavy icebreaker, as partially provided in the Department of Defense Appropriations Act for Fiscal Year 2017. Accordingly, the conferees support the authorization of this icebreaker in this Act.

The conferees note the Undersecretary of Management in the Department of Homeland Security (DHS) serves as the Acquisition Decision Authority for the Polar Icebreaker Program and that this program is governed in accordance with DHS Acquisition Management Directive 102–01 and Instruction 102–01–001.

The conferees believe maintaining clear lines of authority, responsibility, accountability, and resources with the Secretary and Acquisition Decision Authority of the department in which the U.S. Coast Guard is operating are essential to delivering icebreakers on cost and schedule.

Accordingly, the conferees believe the Secretary of the Department of Homeland Security and the Undersecretary of Management in the DHS should be the officials provided with authorities and resources related to the Polar Icebreaker Program.

Therefore, the conferees expect subsequent icebreakers to be authorized by the congressional committees with jurisdiction over the Coast Guard and funded using Coast Guard appropriations. (Pages 765-766)

---

[54] Section 122 also includes a subsection (d) that requires a GAO report assessing the cost of, and schedule for, the procurement of new icebreakers.

Technical, Schedule, and Cost Risk for PSC Program

Another potential issue for Congress concerns technical, schedule, and cost risk in the PSC program. A September 2018 GAO report on the PSC program states that the Coast Guard did not have a sound business case in March 2018, when it established the cost, schedule, and performance baselines for its heavy polar icebreaker acquisition program, because of risks in four key areas:

> *Design.* The Coast Guard set program baselines before conducting a preliminary design review, which puts the program at risk of having an unstable design, thereby increasing the program's cost and schedule risks. While setting baselines without a preliminary design review is consistent with DHS's current acquisition policy, it is inconsistent with acquisition best practices. Based on GAO's prior recommendation, DHS is currently evaluating its policy to better align technical reviews and acquisition decisions.
>
> *Technology.* The Coast Guard intends to use proven technologies for the program, but did not conduct a technology readiness assessment to determine the maturity of key technologies prior to setting baselines. Coast Guard officials indicated such an assessment was not necessary because the technologies the program plans to employ have been proven on other icebreaker ships. However, according to best practices, such technologies can still pose risks when applied to a different program or operational environment, as in this case. Without such an assessment, the program's technical risk is underrepresented.
>
> *Cost.* The lifecycle cost estimate that informed the program's $9.8 billion cost baseline substantially met GAO's best practices for being comprehensive, well-documented, and accurate, but only partially met best practices for being credible. The cost estimate did not quantify the range of possible costs over the entire life of the program. As a result, the cost estimate was not fully reliable and may underestimate the total funding needed for the program.
>
> *Schedule.* The Coast Guard's planned delivery dates were not informed by a realistic assessment of shipbuilding activities, but rather driven by the potential gap in icebreaking capabilities once the Coast Guard's only operating heavy polar icebreaker—the Polar Star—reaches the end of its service life....
>
> GAO's analysis of selected lead ships for other shipbuilding programs found the icebreaker program's estimated construction time of 3 years is optimistic. As a result, the Coast Guard is at risk of not delivering the icebreakers when promised and the potential gap in icebreaking capabilities could widen.[55]

---

[55] Government Accountability Office, *Coast Guard Acquisitions[:] Polar Icebreaker Program Needs to Address Risks before Committing Resources*, GAO-18-600, summary page.

Common Design for Heavy and Medium Polar Icebreakers

Another potential issue for Congress is whether to procure heavy and medium polar icebreakers to a common basic design. As noted earlier (see "June 2013 DHS Polar Icebreaker Mission Need Statement"), the DHS polar icebreaker mission need statement (MNS) states that "current requirements and future projections ... indicate the Coast Guard will need to expand its icebreaking capacity, potentially requiring a fleet of up to six icebreakers (3 heavy and 3 medium) to adequately meet mission demands in the high latitudes...." Consistent with this statement, the Coast Guard envisages procuring up to three new medium icebreakers after it procures three new heavy polar icebreakers. The question is whether to develop a separate design for the medium polar icebreakers, or instead build the medium polar icebreakers to the same basic design as the heavy polar icebreakers.

A congressionally mandated July 2017 report from the National Academies of Sciences, Engineering, and Medicine (NASEM) on the acquisition and operation of polar icebreakers concluded that notional operational requirements for new medium polar icebreakers would result in ships that would not be too different in size from new heavy polar icebreakers. (As shown in *Table 1*, the Coast Guard's current medium polar icebreaker, *Healy*, is actually somewhat larger than the Coast Guard's heavy polar icebreaker, *Polar Star*.) Given what it concluded as the probable similarity in size between future U.S. heavy and medium polar icebreakers, the NASEM report recommended building a single medium polar icebreaker to the same common design as three new heavy polar icebreakers. This approach, the report concluded, would reduce the cost of the medium icebreaker by avoiding the cost of developing a new design and by making the medium polar icebreaker the fourth ship on an existing production learning curve rather than the first ship on a new production learning curve. The NASEM report stated (emphasis as in original):

> 2. Recommendation: The United States Congress should fund the construction of four polar icebreakers of common design that would be owned and operated by the United States Coast Guard (USCG).
>
> The current Department of Homeland Security (DHS) Mission Need Statement... contemplates a combination of medium and heavy icebreakers. The committee's recommendation is for a single class of polar icebreaker with heavy icebreaking capability. Proceeding with a single class means that only one design will be needed, which will provide cost savings. The committee has found that the fourth heavy icebreaker could be built for a lower cost than the lead ship of a medium icebreaker class....
>
> The DHS Mission Need Statement contemplated a total fleet of "potentially" up to six ships of two classes—three heavy and three medium icebreakers. Details appear in the High Latitude Mission Analysis Report. The Mission Need Statement indicated that to

fulfill its statutory missions, USCG required three heavy and three medium icebreakers; each vessel would have a single crew and would homeport in Seattle. The committee's analysis indicated that four heavy icebreakers will meet the statutory mission needs gap identified by DHS for the lowest cost....

4. Finding: In developing its independent concept designs and cost estimates, the committee determined that the costs estimated by USCG for the heavy icebreaker are reasonable. However, the committee believes that the costs of medium icebreakers identified in the High Latitude Mission Analysis Report are significantly underestimated....

Although USCG has not yet developed the operational requirements document for a medium polar icebreaker, the committee was able to apply the known principal characteristics of the USCG Cutter Healy to estimate the scope of work and cost of a similar medium icebreaker. The committee estimates that a first-of-class medium icebreaker will cost approximately $786 million. The fourth ship of the heavy icebreaker series is estimated to cost $692 million. Designing a medium-class polar icebreaker in a second shipyard would incur the estimated engineering, design, and planning costs of $126 million and would forgo learning from the first three ships; the learning curve would be restarted with the first medium design. Costs of building the fourth heavy icebreaker would be less than the costs of designing and building a first-of-class medium icebreaker....

6. Recommendation: USCG should ensure that the common polar icebreaker design is science-ready and that one of the ships has full science capability.

All four proposed ships would be designed as "science-ready," which will be more cost-effective when one of the four ships—most likely the fourth—is made fully science capable. Including science readiness in the common polar icebreaker design is the most cost-effective way of fulfilling both the USCG's polar missions and the nation's scientific research polar icebreaker needs.... The incremental costs of a science-ready design for each of the four ships ($10 million to $20 million per ship) and of full science capability for one of the ships at the initial build (an additional $20 million to $30 million) are less than the independent design and build cost of a dedicated research medium icebreaker.... In briefings at its first meeting, the committee learned that the National Science Foundation and other agencies do not have budgets to support full-time heavy icebreaker access or the incremental cost of design, even though their science programs may require this capability. Given the small incremental cost, the committee believes that the science capability cited above should be included in the acquisition costs.

Science-ready design includes critical elements that cannot be retrofitted cost-effectively into an existing ship and that should be incorporated in the initial design and build. Among these elements are structural supports, appropriate interior and exterior spaces, flexible accommodation spaces that can embark up to 50 science personnel, a hull design that accommodates multiple transducers and minimizes bubble sweep while optimizing icebreaking capability, machinery arrangements and noise dampening to mitigate interference with sonar transducers, and weight and stability latitudes to allow installation of scientific equipment. Such a design will enable any of the ships to be retrofitted for full science capability in the future, if necessary....

> Within the time frame of the recommended build sequence, the United States will require a science-capable polar icebreaker to replace the science capabilities of the *Healy* upon her retirement. To fulfill this need, one of the heavy polar icebreakers would be procured at the initial build with *full* science capability; the ability to fulfill other USCG missions would be retained. The ship would be outfitted with oceanographic overboarding equipment and instrumentation and facilities comparable with those of modern oceanographic research vessels. Some basic scientific capability, such as hydrographic mapping sonar, should be acquired at the time of the build of each ship so that environmental data that are essential in fulfilling USCG polar missions can be collected.[56]

If policymakers decide to procure a second new medium polar icebreaker or a third new medium polar icebreaker, the same general approach recommended by the NASEM report could be followed—a second medium polar icebreaker and third medium polar icebreaker could be built to the same common design used for the three new heavy polar icebreakers and the first new medium polar icebreaker.

An April 12, 2018, press report states:

> As the Coast Guard prepares to review industry bids for a new heavy polar icebreaker, the service is keeping its options open for the right number and mix of polar icebreakers it will need in the future, Adm. Paul Zukunft, the [then-]commandant of the Coast Guard, said on Wednesday [April 11].
>
> The Coast Guard's program of record is for three heavy and three medium polar icebreakers but Zukunft said the "jury is still out" whether that will remain so. Right now, the service is aiming toward building three new heavy icebreakers, but it might make sense just to keep building these ships, he told reporters at a Defense Writers Group breakfast in Washington, D.C.
>
> Zukunft said that "when you start looking at the business case after you build three, and then you need to look at what is the economy of scale when you start building heavy icebreakers, and would it be less expensive to continue to build heavies and not mediums." He added that the heavy icebreakers provide more capability, and if the price is "affordable" and in "the same range" as building medium icebreakers, then "maybe you end up with one class of heavy icebreakers."
>
> Building only one class of ships has a number of advantages in terms of maintenance, crew familiarity, configuration management, and more, he said. A decision on what the future icebreaker fleet will consist of is "still probably several years out .... but that's one option that we want to keep open going forward," Zukunft said.[57]

---

[56] National Academies of Sciences, Engineering, and Medicine, Division on Earth and Life Studies and Transportation Research Board, *Acquisition and Operation of Polar Icebreakers: Fulfilling the Nation's Needs*, Letter Report, with cover letter dated July 11, 2017, pp. 2, 4-6.

[57] Calvin Biesecker, "Coast Guard Leaving Options Open For Future Polar Icebreaker Fleet Type," *Defense Daily*, April 12, 2018. Ellipse as in original.

Building Polar Icebreakers in Foreign Shipyards

Another potential issue for Congress concerns the possibility of building polar icebreakers for the U.S. Coast guard in foreign shipyards. Some observers believe the acquisition cost of U.S. Coast Guard polar icebreakers could be reduced, perhaps substantially, by building them in a foreign shipyard, such as a yard in one of the Nordic countries that is experienced in building icebreakers. Shipyards in Finland reportedly are interested in building polar icebreakers for the U.S. Coast Guard.[58]

Laws Relating to Building Ships in Foreign Shipyards

Some observers have suggested that a U.S. law known as the Jones Act prevents the U.S. Coast Guard from buying or operating a foreign-built polar icebreaker. The Jones Act, however, does not prevent the U.S. Coast Guard from buying or operating a foreign-built polar icebreaker.[59] Two other laws, however, are of note in connection with the idea of building a U.S. Coast Guard polar icebreaker in a foreign shipyard. One is 14 U.S.C. 665, which states the following:

> §665. Restriction on construction of vessels in foreign shipyards
>
> (a) Except as provided in subsection (b), no Coast Guard vessel, and no major component of the hull or superstructure of a Coast Guard vessel, may be constructed in a foreign shipyard.
>
> (b) The President may authorize exceptions to the prohibition in subsection (a) when the President determines that it is in the national security interest of the United States to do so. The President shall transmit notice to Congress of any such determination, and no contract may be made pursuant to the exception authorized until the end of the

---

[58] See, for example, Yereth Rosen, "Can the U.S. Benefit from Finland and Russia's Icebreaker Expertise?" *Arctic Now*, October 9, 2017. See also Jim Paulin, "Finland Wants In On US Icebreaker Investment," *Alaska Dispatch News*, September 8, 2015.

[59] The Jones Act (Section 27 of the Merchant Marine Act of 1920, P.L. 66-261) applies to vessels transporting "merchandise" from one U.S. point to another U.S. point. It requires that such transportation be performed in U.S.-built vessels owned by U.S. citizens and registered in the United States; U.S. registration, in turn, requires that crew members be U.S. citizens. Merchandise is defined to include "merchandise owned by the U.S. Government, a State, or a subdivision of a State; and valueless material" (46 U.S.C. §55102). Merchandise is further defined at 19 U.S.C. §1401(c) to mean "goods, wares, and chattels of every description." It is the waterborne transportation of merchandise domestically that triggers the Jones Act. A vessel wishing to engage in such transportation would apply to the U.S. Coast Guard for a "coastwise endorsement." Thus, an icebreaker strictly performing the task it is designed for and not transporting cargo from one U.S. point to another would not be subject to the Jones Act.

The federal agency in charge of deciding what kind of maritime activity must comply with the Jones Act, U.S. Customs and Border Protection (CBP), has confirmed that icebreaking is not one of those activities. In a 2006 ruling, which appears to be its most recent ruling on the subject, CPB informed Alcoa, Inc. that it could use foreign-built and foreign-flagged vessels for icebreaking on the Hudson River in New York State. CBP reasoned that the transporting of equipment, supplies, and materials used on or from the vessel in effecting its service is not coastwise trade, provided that these articles are necessary for the accomplishment of the vessel's mission and are usually carried aboard the vessel as a matter of course. The 2006 ruling cited earlier rulings in 1974, 1985, and 2000 as precedent.

30-day period beginning on the date the notice of such determination is received by Congress.

The other is 10 U.S.C. 7309, which states the following:

§7309. Construction of vessels in foreign shipyards: prohibition

(a) Prohibition.-Except as provided in subsection (b), no vessel to be constructed for any of the armed forces,[60] and no major component of the hull or superstructure of any such vessel, may be constructed in a foreign shipyard.

(b) Presidential Waiver for National Security Interest.-(1) The President may authorize exceptions to the prohibition in subsection (a) when the President determines that it is in the national security interest of the United States to do so.

(2) The President shall transmit notice to Congress of any such determination, and no contract may be made pursuant to the exception authorized until the end of the 30-day period beginning on the date on which the notice of the determination is received by Congress.

(a)Exception for Inflatable Boats.-An inflatable boat or a rigid inflatable boat, as defined by the Secretary of the Navy, is not a vessel for the purpose of the restriction in subsection (a).

October 2017 Press Report

An October 9, 2017, press report states the following:

Finland, the world leader in icebreaker design and construction, could help pull the United States out of its icebreaker crisis, a diplomat said at a business conference in Anchorage last week.

"The U.S. is now in dire straits about its own icebreaker fleet. They only have two and they are both seriously outdated. We can help," Stefan Lindstrom, Finland's Los Angeles-based consul general, said in a presentation at last week's Arctic Ambitions conference held by the World Trade Center of Alaska....

If the U.S. makes a decision to buy a replacement from overseas, Finnish shipbuilders could respond quickly, Lindstrom said.

In Finland, a shipyard can build and deliver a polar-class icebreaker within 24 months after a contract is signed—a sharp contrast, Lindstrom said, to the extended discussions that the U.S. Coast Guard and Congress have had over planning for potential new icebreakers.

And the costs for a Finnish-designed and Finnish-built polar-class icebreaker is about 200 million to 220 million Euros ($235 [million] to [$]258 million), he said. That's far lower than the price tag being discussed in the US.

[60] 14 U.S.C. 1, which establishes the Coast Guard, states the following: "The Coast Guard, established January 28, 1915, shall be a military service and a branch of the armed forces of the United States at all times."

"I have serious difficulties, however, understanding how you can pay a billion for an icebreaker that costs one-fifth of it if you order it from abroad," Lindstrom said. "But I'm not going to go into those political situations."[61]

It is unclear from the above-quoted remarks whether the €220-million polar-class icebreaker being referred to would qualify as a heavy, medium, or light polar icebreaker, or to what degree it would meet the Coast Guard's desired capabilities for a heavy polar icebreaker (see "Desired Capabilities for New Polar Icebreaker"). Of the six Russian heavy polar icebreakers shown in *Table 2* (all of which are nuclear-powered), four were built in Russia, while the other two—sister ships named Taymyr and Vaygach that entered service around 1989 and 1990—were mostly built in Finland and then moved to a Russian shipyard for the installation of their nuclear reactors. All other Finnish-built icebreakers shown in *Table 2* (whether operated by Finland or other countries) could be considered, based on their brake horsepower (BHP), to be medium or light polar icebreakers.

Short-Term Bridge to One or More New Polar Icebreakers

Overview: Two Basic Options

As mentioned earlier, a new heavy polar icebreaker that begins construction in FY2019 might enter service in 2023, while *Polar Star* was refurbished and reentered service in December 2012 for an intended period of 7 to 10 years—a period that will end between December 2019 and December 2022.[62] Consequently, another potential issue for Congress concerns how to bridge a potential gap in time between the end of *Polar Star's* current intended service life and the entry into service of one or more new heavy polar icebreakers.

As testified by CRS on July 21, 2016,[63] there are at least two options for bridging this time period: One would be to further extend the service life of *Polar Star*. The other would be to charter (i.e., lease) one or more other icebreakers (perhaps foreign-owned ones), if such ships are available for charter and have capabilities for performing missions performed by U.S. heavy polar icebreakers. The United States has used both of these approaches in the past to mitigate polar icebreaking capacity gaps.[64]

[61] Yereth Rosen, "Can the U.S. Benefit from Finland and Russia's Icebreaker Expertise?" *Arctic Now*, October 9, 2017.

[62] The September 25, 2017, GAO report on polar icebreakers states the following: "According to Coast Guard planning documents, the *Polar Star's* useful service life will end between fiscal years 2020 and 2023." (Government Accountability Office, *Coast Guard: Status of Polar Icebreaking Fleet Capability and Recapitalization Plan*, GAO-17- 698R, September 25, 2017, p. 6.

[63] See CRS Testimony TE10012, *Coast Guard Arctic Implementation Capabilities*, by Ronald O'Rourke.

[64] Regarding the first option, the Coast Guard, in addition to the work done to extend the service life of *Polar Star* by an additional 7 to 10 years, also mitigated a polar icebreaking capacity gap in the 1970s by putting two of its older Wind-class icebreakers through a vessel rehabilitation and modernization (VRAM) program. (See National Research Council, Polar Icebreakers in a Changing World: An Assessment of U.S. Needs, Washington, 2007, p. 55. See also Donald L. Canney, "*Icebreakers and the U.S. Coast Guard*," accessed June 28, 2016, at http://www.uscg.mil/history/ webcutters/Icebreakers.asp.)

Coast Guard Plan is to Further Extend Life of Polar Star

The Coast Guard plans to pursue the first of the two options outlined above—further extend the service life of *Polar Star*—and has requested funding in its FY2019 budget for service life extension work on *Polar Star.* A September 25, 2017, GAO report on polar icebreakers states the following:

> While the Coast Guard considered various options to bridge this potential heavy icebreaker gap, in a January 2017 study the Coast Guard reported that it was planning for a limited service life extension of the Polar Star to keep it operational until fiscal year 2025, at an initial cost estimate of $75 million. However, the Coast Guard has not completed a formal cost estimate for this effort and we have previously reported that the $75 million estimate may be unrealistic....
>
> The Coast Guard's Capital Investment Plan for fiscal years 2018-2022 includes $60 million of a planned $75 million for polar icebreaker sustainment, which officials reported as being the rough estimate for the Polar Star's limited service life extension. Coast Guard officials stated that the $75 million rough estimate is based on the cost of the Polar Star's prior 7- 10 year service life extension which was completed in fiscal year 2013. However, in July 2017 we reported that the Coast Guard has not completed a cost estimate for this effort, and that the $75 million estimate may be unrealistic based on the assumptions the Coast Guard used, such as continuing to use parts from the Polar Sea as has been done in previous maintenance events.[65]

A July 2018 GAO report states:

> The Coast Guard is planning a SLEP on the Polar Star to keep it operational until the first and second new heavy polar icebreakers are delivered (planned for 2023 and 2025, according to current acquisition plans) in order to bridge a potential operational gap. This approach would allow the Coast Guard to operate a minimum of two heavy icebreakers once the first polar icebreaker is delivered. The approach would also provide the Coast Guard with a self-rescue capability—the ability for one icebreaker to rescue the other if it became incapacitated while performing icebreaking operations.
>
> The Coast Guard's plan to conduct the Polar Star SLEP during its existing annual depot-level maintenance periods may not be feasible given the amount of maintenance already required on the cutter. The Polar Star's mission capable rating has been decreasing in recent years and reached a low point of 29 percent—well below the target of 41 percent—from October 2016 to September 2017. Based on mission capable data, we found this is mostly due to additional time spent in depot-level maintenance, which has increased in recent years from about 6 months in 2015 to more than 8 months in 2017.
>
> Additionally, the Polar Star has required extensions of about 3 months for its annual dry dock periods—the period of time when a cutter is removed from the water so that

---

[65] Government Accountability Office, *Coast Guard: Status of Polar Icebreaking Fleet Capability and Recapitalization Plan*, GAO-17-698R, September 25, 2017, pp. 3, 8.

maintenance can be conducted—in 2016 and 2017 to complete required maintenance activities. These dry docks were originally planned to last between 2-1/2 months and 4 months. These extensions also compressed the amount of time that the crew had to prepare for its annual mission to Antarctica, which, according to members of the Polar Star crew, placed a large stress on the crew, risked the quality of work, and reduced or eliminated the crews' planned rest and personal preparation for their roughly 4-month deployment. Based on our analysis, these delays and extensions are likely to continue as the cutter ages. According to Coast Guard officials, the Polar Star's SLEP work will be conducted during the annual dry dock periods by adding an additional 1 or 2 months to the annual dry docks. However, if the work is unable to be completed during this time frame, it could force the Coast Guard to miss its commitment to conduct the annual Antarctica mission. Coast Guard maintenance officials stated that until the Polar Star completes the SLEP, its repairs will likely continue to get more expensive and time consuming. We will continue to monitor the Polar Star's SLEP through our annual review of DHS programs.

As we found in July 2017, the Polar Star SLEP effort has a rough order cost estimate of $75 million, which is based on the reactivation work completed in 2013.41 However, this estimate may be unrealistic based on assumptions the Coast Guard used, such as that it would continue to use parts from the Coast Guard's other heavy polar icebreaker, the Polar Sea, which has been inactive since 2010.42 The Coast Guard's recent assessment of the Polar Star's material condition—the physical condition of the cutter, which includes the hull structure, habitability, major equipment systems, and spare parts availability—was completed in January 2018.43 The material assessment stated that many of the available parts from the Polar Sea have already been removed and installed on the Polar Star. As a result of the finite parts available from the Polar Sea, the Coast Guard may have to acquire new parts for the Polar Star that could increase the $75 million SLEP estimate. The Polar Star's recent material assessment will form the basis to determine which systems will be overhauled during the SLEP and for a more detailed cost estimate. The Coast Guard expects the program to reach the obtain phase of the acquisition life cycle by December 2019, at which time the Polar Star could reach the end of its current useful service life (currently projected to be between 2020 to 2023). This timeline contains risk that the Polar Star could be rendered inoperable before the cutter is able to undergo a SLEP.[66]

Another Option: Chartering an Icebreaker

Overview

The feasibility of the second of the two options outlined above— charter (i.e., lease) one or more other icebreakers—would depend on whether an icebreaker was available for charter at the time of the year when the United States would need it to perform desired missions in the Arctic or Antarctic. Foreign polar icebreakers are used by their own

[66] Government Accountability Office, *Coast Guard Acquisitions[:] Actions Needed to Address Longstanding Portfolio Management Challenges*, GAO-18-454, July 2018, pp. 29-31.

countries for icebreaking operations, and may not always be available for charter when the United States might want to use them. If an icebreaker were available for charter, the potential cost effectiveness of this option would then depend on the cost of the charter, the ability of the ship to perform U.S. polar icebreaker missions, and how these costs and capabilities compare to the option of extending the service life of *Polar Star*.

Aiviq Being Offered for Lease

One ship that is being offered for lease to the Coast Guard as an interim polar icebreaker is *Aiviq* (*Figure 4*), an Arctic oil-exploration support ship owned by Edison Chouest Offshore (ECO). The 361-foot-long ship was ordered in 2009, completed in 2012, and chartered by Royal Dutch Shell to support that company's effort (now ended) to explore for oil in Arctic waters. Following Shell's decision to end that effort, alternative uses for *Aiviq* have been sought. The ship has been modified to serve as a polar icebreaker, and it is being offered to the Coast Guard for lease as an interim polar icebreaker. It reportedly has also been offered for use as an icebreaker to the Canadian government.[67]

The possibility of leasing *Aiviq* as an interim polar icebreaker has been discussed at certain recent hearings about the Coast Guard. For example, at a July 25, 2017, hearing on Coast Guard capabilities before the Coast Guard and Maritime transportation subcommittee of the House Transportation and Infrastructure Committee, the following exchange occurred:

> REPRESENTATIVE DON YOUNG (continuing):
>
> Have you looked at, Admiral, I know this has been an ongoing battle with me and the Coast Guard over the years, the other possibility of getting an ice breaker into the arena quicker than having one constructed like leasing from another outfit? You know, I've been talking about this a long time. Have you analyzed this again?
>
> I know the last time we had a study, it was 1980. That's a long time ago. So is there a way we can put metal on the water, especially for the new shipping through and the—and the cruise ships, because that Healy is old, and—is—have you looked at that at all?
>
> ADMIRAL PAUL ZUKUNFT, [THEN-]COMMANDANT, U.S. COAST GUARD:
>
> We have. In fact, one potential vendor, we've had multiple interactions. They have a platform that has yet to complete ice trials. We—we would not want to lease something they can't demonstrate its ability to actually operate in the ice that—that Healy sees. Healy was actually beset in ice for 36 hours last year, so it's not ice free up there, and that's a medium ice breaker. This particular platform doesn't have the capability of Healy.

[67] See for example, "More Sparks Fly in Canada Shipbuilding Controversy," *Marine Log*, March 18, 2016; Pierre Leblanc, "An Out-of-the-Blue Icebreaker Opportunity," *Maritime Executive*, January 2, 2018.

But we would at least want to make sure that ice trials were completed. That we could actually be a good steward of taxpayer dollars, so at least a platform that would meet our requirements. So we've had multiple interactions, the last one was probably in May, and the issue of ice trials is still on the table right now.[68]

Source: "Arctic Supply Vessel Aiviq," accessed September 14, 2017, at http://www.mascoat.com/arctic-supplyvessel-aiviq1.html.

Figure 4. Aiviq.

Later in the same hearing, the following exchange occurred:

REPRESENATIVE DUNCAN HUNTER, CHAIRMAN:

Going back to Mr. Young's question. too, about leasing. You said you—you're—you're waiting for—I'm—I'm guessing money for ice trials. That's what you said.

ZUKUNFT:

No real dollars have been negotiated in any of this. So...

HUNTER:

But in—in real terms, you're only paying for gas? I mean what—what does it cost to do ice trials. It's gas, right? You're not going to hire more Coast Guardsmen to come in and— and do it. I mean so that's a figure—your—your overhead's fixed. So what is the cost to— to go do ice trials with the (inaudible)?

ZUKUNFT:

That would really be for the...

[68] Source: Transcript of hearing.

HUNTER:
The ice—once again the only...

ZUKUNFT:
... vendor to decide.

HUNTER:
...existing U.S. made ice breaker in America.

ZUKUNFT:
Yeah. So this—this is a ship that is built with direct drive diesel. Ice breakers are typically diesel electric, which means the generators push the shaft, and they absorb that shock load every time you collide with ice.

A reduction gear, fixed gear is going to that—that gear box is going to absorb all that shock. So if you're going to do ice trials, there's a likelihood you might have to replace a reduction gear. There might be real hidden costs of doing ice trials. So if I'm a vendor, I might want to protect myself from some of that risk.

Now I'm not the vendor but those would be some of my thoughts of, OK, if you're really serious about this and I do ice trials and now I've just caused X number of dollars that I am now going to have to fit. And oh, by the way, you're not going to lease it because it didn't meet your requirements. I think those are some of the issues that we still have to negotiate.[69]

At a June 14, 2016, hearing on Coast Guard mission needs and resource allocation before the Coast Guard and Maritime Transportation subcommittee of the House Transportation and Infrastructure Committee, the following exchange occurred:

REPRESENTATIVE HUNTER (Chairman):
How do you plan on—on filling the capability gap until you get a heavy icebreaker, which is 10 years at the least based on the best projections of Congress and everybody working together? You still haven't answered that one.

ADMIRAL MICHEL:
Well, right—the alternatives now, since we'll provide the answer to that, and it's probably going to be either a rolling recapitalization of the *Polar Star* or to try to bring—let *Polar Star* taper off and then try to bring *Polar Sea* back on and bridge out to the new icebreaker.

I do not know which one at this point, which path we would want to take. I'm not aware of any other—we've looked out there for vessels to lease for heavy icebreaking capabilities. There's nothing out there on planet earth that you can lease in the heavy icebreaking area. So that's kind of where we are, sir.

---

[69] Source: Transcript of hearing.

HUNTER:
Was it the—the Finns that came into my office?

(UNKNOWN)
Mm-hmm.

HUNTER:
Can't remember whether we had the Norwegians or the Finns. I mean, they—have you— you've obviously looked at that, right?

MICHEL:
Yes. As a matter of fact I—I traveled to Sweden and Finland...

HUNTER:
Yeah.

MICHEL:
... and talked to them. And they do not have heavy icebreaking capability that will meet the needs as in the FedBizOpps. As a matter of fact, in—when I'm talking FedBizOpps [I mean] there's a technical package that the Coast Guard put out for our [new] heavy icebreaker [i.e., the one that the Obama Administration wanted to begin building in 2020].

It kind of lays out our basic requirements including the long pole in the tent which is the icebreaking requirement, which is six foot minimum at three knots, desirable eight-foot minimum at three knots and then 21 feet backing and ramming.

When I talked to the shipbuilders over there, they said there is not a vessel like that that currently exists that will meet those requirements in the—in the FedBizOpps technical package. So you'd have to build a vessel like that. And that's the type of vessel that we're looking for.[70]

The Coast Guard stated in July 2016 that

NSF [the National Science Foundation] leased the icebreaker KRASIN from Russia from 2005-2006, ODEN from the Swedish government from 2007-2010, and VLADIMIR IGNATYUK from Russia in 2012 to support the McMurdo resupply mission. All leases were time charters, and crews were supplied with the leases. As a contingency measure, NSF obtained assurances of assistance from other vessels in the area, such as the Chinese flagged [icebreaking] vessel XUE LONG, in the event they encountered difficulty. They also hired icebreaker captains with previous McMurdo experience to supplement the crew. NSF acquired these leases through a RFP process,

[70] Transcript of hearing.

and had no assurances that icebreakers would be available to perform the mission, or what price would be quoted.

This process came with risks, as there was no way to gauge icebreaker availability until NSF received responses to their RFP. Additionally, a foreign-flagged commercial or state vessel can become unavailable for a variety of environmental and political reasons. For example, the Swedish government abruptly terminated their contract during the spring/summer of 2011, and NSF was left without a platform to conduct its mission. NSF requested support from CGC [Coast Guard cutter] HEALY, but it was employed in the Arctic. NSF ultimately leased the Russian icebreaker VLADIMIR IGNATYUK. After that incident, NSF decided to utilize CGC POLAR STAR to support the McMurdo mission, which it has been doing since 2013.[71]

Acquisition vs. Leasing

In addition to the issues for Congress discussed earlier in this report, another potential issue for Congress that arises from time to time is whether future polar icebreakers should be acquired through a traditional acquisition (i.e., the government procuring the ship and owning it throughout its service life) or through a leasing arrangement (under which the icebreakers would be privately built and privately owned, leased to the Coast Guard, and crewed by an all-Coast Guard crew or a mix of Coast Guard personnel and civilian mariners). For additional information on this issue, see *Appendix E*.

Legislative Activity for FY2019

Summary of Appropriation Action on FY2019 Funding Request

The Coast Guard's proposed FY2019 budget requests $750 million in Coast Guard acquisition funding for the PSC program. *Table 3* summarizes congressional appropriation action on the program's FY2019 funding request.

FY2019 DHS Appropriations Act (H.R. XXXX/S. 3109)

House

The House Appropriations Committee marked up the FY2019 DHS Appropriations Act (referred to here as H.R. XXXX) on July 25, 2018. The text of the bill as marked up,

[71] Source: Email from Guard Office of Congressional Affairs to CRS, July 8, 2016.

and the committee's report reflecting the markup, were not available as of August 3, 2018. The figures shown in the HAC column of *Table 3* and the discussion below are based on the bill text and draft committee report (referred to here as H.Rept. 115-XXX) going into July 25, 2018, markup meeting, combined with a summary of the amendments adopted at the markup meeting, which were posted on the committee's website in conjunction with the pre-markup bill text and draft committee report.[72]

Table 3. Summary of Congressional Appropriations Action on FY2019 Funding Request (millions of dollars)

Polar icebreaker	**Request**	**HAC**	**SAC**	**Conf.**
New polar icebreaker				
Coast Guard acquisition account	750	0	750	
Navy shipbuilding account	0	0	0	
Subtotal	750	0	750	
Polar sustainment (service life extension of Polar Star)				
Coast Guard acquisition account	15	15	15	
Total	765	15	765	

Source: Table prepared by CRS, based on Coast Guard's FY2019 budget submission and HAC committee report, and SAC chairman's recommendation and explanatory statement on FY2019 DHS Appropriations Act and FY2019 DOD Appropriations Act. HAC is House Appropriations Committee; SAC is Senate Appropriations Committee; Conf. is conference agreement.

H.Rept. 115-XXX recommends the funding levels shown in the HAC column of *Table 3*. H.Rept. 115-XXX states:

> *Polar Ice Breaking Vessel.* The Committee recognizes that Polar icebreakers are essential to securing the nation's security and economic interests in the Polar regions. While the recommendation does not include the requested funding for the construction of a new Polar Icebreaker, the Committee plans to work with the Coast Guard to determine what is needed in fiscal year 2019 to advance this program. The Coast Guard's existing operational icebreaking fleet consists of just one heavy icebreaker, the *POLAR STAR*, which entered into service in 1976. The Coast Guard has testified that it will need to sustain the *POLAR STAR* beyond two years after delivery of the first of the new class of icebreakers to ensure mission readiness, thus the Committee recommends $15,000,000, as requested. The Committee looks forward to the updated cost estimate for the *POLAR STAR*'s service life extension project (SLEP) that is anticipated this summer. (Pages 38-39)

[72] Houser Appropriations Committee, "*Appropriations Committee Approves Fiscal Year 2019 Homeland Security Funding Bill*," July 25, 2019, accessed August 3, 2018, at https://appropriations.house.gov/news/document single.aspx?DocumentID=395388.

H.Rept. 115-XXX also states:

> The Coast Guard is directed to brief the Committee not later than 90 days after the date of enactment of this Act with an update on the results of the examination, for which up to $5,000,000 was provided in the fiscal year 2018 DHS Appropriations Act, on whether the Coast Guard's heavy icebreaking requirements can be met by existing vessels using short-term procurement strategies. (Page 40)

Senate

The Senate Appropriations Committee, in its report (S.Rept. 115-283 of June 21, 2018) on S. 3109, recommends the funding levels shown in the SAC column of *Table 3*. S.Rept. 115-283 states:

> *Polar Ice Breaking Vessel.*—The recommendation includes $750,000,000, as requested, to maintain the accelerated acquisition schedule for a new class of heavy polar icebreakers that was established in fiscal year 2018. These funds will be used to request proposals and award contracts for detail design and construction near the beginning of fiscal year 2020. Heavy polar icebreakers are essential to securing the national security and economic interests of the United States in the Arctic and Antarctic. To ensure the United States is able to achieve this objective in the most expeditious and efficient manner possible, the Coast Guard should explore block buy pricing for the heavy polar icebreakers. Not later than 180 days after the date of the enactment of this act, the Coast Guard shall submit a report to the Committee on the feasibility of block buy contracting for the acquisition of heavy polar icebreakers.
>
> *Polar Star.*—The recommendation includes $15,000,000 to carry out a service life extension program [SLEP] for the POLAR STAR to extend its service life so that it remains operational until the delivery of the second new heavy polar icebreaker. The Coast Guard's two existing heavy polar icebreakers are over 45 years old and well past their planned service life. Currently, only one heavy polar icebreaker is in active service [POLAR STAR], and the other vessel [POLAR SEA] is in an inactive status serving to provide specialty parts to help sustain POLAR STAR. Continued funding for its SLEP will ensure the POLAR STAR can meet and support national interests and provide assured surface presence in the Arctic and Antarctic. (Pages 68-69)

S.Rept. 115-283 also states (emphasis added):

> *Full-Funding Policy.*—The Committee again directs an exception to the administration's current acquisition policy that requires the Coast Guard to attain the total acquisition cost for a vessel, including long lead time materials [LLTM], production costs, and postproduction costs, before a production contract can be awarded. This policy has the potential to make shipbuilding less efficient, to force delayed obligation of production funds, and to require post-production funds far in advance of when they will be used. The Department should position itself to acquire vessels in the most efficient

manner within the guidelines of strict governance measures. The Committee expects the administration to adopt a similar policy for the acquisition of the Offshore Patrol Cutter [OPC] and *heavy polar icebreaker*. (Page 67)

Frank LoBiondo Coast Guard Authorization Act of 2018 (S. 140/P.L. 115-282)

S. 140 was introduced in the Senate on January 12, 2017, as a bill to amend the White Mountain Apache Tribe [WMAT] Water Rights Quantification Act of 2010 to clarify the use of amounts in the WMAT Settlement Fund. The bill retained that purpose and was acted on by Congress into 2018. Later in 2018, S. 140 became a bill to authorize appropriations for the Coast Guard, and for other purposes. On November 14, 2018, the Senate concurred, 94-6, in the House amendment to S. 140 with an amendment (S.Amdt. 4054 as modified). On November 27, 2018, the House agreed to by voice vote a motion that the House suspend the rules and agree to the Senate amendment to the House amendment. S. 140 was signed into law as P.L. 115-282 on December 4, 2018.

Section 311 of S.140/P.L. 115-282 states:

> SEC. 311. Contracting for major acquisitions programs.
>
> (a) General acquisition authority.—Section 501(d) of title 14, United States Code, is amended by inserting "aircraft, and systems," after "vessels,".
>
> (b) Contracting authority.—Chapter 11 of title 14, United States Code, as amended by this Act, is further amended by inserting after section 1136 the following:
>
> "§ 1137. Contracting for major acquisitions programs
>
> "(a) In general.—In carrying out authorities provided to the Secretary to design, construct, accept, or otherwise acquire assets and systems under section 501(d), the Secretary, acting through the Commandant or the head of an integrated program office established for a major acquisition program, may enter into contracts for a major acquisition program.
>
> "(b) Authorized methods.—Contracts entered into under subsection (a)—
>
> "(1) may be block buy contracts;
>
> "(2) may be incrementally funded;
>
> "(3) may include combined purchases, also known as economic order quantity purchases, of—
>
> (A) materials and components; and
>
> (B) long lead time materials; and
>
> "(4) as provided in section 2306b of title 10, may be multiyear contracts.
>
> "(c) Subject to appropriations.—Any contract entered into under subsection (a) shall provide that any obligation of the United States to make a payment under the contract is subject to the availability of amounts specifically provided in advance for that purpose in subsequent appropriations Acts.".

(c) Clerical amendment.—The analysis for chapter 11 of title 14, United States Code, as amended by this Act, is further amended by inserting after the item relating to section 1136 the following:

"1137. Contracting for major acquisitions programs.".

(d) Conforming amendments.—The following provisions are repealed:

(1) Section 223 of the Howard Coble Coast Guard and Maritime Transportation Act of 2014 (14 U.S.C. 1152 note), and the item relating to that section in the table of contents in section 2 of such Act.

(2) Section 221(a) of the Coast Guard and Maritime Transportation Act of 2012 (14 U.S.C. 1133 note).

(3) Section 207(a) of the Coast Guard Authorization Act of 2016 (14 U.S.C. 561 note).

(e) Internal regulations and policy.—Not later than 180 days after the date of enactment of this Act, the Secretary of the department in which the Coast Guard is operating shall establish the internal regulations and policies necessary to exercise the authorities provided under this section, including the amendments made in this section.

(f) Multiyear contracts.—The Secretary of the department in which the Coast Guard is operating is authorized to enter into a multiyear contract for the procurement of a tenth, eleventh, and twelfth National Security Cutter and associated government-furnished equipment.

Section 821 of S. 140/P.L. 115-282 states:

SEC. 821. Polar icebreakers.

(a) Enhanced maintenance program for the Polar Star.—

(1) IN GENERAL.—Subject to the availability of appropriations, the Commandant of the Coast Guard shall conduct an enhanced maintenance program on Coast Guard Cutter Polar Star (WAGB–10) to extend the service life of such vessel until at least December 31, 2025.

(2) REQUIREMENT FOR REPORT.—Not later than 180 days after the date of the enactment of the Coast Guard Authorization Act of 2017, the Secretary of the department in which the Coast Guard is operating, in consultation with Naval Sea Systems Command, shall submit to the Committee on Commerce, Science, and Transportation and the Committee on Armed Services of the Senate and the Committee on Transportation and Infrastructure and the Committee on Armed Services of the House of Representatives a detailed report describing a plan to extend the service life of the Coast Guard Cutter Polar Star (WAGB–10) until at least December 31, 2025, through an enhanced maintenance program.

(3) CONTENT.—The report required by paragraph (2) shall include the following:

(A) An assessment and discussion of the enhanced maintenance program recommended by the National Academies of Sciences, Engineering, and Medicine's Committee on Polar Icebreaker Cost Assessment in the letter report "Acquisition and Operation of Polar Icebreakers: Fulfilling the Nation's Needs".

(B) An assessment and discussion of the Government Accountability Office's concerns and recommendations regarding service life extension work on Coast Guard Cutter Polar Star (WAGB–10) in the report "Status of the Coast Guard's Polar Icebreaking Fleet Capability and Recapitalization Plan".

(C) Based upon a materiel condition assessment of the Coast Guard Cutter Polar Star (WAGB–10)—

(i) a description of the service life extension needs of the vessel;

(ii) detailed information regarding planned shipyard work for each fiscal year to meet such needs; and

(iii) an estimate of the amount needed to be appropriated to complete the enhanced maintenance program.

(D) A plan to ensure the vessel will maintain seasonally operational status during the enhanced maintenance program.

(4) AUTHORIZATION OF APPROPRIATIONS.—The Commandant of the Coast Guard may use funds made available pursuant to section 4902 of title 14, United States Code, as amended by section 202 of this Act, for the enhanced maintenance program described in the report required by subsection (a).

(b) Coast Guard and Maritime Transportation Act of 2012; amendment.—Section 222 of the Coast Guard and Maritime Transportation Act of 2012 (Public Law 112–213), as amended, is further amended as follows:

(1) by striking subsections (a) through (d);

(2) by redesignating subsections (e) through (g) as subsections (a) through (c), respectively;

(3) in subsection (a), as redesignated—

(A) in the matter preceding paragraph (1), by striking "Except as provided in subsection (c), the Commandant" and inserting "The Commandant";

(B) in paragraph (1) by striking "Polar Sea or";

(C) in paragraph (2) by striking "either of the vessels" and inserting "the Polar Star or the Polar Sea"; and

(D) in paragraph (3) by striking "either of the vessels" each place it appears and inserting "the Polar Star".

National Defense Authorization Act for Fiscal Year 2019/John S. McCain National Defense Authorization Act for Fiscal Year 2019 (H.R. 5515/S. 2987)

House Committee Report

In H.R. 5515 as reported by the House Armed Services Committee (H.Rept. 115-676 of May 15, 2018), *Section 841* states (emphasis added):

SEC. 841. Requirement that certain ship components be manufactured in the national technology and industrial base.

(a) Additional procurement limitation.—Section 2534(a) of title 10, United States Code, is amended by adding at the end the following new paragraph:

"(6) COMPONENTS FOR AUXILIARY SHIPS.—Subject to subsection (k), the following components:

(A) Auxiliary equipment, including pumps, for all shipboard services.

(A) Propulsion system components, including engines, reduction gears, and propellers.

Shipboard cranes.

Spreaders for shipboard cranes.".

(b) Implementation.—Such section is further amended by adding at the end the following new subsection:

"(k) Implementation of auxiliary ship component limitation.—Subsection (a)(6) applies only with respect to contracts awarded by the Secretary of a military department for new construction of an auxiliary ship after the date of the enactment of the National Defense Authorization Act for Fiscal Year 2019 using funds available for National Defense Sealift Fund programs or Shipbuilding and Conversion, Navy. *For purposes of this subsection, the term 'auxiliary ship' does not include an icebreaker."*

House Floor Action (Adding Coast Guard Authorization Act of 2017)

On May 22, 2018, as part of its consideration of the National Defense Authorization Act for Fiscal Year 2019 (H.R. 5515), the House agreed to by voice vote H.Amdt. 641, an en bloc amendment that included, inter alia, amendment number 52 as printed in H.Rept. 115-702 of May 22, 2018, on H.Res. 908, providing for the further consideration of H.R. 5515. Amendment number 52 added to H.R. 5515, as a new Division D, the *Coast Guard Authorization Act of 2017.*

Section 4311 within Division D states:

SEC. 4311. Contracting for major acquisitions programs.

(a) General acquisition authority.—Section 501(d) of title 14, United States Code, is amended by inserting "aircraft, and systems," after "vessels,".

(b) Contracting authority.—Chapter 11 of title 14, United States Code, as amended by this division, is further amended by inserting after section 1136 the following:

"§ 1137. Contracting for major acquisitions programs

"(a) In general.—In carrying out authorities provided to the Secretary to design, construct, accept, or otherwise acquire assets and systems under section 501(d), the Secretary, acting through the Commandant or the head of an integrated program office established for a major acquisition program, may enter into contracts for a major acquisition program.

"(b) Authorized methods.—Contracts entered into under subsection (a)—

"(1) may be block buy contracts;

"(2) may be incrementally funded;

"(3) may include combined purchases, also known as economic order quantity purchases, of—

(A) materials and components; and

(B) long lead time materials; and

"(4) as provided in section 2306b of title 10, may be multiyear contracts.

"(c) Subject to appropriations.—Any contract entered into under subsection (a) shall provide that any obligation of the United States to make a payment under the contract is subject to the availability of amounts specifically provided in advance for that purpose in subsequent appropriations Acts.".

(c) Clerical amendment.—The analysis for chapter 11 of title 14, United States Code, as amended by this division, is further amended by inserting after the item relating to section 1136 the following:

"1137. Contracting for major acquisitions programs.".

(d) Conforming amendments.—The following provisions are repealed:

(1)Section 223 of the Howard Coble Coast Guard and Maritime Transportation Act of 2014 (14 U.S.C. 1152 note), and the item relating to that section in the table of contents in section 2 of such Act.

(2) Section 221(a) of the Coast Guard and Maritime Transportation Act of 2012 (14 U.S.C. 1133 note).

Section 207(a) of the Coast Guard Authorization Act of 2016 (14 U.S.C. 561 note).

(e) Internal regulations and policy.—Not later than 180 days after the date of enactment of this Act, the Secretary of the department in which the Coast Guard is operating shall establish the internal regulations and policies necessary to exercise the authorities provided under this section, including the amendments made in this section.

(f) Multiyear contracts.—The Secretary of the department in which the Coast Guard is operating is authorized to enter into a multiyear contract for the procurement of a tenth, eleventh, and twelfth National Security Cutter and associated government-furnished equipment.

Section 4822 within Division D states:

SEC. 4821. Polar icebreakers.

(a) Enhanced maintenance program for the Polar Star.—

(1) IN GENERAL.—Subject to the availability of appropriations, the Commandant of the Coast Guard shall conduct an enhanced maintenance program on Coast Guard Cutter Polar Star (WAGB–10) to extend the service life of such vessel until at least December 31, 2025.

(2) REQUIREMENT FOR REPORT.—Not later than 180 days after the date of the enactment of the Coast Guard Authorization Act of 2017, the Secretary of the department in which the Coast Guard is operating, in consultation with Naval Sea Systems Command, shall submit to the Committee on Commerce, Science, and Transportation of the Senate and the Committee on Transportation and Infrastructure of the House of Representatives a detailed report describing a plan to extend the service life of the Coast Guard Cutter Polar Star (WAGB–10) until at least December 31, 2025, through an enhanced maintenance program.

(3) CONTENT.—The report required by paragraph (2) shall include the following:

(A) An assessment and discussion of the enhanced maintenance program recommended by the National Academies of Sciences, Engineering, and Medicine's

Committee on Polar Icebreaker Cost Assessment in the letter report "Acquisition and Operation of Polar Icebreakers: Fulfilling the Nation's Needs".

(B) An assessment and discussion of the Government Accountability Office's concerns and recommendations regarding service life extension work on Coast Guard Cutter Polar Star (WAGB–10) in the report "Status of the Coast Guard's Polar Icebreaking Fleet Capability and Recapitalization Plan".

(C) Based upon a materiel condition assessment of the Coast Guard Cutter Polar Star (WAGB–10)—

(i) a description of the service life extension needs of the vessel;

(ii) detailed information regarding planned shipyard work for each fiscal year to meet such needs; and

(iii) an estimate of the amount needed to be appropriated to complete the enhanced maintenance program.

(D) A plan to ensure the vessel will maintain seasonally operational status during the enhanced maintenance program.

(4) AUTHORIZATION OF APPROPRIATIONS.—The Commandant of the Coast Guard may use funds made available pursuant to section 4902 of title 14, United States Code, as amended by section 4202 of this division, for the enhanced maintenance program described in the report required by subsection (a).

(b) Overdue report.—Upon the date of enactment of the Coast Guard Authorization Act of 2017, the Secretary of the department in which the Coast Guard is operating shall submit to the Committee on Commerce, Science, and Transportation of the Senate and the Committee on Transportation and Infrastructure of the House of Representatives the polar icebreaker recapitalization plan required under section 3523 of the National Defense Authorization Act for Fiscal Year 2017 (Public Law 114–328).

(c) Coast Guard and Maritime Transportation Act of 2012; amendment.—Section 222 of the Coast Guard and Maritime Transportation Act of 2012 (Public Law 112–213), as amended, is further amended as follows:

(i) by striking subsections (a) through (d);

(2) by redesignating subsections (e) through (g) as subsections (a) through (c), respectively;

(3) in subsection (a), as redesignated—

(A) in the matter preceding paragraph (1), by striking "Except as provided in subsection (c), the Commandant" and inserting "The Commandant";

(B) in paragraph (1) by striking "Polar Sea or";

(C) in paragraph (2) by striking "either of the vessels" and inserting "the Polar Star or the Polar Sea"; and

(D) in paragraph (3) by striking "either of the vessels" each place it appears and inserting "the Polar Star".

Senate

In S. 2987 as reported by the Senate Armed Services Committee (S.Rept. 115-262 of June 5, 2018), *Section 153* states:

SEC. 153. Authority to procure additional polar-class icebreakers.

Section 122 of the National Defense Authorization Act for Fiscal Year 2018 (Public Law 115–91) is amended—

(1) in the section heading, by striking "Icebreaker vessel" and inserting "Authorization to procure up to six polar-class icebreakers";

(2) by striking subsections (a) and (b);

(3) by inserting before subsection (c) the following new subsection:

"(a) Authority To procure icebreakers.—The Secretary of the department in which the Coast Guard is operating may, in consultation with the Secretary of the Navy, enter into a contract or contracts for the procurement of up to six polar-class icebreakers, including—

(1) polar-class heavy icebreakers; and

(2) polar-class medium icebreakers.";

(4) by redesignating subsections (c) and (d) as subsections (b) and (c), respectively; and

(5) in paragraph (1) of subsection (b), as redesignated by paragraph (4) of this section, by striking "subsection (a)(1)" and inserting "subsection (a)".

S.Rept. 115-262 states:

Navy equipment for the Heavy Polar Icebreaker program

The committee notes the Government Accountability Office (GAO) published a report on April 13, 2017, titled "Status of Coast Guard's Heavy Polar Icebreaker Acquisition" (GAO–18–385R), which noted added space, weight, and power reservations for Navy equipment, such as a multi-mode radar and minor caliber weapons, were incorporated in the Department of Homeland Security-approved Operational Requirements Document forthe Heavy Polar Icebreaker (HPIB) in January 2018. The committee is interested in better understanding the plan for Navy equipment to be incorporated on HPIBs.

Accordingly, not later than December 1, 2018, the Secretary of the Navy, in consultation with the Under Secretary of Homeland Security for Management, shall submit to the Committees on Armed Services of the Senate and the House of Representatives an unclassified report, which may include a classified annex, containing the following: (1) A detailed description of Navy equipment planned to be included in HPIBs, including Navy-Type, Navy-Owned equipment; (2) The estimated space, weight, power, and cost for the equipment described in paragraph (1); (3) A description of Navy equipment under consideration to be included in HPIBs; (4) The estimated space, weight, power, and cost for the equipment described in paragraph (3); (5) An explanation of the capability of the equipment listed in paragraphs (1) and (3) to assist or augment the missions of the Combatant Commanders and the execution of the Department of Defense's 2016 Arctic Strategy; and (6) A description of how the equipment listed in paragraphs (1) and (3) will meet a modular open systems approach to allow for future mission expansion. (Page 47)

Conference

In the conference report (H.Rept. 115-874 of July 25, 2018) on H.R. 5515, *Section 151* states:

SEC. 151. PROCUREMENT AUTHORITY FOR ADDITIONAL ICEBREAKER VESSELS.

(a) PROCUREMENT AUTHORITY.—

(1) IN GENERAL.—In addition to the icebreaker vessel authorized to be procured under section 122(a) of the National Defense Authorization Act for Fiscal Year 2018 (Public Law 115–91), the Secretary of the department in which the Coast Guard is operating may enter into one or more contracts for the procurement of up to five additional polar-class icebreaker vessels.

(2) CONDITION FOR OUT-YEAR CONTRACT PAYMENTS.—A contract entered into under paragraph (1) shall provide that any obligation of the United States to make a payment under the contract for a fiscal year after fiscal year 2019 is subject to the availability of appropriations or funds for that purpose for such later fiscal year.

(b) SENSE OF CONGRESS.—It is the sense of Congress that the Coast Guard should maintain an inventory of not fewer than six polar-class icebreaker vessels beginning not later than fiscal year 2029 and, to achieve such inventory, should—

(1) award a contract for the first new polar-class icebreaker not later than fiscal year 2019;

(2) deliver the first new polar-class icebreaker not later than fiscal year 2023;

(3) start construction on the second through sixth new polar-class icebreakers at a rate of one vessel per year in fiscal years 2022 through 2026; and

(4) accept delivery of the second through sixth new polar-class icebreakers at a rate of one vessel per year in fiscal years 2025 through 2029.

Regarding Section 151, H.Rept. 115-874 states:

Procurement authority for additional icebreaker vessels (sec. 151)

The Senate amendment contained a provision (sec. 153) that would amend section 122 of the National Defense Authorization Act for Fiscal Year 2018 (Public Law 115–91) by striking subsections (a) and (b), as well as providing authority to enter into a contract or contracts for up to six polar-class icebreakers.

The House bill contained no similar provision.

The House recedes with an amendment that would provide the secretary of the department in which the Coast Guard is operating the authority to enter into a contract or contracts for the procurement of up to five additional polar-class icebreakers and express the sense of Congress regarding polar-class icebreakers.

The conferees note that section 207 of the Coast Guard Authorization Act of 2015 (Public Law 114–120) provided authority for the Commandant of the Coast Guard to enter into a contract or contracts for the acquisition of polar icebreakers and associated equipment using incremental funding. The conferees further note the Fiscal Years 2019 through 2023 Future Years Homeland Security Program includes $1.8 billion to fully fund 3 icebreakers. The conferees understand that additional Department of Defense funds are not required to procure icebreakers for the foreseeable future. The conferees support the Coast Guard's stated goal of building six icebreakers and believe achieving this objective should be accomplished as expeditiously as possible. (Page 806)

H.Rept. 115-874 also states:

Coast Guard Authorization Act of 2018.

The House bill contained a division (Division D) that would authorize certain aspects of the Coast Guard.

The Senate amendment contained no similar provisions. The House recedes. (Page 1137)

Appendix A. Studies in Recent Years Relating to Coast Guard Polar Icebreakers

A number of studies have been conducted in recent years to assess U.S. requirements for polar icebreakers and options for sustaining and modernizing the Coast Guard's polar icebreaker fleet. This appendix presents the findings of some of these studies, with the most recent study on top.

July 2017 National Academies (NASEM) Report

A July 2017 report on the acquisition and operation of polar icebreakers by the National Academies of Sciences, Engineering, and Medicine (NASEM) that was directed by Congress in Section 604 of the Coast Guard Authorization Act of 2015 (H.R. 4188/P.L. 114-120 of February 8, 2016) concluded the following:

INTRODUCTION

The United States has strategic national interests in the polar regions. In the Arctic, the nation must protect its citizens, natural resources, and economic interests; assure sovereignty, defense readiness, and maritime mobility; and engage in discovery and research. In the Antarctic, the United States must maintain an active presence that includes access to its research stations for the peaceful conduct of science and the ability to participate in inspections as specified in the Antarctic Treaty. The committee's charge... was to advise the U.S. House of Representatives and the U.S. Senate on an assessment of the costs incurred by the federal government in carrying out polar icebreaking missions and on options that could minimize lifecycle costs. The committee's consensus findings and recommendations are presented below. Unless otherwise specified, all estimated costs and prices for the future U.S. icebreakers are expressed in 2019 dollars, since that is the year in which the contracts are scheduled to be made. Supporting material is found in the appendices.

FINDINGS AND RECOMMENDATIONS

1. Finding: The United States has insufficient assets to protect its interests, implement U.S. policy, execute its laws, and meet its obligations in the Arctic and Antarctic because it lacks adequate icebreaking capability.

For more than 30 years, studies have emphasized the need for U.S. icebreakers to maintain presence, sovereignty, leadership, and research capacity—but the nation has failed to respond....The strong warming and related environmental changes occurring in both the Arctic and the Antarctic have made this failure more critical. In the Arctic, changing sea ice conditions will create greater navigation hazards for much of the year, and expanding human industrial and economic activity will magnify the need for national presence in the region. In the Antarctic, sea ice trends have varied greatly from year to year, but the annual requirements for access into McMurdo Station have not changed. The nation is ill-equipped to protect its interests and maintain leadership in these regions and has fallen behind other Arctic nations, which have mobilized to expand their access to ice-covered regions. The United States now has the opportunity to move forward and acquire the capability to fulfill these needs....

2. Recommendation: The United States Congress should fund the construction of four polar icebreakers of common design that would be owned and operated by the United States Coast Guard (USCG).

The current Department of Homeland Security (DHS) Mission Need Statement (DHS 2013) contemplates a combination of medium and heavy icebreakers. The committee's recommendation is for a single class of polar icebreaker with heavy icebreaking capability.

Proceeding with a single class means that only one design will be needed, which will provide cost savings. The committee has found that the fourth heavy icebreaker could be built for a lower cost than the lead ship of a medium icebreaker class....

The DHS Mission Need Statement contemplated a total fleet of "potentially" up to six ships of two classes—three heavy and three medium icebreakers. Details appear in the High Latitude Mission Analysis Report. The Mission Need Statement indicated that to fulfill its statutory missions, USCG required three heavy and three medium icebreakers; each vessel would have a single crew and would homeport in Seattle. The committee's analysis indicated that four heavy icebreakers will meet the statutory mission needs gap identified by DHS for the lowest cost. Three of the ships would allow continuous presence in the Arctic, and one would service the Antarctic.

As noted in the High Latitude Report, USCG's employment standard is 185 days away from home port (DAFHP) for a single crew. Three heavy icebreakers in the Arctic provide 555 DAFHP, sufficient for continuous presence. In addition, the medium icebreaker USCG Cutter Healy's design service life runs through 2030. If greater capacity is required, USCG could consider operating three ships with four crews, which would provide 740 DAFHP. The use of multiple crews in the Arctic could require fewer ships while providing a comparable number of DAFHP. For example, two ships (instead of the recommended three) operating in the Arctic with multiple crews could provide a similar number of annual operating days at a lower cost, but such an arrangement may not permit simultaneous operations in both polar regions and may not provide adequate redundancy in capability. More important, an arrangement under which fewer boats are operated more often would require more major maintenance during shorter time in port, often at increasing cost. In addition, if further military presence is desired in the Arctic, USCG could consider ice-strengthening the ninth national security cutter.

One heavy icebreaker servicing the Antarctic provides for the McMurdo breakout and international treaty verification. The availability of the vessel could be extended by homeporting in the Southern Hemisphere. If the single vessel dedicated to the Antarctic is rendered inoperable, USCG could redirect an icebreaker from the Arctic, or it could rely on support from other nations. The committee considers both options to be viable and believes it difficult to justify a standby (fifth) vessel for the Antarctic mission when the total acquisition and lifetime operating costs of a single icebreaker are projected to exceed $1.6 billion. Once the four new icebreakers are operational, USCG can reasonably be expected to plan for more distant time horizons. USCG could assess the performance of the early ships once they are operational and determine whether additional capacity is needed.

USCG is the only agency of the U.S. government that is simultaneously a military service, a law enforcement agency, a marine safety and rescue agency, and an environmental protection agency. All of these roles are required in the mission need statement for a polar icebreaker. USCG, in contrast to a civilian company, has the authorities, mandates, and competencies to conduct the missions contemplated for the polar icebreakers. Having one agency with a multimission capability performing the range of services needed would be more efficient than potentially duplicating effort by splitting polar icebreaker operations among other agencies.

The requirement for national presence is best accomplished with a military vessel. In addition, USCG is fully interoperable with the U.S. Navy and the nation's North Atlantic Treaty Organization partners. USCG is already mandated to operate the nation's domestic and polar icebreakers. Continuing to focus this expertise in one agency remains the logical approach....

Government ownership of new polar icebreakers would be less costly than the use of lease financing (see Appendix C). The government has a lower borrowing cost than any U.S.- based leasing firm or lessor. In addition, the lessor would use higher-cost equity (on which it would expect to make a profit) to cover a portion of the lease financing. The committee's analysis shows that direct purchase by the government would cost, at a minimum, 19 percent less than leasing on a net present value basis (after tax). There is also the risk of the lessor going bankrupt and compromising the availability of the polar icebreaker to USCG. For its analysis, the committee not only relied on its extensive experience with leveraged lease financing but also reviewed available Government

Accountability Office reports and Office of Management and Budget rules, examined commercial leasing economics and current interest rates, and validated its analysis by consulting an outside expert on the issue....

Chartering (an operating lease) is not a viable option.... The availability of polar icebreakers on the open market is extremely limited. (The committee is aware of the sale of only one heavy icebreaker since 2010.) U.S. experience with chartering a polar icebreaker for the McMurdo resupply mission has been problematic on two prior charter attempts. Chartering is workable only if the need is short term and mission specific. The committee notes that chartering may preclude USCG from performing its multiple missions....

In the committee's judgment, an enlarged icebreaker fleet will provide opportunities for USCG to strengthen its icebreaking program and mission. Although the number of billets that require an expert is small compared with the overall number of billets assigned to these icebreakers, more people performing this mission will increase the pool of experienced candidates. This will provide personnel assignment officers with a larger pool of candidates when the more senior positions aboard icebreakers are designated, which will make icebreaking more attractive as a career path and increase the overall level of icebreaking expertise within USCG. Importantly, the commonality of design of the four recommended heavy icebreakers will reduce operating and maintenance costs over the service life of these vessels through efficiencies in supporting and crewing them. Having vessels of common design will likely improve continuity of service, build icebreaking competency, improve operational effectiveness, and be more cost-efficient....

3. Recommendation: USCG should follow an acquisition strategy that includes block buy contracting with a fixed price incentive fee contract and take other measures to ensure best value for investment of public funds.

Icebreaker design and construction costs can be clearly defined, and a fixed price incentive fee construction contract is the most reliable mechanism for controlling costs for a program of this complexity. This technique is widely used by the U.S. Navy. To help ensure best long-term value, the criteria for evaluating shipyard proposals should incorporate explicitly defined lifecycle cost metrics....

A block buy authority for this program will need to contain specific language for economic order quantity purchases for materials, advanced design, and construction activities. A block buy contracting program3 with economic order quantity purchases enables series construction, motivates competitive bidding, and allows for volume purchase and for the timely acquisition of material with long lead times. It would enable continuous production, give the program the maximum benefit from the learning curve, and thus reduce labor hours on subsequent vessels.

The acquisition strategy would incorporate (a) technology transfer from icebreaker designers and builders with recent experience, including international expertise in design, construction, and equipment manufacture; (b) a design that maximizes use of commercial off-the-shelf (COTS) equipment, applies Polar Codes and international standards, and only applies military specifications (MIL-SPEC) to the armament, aviation, communications, and navigation equipment; (c) reduction of any "buy American" provisions to allow the sourcing of the most suitable and reliable machinery available on the market; and (d) a program schedule that allows for completion of design and planning before the start of construction. These strategies will allow for optimization of design, reduce construction costs, and enhance reliability and maintainability....

4. Finding: In developing its independent concept designs and cost estimates, the committee determined that the costs estimated by USCG for the heavy icebreaker are reasonable. However, the committee believes that the costs of medium icebreakers identified in the High Latitude Mission Analysis Report are significantly underestimated.

The committee estimates the rough order-of-magnitude (ROM) cost of the first heavy icebreaker to be $983 million. (See Appendix D, Table D-6.) Of these all-in costs, 75 to 80 percent are shipyard design and construction costs; the remaining 20 to 25 percent cover government-incurred costs such as government-furnished equipment and government-incurred program expenses. If advantage is taken of learning and quantity discounts available through the recommended block buy contracting acquisition strategy, the average cost per heavy icebreaker is approximately $791 million, on the basis of the acquisition of four ships. The committee's analysis of the ship size to incorporate the required components (stack-up length) suggests an overall length of 132 meters (433 feet) and a beam of 27 meters (89 feet). This is consistent with USCG concepts for the vessel.

Costs can be significantly reduced by following the committee's recommendations. Reduction of MIL-SPEC requirements can lower costs by up to $100 million per ship with no loss of mission capability.... The other recommended acquisition, design, and construction strategies will control possible cost overruns and provide significant savings in overall life-cycle costs for the program.

Although USCG has not yet developed the operational requirements document for a medium polar icebreaker, the committee was able to apply the known principal characteristics of the USCG Cutter Healy to estimate the scope of work and cost of a similar medium icebreaker. The committee estimates that a first-of-class medium icebreaker will cost approximately $786 million. The fourth ship of the heavy icebreaker series is estimated to cost $692 million. Designing a medium-class polar icebreaker in a second shipyard would incur the estimated engineering, design, and planning costs of $126 million and would forgo learning from the first three ships; the learning curve would be restarted with the first medium design. Costs of building the fourth heavy icebreaker would be less than the costs of designing and building a first-of-class medium icebreaker... . In developing its ROM cost estimate, the committee agreed on a common notional design and basic assumptions.... Two committee members then independently developed cost estimating models, which were validated internally by other committee members. These analyses were then used to establish the committee's primary cost estimate....

5. Finding: Operating costs of new polar icebreakers are expected to be lower than those of the vessels they replace.

The committee expects the operating costs for the new heavy polar icebreakers to be lower than those of USCG's Polar Star. While USCG's previous experience is that operating costs of new cutters are significantly higher than those of the vessels they replace, the committee does not believe this historical experience applies in this case. There is good reason to believe that operating costs for new ships using commercially available modern technology will be lower than costs for existing ships.... The more efficient hull forms and modern engines will reduce fuel consumption, and a well-designed automation plant will require fewer operation and maintenance personnel, which will allow manning to be reduced or freed up for alternative tasks. The use of COTS technology and the minimization of MIL-SPEC, as recommended, will also reduce long-term maintenance costs, since use of customized equipment to meet MIL-SPEC requirements can reduce reliability and increase costs. A new vessel, especially over the first 10 years, typically has significantly reduced major repair and overhaul costs, particularly during dry-dock periods, compared with existing icebreakers—such as the Polar Star—that are near or at the end of their service life.... The Polar Star has many age-related issues that require it to be extensively repaired at an annual dry-docking. These issues will be avoided in the early years of a new ship. However, the committee recognizes that new ship operating costs can be higher than those of older ships if the new ship has more complexity to afford more capabilities. Therefore, any direct comparisons of operating costs of newer versus older ships would need to take into account the benefits of the additional capabilities provided by the newer ship.

USCG will have an opportunity to evaluate the manning levels of the icebreaker in light of the benefits of modern technology to identify reductions that can be made in operating costs....

6. Recommendation: USCG should ensure that the common polar icebreaker design is science-ready and that one of the ships has full science capability.

All four proposed ships would be designed as "science-ready," which will be more cost-effective when one of the four ships—most likely the fourth—is made fully science capable. Including science readiness in the common polar icebreaker design is the most cost-effective way of fulfilling both the USCG's polar missions and the nation's scientific research polar icebreaker needs.... The incremental costs of a science-ready design for each of the four ships ($10 million to $20 million per ship) and of full science capability for one of the ships at the initial build (an additional $20 million to $30 million) are less than the independent design and build cost of a dedicated research medium icebreaker.... In briefings at its first meeting, the committee learned that the National Science Foundation and other agencies do not have budgets to support full-time heavy icebreaker access or the incremental cost of design, even though their science programs may require this capability. Given the small incremental cost, the committee believes that the science capability cited above should be included in the acquisition costs.

Science-ready design includes critical elements that cannot be retrofitted cost-effectively into an existing ship and that should be incorporated in the initial design and build. Among these elements are structural supports, appropriate interior and exterior spaces, flexible accommodation spaces that can embark up to 50 science personnel, a hull design that accommodates multiple transducers and minimizes bubble sweep while optimizing icebreaking capability, machinery arrangements and noise dampening to mitigate interference with sonar transducers, and weight and stability latitudes to allow installation of scientific equipment. Such a design will enable any of the ships to be retrofitted for full science capability in the future, if necessary....

Within the time frame of the recommended build sequence, the United States will require a science-capable polar icebreaker to replace the science capabilities of the Healy upon her retirement. To fulfill this need, one of the heavy polar icebreakers would be procured at the initial build with full science capability; the ability to fulfill other USCG missions would be retained. The ship would be outfitted with oceanographic overboarding equipment and instrumentation and facilities comparable with those of modern oceanographic research vessels. Some basic scientific capability, such as hydrographic mapping sonar, should be acquired at the time of the build of each ship so that environmental data that are essential in fulfilling USCG polar missions can be collected.

7. Finding: The nation is at risk of losing its heavy polar icebreaking capability—experiencing a critical capacity gap—as the Polar Star approaches the end of its extended service life, currently estimated at 3 to 7 years.

The Polar Star, built in 1976, is well past its 30-year design life. Its reliability will continue to decline, and its maintenance costs will continue to escalate. Although the ship went through an extensive life-extending refit in 2011–2012, the Polar Star's useful life is estimated to end between 2020 and 2024. As USCG has recognized, the evaluation of alternative arrangements to secure polar icebreaking capacity is important, given the growing risks of the Polar Star losing its capability to fulfill its mission....

8. Recommendation: USCG should keep the Polar Star operational by implementing an enhanced maintenance program (EMP) until at least two new polar icebreakers are commissioned.

Even if the committee's notional schedule for new polar icebreakers is met, the second polar icebreaker would not be ready until July 2025.... The committee's proposed EMP could be designed with planned—and targeted—upgrades that allow the Polar Star to operate every year for its Antarctic mission. The necessary repairs could be performed in conjunction with the ship's current yearly dry-docking schedule within existing annual expenditures, estimated to average $5 million. In particular, the EMP would require improvements in the ship's operating systems, sanitary system, evaporators, main propulsion systems, and controllable pitch propellers. In the committee's judgment, the EMP could be accomplished within USCG's average annual repair expenditures for the Polar Star, which currently range between $2 million and $9 million.[73]

Coast Guard High Latitude Study Provided to Congress in July 2011

In July 2011, the Coast Guard provided to Congress a study on the Coast Guard's missions and capabilities for operations in high-latitude (i.e., polar) areas. The study,

[73] National Academies of Sciences, Engineering, and Medicine, Division on Earth and Life Studies and Transportation Research Board, *Acquisition and Operation of Polar Icebreakers: Fulfilling the Nation's Needs*, Letter Report, with cover letter dated July 11, 2017, pp. 9-20.

commonly known as the High Latitude Study, is dated July 2010 on its cover. The High Latitude Study concluded the following:

> [The study] concludes that future capability and capacity gaps will significantly impact four [Coast Guard] mission areas in the Arctic: Defense Readiness, Ice Operations, Marine Environmental Protection, and Ports, Waterways, and Coastal Security. These mission areas address the protection of important national interests in a geographic area where other nations are actively pursuing their own national goals....
>
> The common and dominant contributor to these significant mission impacts is the gap in polar icebreaking capability. The increasing obsolescence of the Coast Guard's icebreaker fleet will further exacerbate mission performance gaps in the coming years....
>
> The gap in polar icebreaking capacity has resulted in a lack of at-sea time for crews and senior personnel and a corresponding gap in training and leadership. In addition to providing multi-mission capability and intrinsic mobility, a helicopter-capable surface unit would eliminate the need for acquiring an expensive shore-based infrastructure that may only be needed on a seasonal or occasional basis. The most capable surface unit would be a polar icebreaker. Polar icebreakers can transit safely in a variety of ice conditions and have the endurance to operate far from logistics bases. The Coast Guard's polar icebreakers have conducted a wide range of planned and unscheduled Coast Guard missions in the past. Polar icebreakers possess the ability to carry large numbers of passengers, cargo, boats, and helicopters. Polar icebreakers also have substantial command, control, and communications capabilities. The flexibility and mobility of polar icebreakers would assist the Coast Guard in closing future mission performance gaps effectively....
>
> Existing capability and capacity gaps are expected to significantly impact future Coast Guard performance in two Antarctic mission areas: Defense Readiness and Ice Operations. Future gaps may involve an inability to carry out probable and easily projected mission requirements, such as the McMurdo resupply, or readiness to respond to less-predictable events. By their nature, contingencies requiring the use of military capabilities often occur quickly. As is the case in the Arctic, the deterioration of the Coast Guard's icebreaker fleet is the primary driver for this significant mission impact. This will further widen mission performance gaps in the coming years. The recently issued Naval Operations Concept 2010 requires a surface presence in both the Arctic and Antarctic. This further exacerbates the capability gap left by the deterioration of the icebreaker fleet....
>
> The significant deterioration of the Coast Guard icebreaker fleet and the emerging mission demands to meet future functional requirements in the high latitude regions dictate that the Coast Guard acquire material solutions to close the capability gaps....
>
> To meet the Coast Guard mission functional requirement, the Coast Guard icebreaking fleet must be capable of supporting the following missions:
>
> - *Arctic North Patrol.* Continuous multimission icebreaker presence in the Arctic.
> - *Arctic West Science.* Spring and summer science support in the Arctic.

- *Antarctic, McMurdo Station resupply.* Planned deployment for break-in, supply ship escort, and science support. This mission, conducted in the Antarctic summer, also requires standby icebreaker support for backup in the event the primary vessel cannot complete the mission.
- *Thule Air Base Resupply and Polar Region Freedom of Navigation Transits.* Provide vessel escort operations in support of the Military Sealift Command's Operation Pacer Goose; then complete any Freedom of Navigation exercises in the region.

In addition, the joint Naval Operations Concept establishes the following mission requirements:

- *Assured access and assertion of U.S. policy in the Polar Regions.* The current demand for this mission requires continuous icebreaker presence in both Polar Regions.

Considering these missions, the analysis yields the following findings:

- *The Coast Guard requires three heavy and three medium icebreakers to fulfill its statutory missions.* These icebreakers are necessary to (1) satisfy Arctic winter and transition season demands and (2) provide sufficient capacity to also execute summer missions. Single-crewed icebreakers have sufficient capacity for all current and expected statutory missions. Multiple crewing provides no advantage because the number of icebreakers required is driven by winter and shoulder season requirements. Future use of multiple or augmented crews could provide additional capacity needed to absorb mission growth.
- The Coast Guard requires six heavy and four medium icebreakers to fulfill its statutory missions and maintain the continuous presence requirements of the Naval Operations Concept. Consistent with current practice, these icebreakers are single-crewed and homeported in Seattle Washington.
- *Applying crewing and home porting alternatives reduces the overall requirement to four heavy and two medium icebreakers.* This assessment of non-material solutions shows that the reduced number of icebreakers can be achieved by having all vessels operate with multiple crews and two of the heavy icebreakers homeporting in the Southern Hemisphere.

Leasing was also considered as a nonmaterial solution. While there is no dispute that the Coast Guard's polar icebreaker fleet is in need of recapitalization, the decision to acquire this capability through purchase of new vessels, reconstruction of existing ships, or commercial lease of suitable vessels must be resolved to provide the best value to the taxpayer. The multi-mission nature of the Coast Guard may provide opportunities to conduct some subset of its missions with non-government-owned vessels. However, serious consideration must be given to the fact that the inherently governmental missions of the Coast Guard must be performed using government-owned and operated vessels.

An interpretation of the national policy is needed to determine the resource level that best supports the nation's interests....

The existing icebreaker capacity, two inoperative heavy icebreakers and an operational medium icebreaker, does not represent a viable capability to the federal government. The time needed to augment this capability is on the order of 10 years. At that point, around 2020, the heavy icebreaking capability bridging strategy expires.[74]

At a July 27, 2011, hearing on U.S. economic interests in the Arctic before the Oceans, Atmosphere, Fisheries, and Coast Guard subcommittee of the Senate Commerce, Science, and Transportation Committee, the following exchange occurred:

SENATOR OLYMPIA J. SNOWE: On the high latitude study, do you agree with—and those—I would like to also hear from you, Admiral Titley, as well, on these requirements in terms of Coast Guard vessels as I understand it, they want to have—I guess, it was a three medium ice breakers. Am in correct in saying that? Three medium ice breakers.

ADMIRAL ROBERT PAPP, COMMANDANT OF THE COAST GUARD: I agree with the mission analysis and as you look at the requirements for the things that we might do up there, if it is in the nation's interest, it identifies a minimum requirement for three heavy ice breakers and three medium ice breakers and then if you want a persistent presence up there, it would require—and also doing things such as breaking out (inaudible) and other responsibilities, then it would take up to a maximum six heavy and four medium.

SNOWE: Right. Do you agree with that?

PAPP: If we were to be charged with carrying out those full responsibilities, yes, ma'am. Those are the numbers that you would need to do it.

SNOWE: Admiral Titley, how would you respond to the high latitude study and has the Navy conducted its own assessment of its capability?

REAR ADMIRAL DAVID TITLEY, OCEANORGRAPHER AND NAVIGATOR OF THE NAVY: Ma'am, we are in the process right now of conducting what we call a capabilities based assessment that will be out in the summer of this year.

We are getting ready to finish that—the Coast Guard has been a key component of the Navy's task force on climate change, literally since day one when the Chief of Naval Operations set this up, that morning, we had the Coast Guard invited as a member of our executive steering committee.

So we have been working very closely with the Coast Guard, with the Department of Homeland Security, and I think Admiral Papp—said it best as far as the specific comments on the high latitude study but we have been working very closely with the Coast Guard.[75]

[74] *United States Coast Guard High Latitude Region Mission Analysis Capstone Summary*, July 2010, pp. 10-13, 15.
[75] Source: Transcript of hearing.

January 2011 DHS Office of Inspector General Report

A January 2011 report on the Coast Guard's polar icebreakers from the DHS Office of the Inspector General stated the following:

> The Coast Guard does not have the necessary budgetary control over its [polar] icebreakers, nor does it have a sufficient number of icebreakers to accomplish its missions in the Polar Regions. Currently, the Coast Guard has only one operational [polar] icebreaker [i.e., Healy], making it necessary for the United States to contract with foreign nations to perform scientific, logistical, and supply activities. Without the necessary budgetary control and a sufficient number of icebreaking assets, the Coast Guard will not have the capability to perform all of its missions, will lose critical icebreaking expertise, and may be beholden to foreign nations to perform its statutory missions. The Coast Guard should improve its strategic approach to ensure that it has the long-term icebreaker capabilities needed to support Coast Guard missions and other national interests in the Arctic and Antarctic regions.[76]

Regarding current polar icebreaking capabilities for performing Arctic missions, the report states the following:

> The Coast Guard's icebreaking resources are unlikely to meet future demands. [The table below] outlines the missions that Coast Guard is unable to meet in the Arctic with its current icebreaking resources.

Arctic Missions Not Being Met

Requesting Agency	Missions Not Being Met
United States Coast Guard	—Fisheries enforcement in Bering Sea to prevent foreign fishing in U.S. waters and overfishing —Capability to conduct search and rescue in Beaufort Sea for cruise line and natural resource exploration ships —Future missions not anticipated to be met: 2010 Arctic Winter Science Deployment
NASA	Winter access to the Arctic to conduct oceanography and study Arctic currents and how they relate to regional ice cover, climate, and biology
NOAA and NSF	Winter research
Department of Defense	Assured access to ice-impacted waters through a persistent icebreaker presence in the Arctic and Antarctic[77]

[76] Department of Homeland Security, Office of Inspector General, *The Coast Guard's Polar Icebreaker Maintenance, Upgrade, and Acquisition Program*, OIG-11-31, January 2011, p. 1 (Executive Summary). Report accessed September 21, 2011, at https://www.oig.dhs.gov/ assets/Mgmt/OIG_11-31_Jan11.pdf.

[77] Department of Homeland Security, *Office of Inspector General, The Coast Guard's Polar Icebreaker Maintenance, Upgrade, and Acquisition Program*, OIG-11-31, January 2011, p. 9.

The report also states the following:

> Should the Coast Guard not obtain funding for new icebreakers or major service life extensions for its existing icebreakers with sufficient lead-time, the United States will have no heavy icebreaking capability beyond 2020 and no polar icebreaking capability of any kind by 2029. Without the continued use of icebreakers, the United States will lose its ability to maintain a presence in the Polar Regions, the Coast Guard's expertise to perform ice operations will continue to diminish, and missions will continue to go unmet.[78]

Regarding current polar icebreaking capabilities for performing Antarctic missions, the report states the following:

> The Coast Guard needs additional icebreakers to accomplish its missions in the Antarctic. The Coast Guard has performed the McMurdo Station resupply in Antarctica for decades, but with increasing difficulty in recent years. The Coast Guard's two heavy-duty icebreakers [i.e., *Polar Star* and *Polar Sea*] are at the end of their service lives, and have become less reliable and increasingly costly to keep in service....
>
> In recent years, the Coast Guard has found that ice conditions in the Antarctic have become more challenging for the resupply of McMurdo Station. The extreme ice conditions have necessitated the use of foreign vessels to perform the McMurdo break-in....
>
> As ice conditions continue to change around the Antarctic, two icebreakers are needed for the McMurdo break-in and resupply mission. Typically, one icebreaker performs the break-in and the other remains on standby. Should the first ship become stuck in the ice or should the ice be too thick for one icebreaker to complete the mission, the Coast Guard deploys the ship on standby. Since the Polar Sea and Polar Star are not currently in service, the Coast Guard has no icebreakers capable of performing this mission. [The table below] outlines the missions that will not be met without operational heavy-duty icebreakers.

Arctic Missions Not Being Met

Requesting Agency	Missions Not Being Met
NSF	Missions not anticipated to be met: 2010-2011 Operation Deep Freeze – McMurdo Station Resupply
Department of State	Additional inspections of foreign facilities in Antarctica to enforce the Antarctic Treaty and ensure facilities' environment compliance[79]

[78] Department of Homeland Security, Office of Inspector General, *The Coast Guard's Polar Icebreaker Maintenance, Upgrade, and Acquisition Program*, OIG-11-31, January 2011, p. 10.

[79] Department of Homeland Security, Office of Inspector General, *The Coast Guard's Polar Icebreaker Maintenance, Upgrade, and Acquisition Program*, OIG-11-31, January 2011, pp. 10-11.

The report's conclusion and recommendations were as follows:

Conclusion

With an aging fleet of three icebreakers, one operational and two beyond their intended 30- year service life, the Coast Guard is at a critical crossroads in its Polar Icebreaker Maintenance, Upgrade, and Acquisition Program. It must clarify its mission requirements, and if the current mission requirements remain, the Coast Guard must determine the best method for meeting these requirements in the short and long term.

Recommendations

We recommend that the Assistant Commandant for Marine Safety, Security, and Stewardship:

Recommendation #1: Request budgetary authority for the operation, maintenance, and upgrade of its icebreakers.

Recommendation #2: In coordination with the Department of Homeland Security, request clarification from Congress to determine whether Arctic missions should be performed by Coast Guard assets or contracted vessels.

Recommendation #3: In coordination with the Department of Homeland Security, request clarification from Congress to determine whether Antarctic missions should be performed by Coast Guard assets or contracted vessels.

Recommendation #4: Conduct the necessary analysis to determine whether the Coast Guard should replace or perform service-life extensions on its two existing heavy-duty icebreaking ships.

Recommendation #5: Request appropriations necessary to meet mission requirements in the Arctic and Antarctic.[80]

The report states that

The Coast Guard concurred with all five of the recommendations and is initiating corrective actions. We consider the recommendations open and unresolved. The Coast Guard provided information on some of its ongoing projects that will address the program needs identified in the report.[81]

2010 U.S. Arctic Research Commission Report

A May 2010 report from the U.S. Arctic Research Commission (USARC) on goals and objectives for Arctic research for 2009-2010 stated the following:

---

[80] Department of Homeland Security, Office of Inspector General, *The Coast Guard's Polar Icebreaker Maintenance, Upgrade, and Acquisition Program*, OIG-11-31, January 2011, p. 12.

[81] Department of Homeland Security, Office of Inspector General, *The Coast Guard's Polar Icebreaker Maintenance, Upgrade, and Acquisition Program*, OIG-11-31, January 2011, p. 13.

To have an effective Arctic research program, the United States must invest in human capital, research platforms, and infrastructure, including new polar class icebreakers, and sustained sea, air, land, space, and social observing systems.... The Commission urges the President and Congress to commit to replacing the nation's two polar class icebreakers.[82]

2007 National Research Council Report

A 2007 National Research Council (NRC) report, *Polar Icebreakers in a Changing World: An Assessment of U.S. Needs*, assessed roles and future needs for Coast Guard polar icebreakers.[83] The study was required by report language accompanying the FY2005 DHS appropriations act (H.R. 4567/P.L. 108-334).[84] The study was completed in 2006 and published in 2007. Some sources refer to the study as the 2006 NRC report. The report made the following conclusions and recommendations:

Based on the current and future needs for icebreaking capabilities, the [study] committee concludes that the nation continues to require a polar icebreaking fleet that includes a minimum of three multimission ships [like the Coast Guard's three current polar icebreakers] and one single-mission [research] ship [like Palmer]. The committee finds that although the demand for icebreaking capability is predicted to increase, a fleet of three multimission and one single-mission icebreakers can meet the nation's future polar icebreaking needs through the application of the latest technology, creative crewing models, wise management of ice conditions, and more efficient use of the icebreaker fleet

---

[82] U.S. Arctic Research Commission, *Report on Goals and Objectives for Arctic Research 2009-2010, May 2010*, p. 4. Accessed online December 5, 2011, at https://storage.googleapis.com/arcticgov-static/publications/goals/usarc_goals_2009-10.pdf.

[83] National Research Council, *Polar Icebreakers in a Changing World, An Assessment of U.S. Needs*, Washington, 2007, 122 pp.

[84] H.R. 4567/P.L. 108-334 of October 18, 2004. The related Senate bill was S. 2537. The Senate report on S. 2537 (S.Rept. 108-280 of June 17, 2004) stated the following:
The Committee expects the Commandant to enter into an arrangement with the National Academy of Sciences to conduct a comprehensive study of the role of Coast Guard icebreakers in supporting United States operations in the Antarctic and the Arctic. The study should include different scenarios for continuing those operations including service life extension or replacement of existing Coast Guard icebreakers and alternative methods that do not use Coast Guard icebreakers. The study should also address changes in the roles and missions of Coast Guard icebreakers in support of future marine operations in the Arctic that may develop due to environmental change, including the amount and kind of icebreaking support that may be required in the future to support marine operations in the Northern Sea Route and the Northwest Passage; the suitability of the Polar Class icebreakers for these new roles; and appropriate changes in existing laws governing Coast Guard icebreaking operations and the potential for new operating regimes. The study should be submitted to the Committee no later than September 30, 2005.
The conference report on H.R. 4567 (H.Rept. 108-774 of October 9, 2004) stated the following:
As discussed in the Senate report and the Coast Guard authorization bill for fiscal year 2005, the conferees require the National Academy of Sciences to study the role of Coast Guard icebreakers.
The earlier House report on H.R. 4567 (H.Rept. 108-541 of June 15, 2004) contained language directing a similar report from the Coast Guard rather than the National Academies. (See the passage in the House report under the header "Icebreaking.")

and other assets. The nation should immediately begin to program, design, and construct two new polar icebreakers to replace the POLAR STAR and POLAR SEA.

Building only one new polar icebreaker is insufficient for several reasons. First, a single ship cannot be in more than one location at a time. No matter how technologically advanced or efficiently operated, a single polar icebreaker can operate in the polar regions for only a portion of any year. An icebreaker requires regular maintenance and technical support from shipyards and industrial facilities, must reprovision regularly, and has to effect periodic crew change outs. A single icebreaker, therefore, could not meet any reasonable standard of active and influential presence and reliable, at-will access throughout the polar regions.

A second consideration is the potential risk of failure in the harsh conditions of polar operations. Despite their intrinsic robustness, damage and system failure are always a risk and the U.S. fleet must have enough depth to provide backup assistance. Having only a single icebreaker would necessarily require the ship to accept a more conservative operating profile, avoiding more challenging ice conditions because reliable assistance would not be available. A second capable icebreaker, either operating elsewhere or in homeport, would provide ensured backup assistance and allow for more robust operations by the other ship.

From a strategic, longer-term perspective, two new Polar class icebreakers will far better position the nation for the increasing challenges emerging in both polar regions. A second new ship would allow the U.S. Coast Guard to reestablish an active patrol presence in U.S. waters north of Alaska to meet statutory responsibilities that will inevitably derive from increased human activity, economic development, and environmental change. It would allow response to emergencies such as search-and-rescue cases, pollution incidents, and assistance to ships threatened with grounding or damage by ice. Moreover, a second new ship will leverage the possibilities for simultaneous operations in widely disparate geographic areas (e.g., concurrent operations in the Arctic and Antarctic), provide more flexibility for conducting Antarctic logistics (as either the primary or the secondary ship for the McMurdo break-in), allow safer multiple-ship operations in the most demanding ice conditions, and increase opportunities for international expeditions. Finally, an up-front decision to build two new polar icebreakers will allow economies in the design and construction process and provide a predictable cost reduction for the second ship....

The [study] committee finds that both operations and maintenance of the polar icebreaker fleet have been underfunded for many years, and the capabilities of the nation's icebreaking fleet have diminished substantially. Deferred long-term maintenance and failure to execute a plan for replacement or refurbishment of the nation's icebreaking ships have placed national interests in the polar regions at risk. The nation needs the capability to operate in both polar regions reliably and at will. Specifically, the committee recommends the following:

The United States should continue to project an active and influential presence in the Arctic to support its interests. This requires U.S. government polar icebreaking capability to ensure year-round access throughout the region.

- The United States should continue to project an active and influential presence in the Antarctic to support its interests. The nation should reliably control sufficient icebreaking capability to break a channel into and ensure the maritime resupply of McMurdo Station.
- The United States should maintain leadership in polar research. This requires icebreaking capability to provide access to the deep Arctic and the ice-covered waters of the Antarctic.
- National interests in the polar regions require that the United States immediately program, budget, design, and construct two new polar icebreakers to be operated by the U.S. Coast Guard.
- To provide continuity of U.S. icebreaking capabilities, the POLAR SEA should remain mission capable and the POLAR STAR should remain available for reactivation until the new polar icebreakers enter service.
- The U.S. Coast Guard should be provided sufficient operations and maintenance budget to support an increased, regular, and influential presence in the Arctic. Other agencies should reimburse incremental costs associated with directed mission tasking.
- Polar icebreakers are essential instruments of U.S. national policy in the changing polar regions. To ensure adequate national icebreaking capability into the future, a Presidential Decision Directive should be issued to clearly align agency responsibilities and budgetary authorities.[85]

The Coast Guard stated in 2008 that it "generally supports" the NRC report, and that the Coast Guard "is working closely with interagency partners to determine a way forward with national polar policy that identifies broad U.S. interests and priorities in the Arctic and Antarctic that will ensure adequate maritime presence to further these interests. Identification and prioritization of U.S. national interests in these regions should drive development of associated USCG [U.S. Coast Guard] capability and resource requirements." The Coast Guard also stated the following: "Until those broad U.S. interests and priorities are identified, the current USG [U.S. Government] polar icebreaking fleet should be maintained in an operational status."[86]

Appendix B. Earlier Estimates of Polar Icebreaker Acquisition Costs

This appendix presents information on earlier estimates of polar icebreaker acquisition costs. These earlier estimates are provided primarily for reference purposes,

---

[85] National Research Council, *Polar Icebreakers in a Changing World: An Assessment of U.S. Needs*, Washington, 2007, pp. 2-3.

[86] Coast Guard point paper provided to CRS on February 12, 2008, and dated with the same date, providing answers to questions from CRS concerning polar icebreaker modernization.

since they can be viewed as having been overtaken by the more-recent cost estimates presented in the main body of this chapter.

2008 Coast Guard Estimate

The Coast Guard estimated in February 2008 that the acquisition of new replacement ships for the *Polar Star* and *Polar Sea* might cost between $800 million and $925 million per ship in 2008 dollars.[87] The Coast Guard said that this estimate

> is based on a ship with integrated electric drive, three propellers, and a combined diesel and gas (electric) propulsion plant. The icebreaking capability would be equivalent to the POLAR Class Icebreakers [i.e., Polar Star and Polar Sea] and research facilities and accommodations equivalent to HEALY. This cost includes all shipyard and government project costs. Total time to procure a new icebreaker [including mission analysis, studies, design, contract award, and construction] is eight to ten years.[88]

The Coast Guard further stated that this notional new ship would be designed for a 30-year service life.

Estimate in Coast Guard High Latitude Study Provided to Congress in July 2011

The High Latitude Study that was provided to Congress in July 2011[89] states that the above figure of $800 million to $925 million in 2008 dollars equates to $900 million to $1,041 million in 2012 dollars. The study provides the following estimates, in 2012 dollars, of the acquisition costs for new polar icebreakers

- $856 million for 1 ship;
- $1,663 million for 2 ships—an average of about $832 million each;
- $2,439 million for 3 ships—an average of $813 million each;
- $3,207 million for 4 ships—an average of about $802 million each;

---

[87] Coast Guard point paper provided to CRS on February 12, 2008, and dated with the same date, providing answers to questions from CRS concerning polar icebreaker modernization.

[88] The Coast Guard states further that the estimate is based on the acquisition cost of the *Mackinaw* (WAGB-30), a Great Lakes icebreaker that was acquired a few years ago and commissioned into service with the Coast Guard in June 2006. The *Mackinaw* is 240 feet long, displaces 3,500 tons, and can break ice up to 2 feet, 8 inches thick at speeds of 3 knots, which is suitable for Great Lakes icebreaking. The Coast Guard says it scaled up the acquisition cost for the *Mackinaw* in proportion to the ship's size compared to that of a polar icebreaker and then adjusted the resulting figure to account for the above-described capabilities of the notional replacement ship and recent construction costs at U.S. Gulf Coast shipyards.

[89] For more on the High Latitude Study, see *Appendix A*.

- $3,961 million for 5 ships—an average of about $792 million each; and
- $4,704 million for 6 ships—an average of $784 million each.

The study refers to the above estimates as "rough order-of-magnitude costs" that "were developed as part of the Coast Guard's independent Polar Platform Business Case Analysis."[90]

2017 NASEM Report

A congressionally mandated July 2017 report from the National Academies of Sciences, Engineering, and Medicine (NASEM) on the acquisition and operation of polar icebreakers estimates that the ship could cost less (and perhaps considerably less) than $1 billion.[91] More specifically, the NASEM study stated the following:

> The committee estimates the rough order-of-magnitude (ROM) cost of the first heavy icebreaker to be $983 million.... If advantage is taken of learning and quantity discounts available through the recommended block buy contracting acquisition strategy, the average cost per heavy icebreaker is approximately $791 million, on the basis of the acquisition of four ships....
>
> Costs can be significantly reduced by following the committee's recommendations. Reduction of MIL-SPEC [military specification] requirements can lower costs by up to $100 million per ship with no loss of mission capability.... The other recommended acquisition, design, and construction strategies will control possible cost overruns and provide significant savings in overall life-cycle costs for the program....
>
> The committee estimates that a first-of-class medium icebreaker will cost approximately $786 million. The fourth ship of the heavy icebreaker series is estimated to cost $692 million. Designing a medium-class polar icebreaker in a second shipyard would incur the estimated engineering, design, and planning costs of $126 million and would forgo learning from the first three ships; the learning curve would be restarted with

---

[90] United States Coast Guard High Latitude Region Mission Analysis Capstone Summary, July 2010, p. 13.

[91] National Academies of Sciences, Engineering, and Medicine, Division on Earth and Life Studies and Transportation Research Board, *Acquisition and Operation of Polar Icebreakers: Fulfilling the Nation's Needs*, Letter Report, with cover letter dated July 11, 2017, 147 pp. For the findings and recommendations of this study, see *Appendix A*. As mentioned earlier, the September 25, 2017, GAO report on polar icebreakers states the following:
According [a January 2017] analysis, the Coast Guard and Navy estimated a preliminary $1.15 billion cost for the lead heavy icebreaker (in fiscal year 2019 dollars). In July 2017, officials said they had reduced the estimated cost to less than $1 billion.
(Government Accountability Office, *Coast Guard: Status of Polar Icebreaking Fleet Capability and Recapitalization Plan*, GAO-17-698R, September 25, 2017, p. 5.)

the first medium design. Costs of building the fourth heavy icebreaker would be less than the costs of designing and building a first-of-class medium icebreaker....[92]

APPENDIX C. PSC PROGRAM FUNDING

This appendix presents additional background information on funding for the PSC program.

Summary of Funding in FY2013-FY2018 Budget Submissions

Table C-1 shows requested and projected funding for the PSC program in the Coast Guard's budget submissions from the initiation of the PSC program in the FY2013 submission through the FY2018 submission.

In addition to the $19 million requested for FY2018, the Coast Guard's acquisition, construction, and improvements FY2018 unfunded priorities list (UPL), dated July 20, 2017, includes, as its first item, a $750 million item for a heavy polar icebreaker. The UPL document states that this "additional funding in FY[20]18 supports construction of the first Heavy Polar Icebreaker and maintains the current strategy to stay on schedule, and maybe even accelerate the acquisition further."[93]

The reduction in programmed five-year funding for a new polar icebreaker during the FY2014- FY2016 budget submissions shown in *Table C-1* appears to have been related to the substantial reduction in the annual funding levels in the Coast Guard's Acquisition, Construction, and Improvements (AC&I) account[94] in those budget submission that is shown in *Table C-2*. Prior to the release of the Administration's September 1, 2015, fact sheet, the Coast Guard testified that if annual funding levels in the AC&I account were not increased from the reduced levels in those budget submissions, the icebreaker would be, essentially, an unfunded requirement. For example, at an April 28, 2015, hearing on Coast Guard resources and priorities before the Oceans, Atmosphere, Fisheries, and Coast Guard subcommittee of the Senate Commerce, Science, and Transportation Committee, Admiral Paul Zukunft, the then-Commandant of the Coast Guard, testified that

---

92 National Academies of Sciences, Engineering, and Medicine, Division on Earth and Life Studies and Transportation Research Board, *Acquisition and Operation of Polar Icebreakers: Fulfilling the Nation's Needs*, Letter Report, with cover letter dated July 11, 2017, pp. 15-16.

93 U.S. Coast Guard, Acquisition, Construction, and Improvements FY2018 Unfunded Priorities List, Submission to Congress, July 20, 2017, p. 2.

94 Prior to FY2019, the PC&I account was called the Acquisition, Construction, and Improvements (AC&I) account.

by reactivating Polar Star, we have purchased up to 10 years of decision space to recapitalize our ice-breaking fleet. Two of those years have expired. And while I'm exploring several options to reconstitute our nation's fleet of icebreakers, I will need topline relief [i.e., an increase] in my acquisition budget to make this requirement a reality.[95]

Table C-1. Funding for Acquisition of New Polar Icebreaker Under FY2013-FY2018 Budget Submissions (millions of then-year dollars)

Budget	FY13	FY14	FY15	FY16	FY17	FY18	FY19	FY20	FY21	FY22	5-year total
FY13	8	120	380	270	82						860
FY14		2	8	100	20	100					230
FY15			6	4	100	20	100				230
FY16				4	10	2	100	50			166
FY17					150	0	50	150	430		780
FY18						19	50	150	430	300	949

Source: Table prepared by CRS based on Coast Guard FY2013-FY2018 budget submissions.

Notes: For each line in the table, the first figure shown (e.g., $8 million in the case of the FY2013 budget) is the amount of funding that was requested for that fiscal year. Actual funding figures for FY2013-FY2017 are as follows: $7.609 million in FY2013; $2.0 million in FY2014; zero in FY2015; $6.0 million in FY2016; and $175 million in FY2017, for a total of $190.609 million for the period FY2013-FY2017. (An additional $30 million in FY2016 funding was subsequently reprogrammed to other uses.)

Table C-2. Funding in Procurement, Construction, and Improvements (PC&I) Account in FY2013-FY2018 Budgets (millions of dollars, rounded to nearest tenth)

Budget	FY 13	FY 14	FY 15	FY 16	FY 17	FY 18	FY 19	FY 20	FY 21	FY 22	Avg.	% change compared to avg. for FY13 budget
FY13	1,217.3	1,429.5	1,619.9	1,643.8	1,722.0						1,526.5	—
FY14		951.1	1,195.7	901.0	1,024.8	1,030.3					1,020.6	-33.1%
FY15			1,084.2	1,103.0	1,128.9	1,180.4	1,228.7				1,145.0	-25.0%
FY16				1,017.3	1,125.3	1,255.7	1,201.0	1,294.6			1,178.8	-22.8%
FY17					1,136.8	1,259.6	1,339.9	1,560.5	1,840.8		1,427.5	-6.5%
FY18						1,203.7	1,360.9	1,602.7	1,810.6	1,687.5	1,533.1	+0.4%

Source: Table prepared by CRS based on Coast Guard FY2013-FY2018 budget submissions. Prior to FY2019, the PC&I account was called the Acquisition, Construction, and Improvements (AC&I) account.

For additional discussion of the issue of the funding level of the Procurement, Construction, and Improvements (PC&I) account, see *Appendix D*. Below are some additional details on each of the budget submissions since the FY2013 submission.

---

[95] Source: Transcript of hearing.

FY2013 Submission

The Administration's FY2013 budget submission initiated a new project for the design and construction of a new polar icebreaker, and included $860 million over five years for the acquisition of the ship (*Table C-1*)—enough or almost enough to fully fund the acquisition of a new polar icebreaker. (Any remaining needed funding might have been projected for FY2018 and perhaps also FY2019, which were beyond the five-year window of the FY2013 budget submission.) The submission stated that DHS anticipated awarding a construction contract for the ship "within the next five years" (i.e., by FY2018) and taking delivery on the ship "within a decade" (i.e., by 2023).[96]

FY2014 Submission

The Administration's FY2014 budget submission reduced the five-year funding for a new polar icebreaker to $230 million (*Table C-1*)—a 73% reduction from the figure in the FY2013 budget submission—but still stated that DHS anticipated awarding a construction contract for the ship "within the next four years" (i.e., by FY2018).[97]

FY2015 Submission

The Administration's FY2015 budget submission maintained five-year funding for a new polar icebreaker at $230 million (*Table C-1*), but did not state when a construction contract for the ship might be awarded, creating uncertainty about the timing of the project.[98]

FY2016 Submission

The Administration's FY2016 budget submission, submitted to Congress in February 2015, reduced five-year funding for a new polar icebreaker further, to $166 million (*Table C-1*)—an 81% reduction from the figure in the FY2013 budget submission—and

---

[96] U.S. Department of Homeland Security, *Annual Performance Report, Fiscal Years 2011-2013*, p. CG-AC&I-40 (PDF page 1,777 of 3,134).

[97] Department of Homeland Security, United States Coast Guard, *Fiscal Year 2014 Congressional Justification*, p. CGAC&I-32 (PDF page 204 of 403).

[98] Department of Homeland Security, United States Coast Guard, *Fiscal Year 2015, Congressional Justification*, p. CGAC&I-42 (PDF page 196 of 474).

again did not state when a construction contract for the ship might be awarded, maintaining the uncertainty about the timing of the project.[99]

On September 1, 2015, the White House issued a fact sheet in conjunction with a visit to Alaska by President Obama indicating that the Administration, in its own internal planning, had at some point over the past two years deferred acquisition of a new polar icebreaker to FY2022, but that this had been changed to FY2020.[100] The newly announced construction start date of FY2020 was a two-year acceleration from the previously unpublicized date of FY2022, and a two-year deferral from the FY2018 date implied in the FY2013 and FY2014 budget submissions. The fact sheet states that the Administration will also "begin planning for construction of additional icebreakers" beyond the one that the Obama Administration proposed to begin building in FY2020.

On January 13, 2016, the Coast Guard announced that it intended to hold an industry day for the PSC program, followed by one-on-one meetings between the Coast Guard and prospective shipbuilders and ship designers, as a part of the Coast Guard's ongoing market research for the program.[101] The industry day was held on March 18, 2016, and the one-on-one meetings between the Coast Guard and industry officials were scheduled for March 28-31, with industry feedback to be submitted to the Coast Guard by April 5, 2016.[102]

---

[99] Department of Homeland Security, United States Coast Guard, *Fiscal Year 2016 Congressional Justification*, p. CGAC&I-36 (PDF page 202 of 518).

[100] The White House, "Fact Sheet: President Obama Announces New Investments to Enhance Safety and Security in the Changing Arctic," September 1, 2015, accessed September 2, 2015, at https://www.whitehouse.gov/the-pressoffice/2015/09/01/fact-sheet-president-obama-announces-new-investments-enhance-safety-and.

Regarding icebreakers, the fact sheet states the following:

Accelerating the acquisition of new Coast Guard icebreakers. After World War II, the United States Coast Guard had seven icebreakers in its fleet—four under the U.S. Navy and three under the U.S. Coast Guard. Today, the United States technically has three icebreakers in its fleet—all under the command of the U.S. Coast Guard. However, when age and reliability are taken into account, the fleet is down to the equivalent of two fully functional icebreakers and only one heavy-duty icebreaker. Russia, on the other hand, has forty icebreakers and another eleven planned or under construction.

The growth of human activity in the Arctic region will require highly engaged stewardship to maintain the open seas necessary for global commerce and scientific research, allow for search and rescue activities, and provide for regional peace and stability. Accordingly, meeting these challenges requires the United States to develop and maintain capacity for year-round access to greater expanses within polar regions.

That is why the Administration will propose to accelerate acquisition of a replacement heavy icebreaker to 2020 from 2022, begin planning for construction of additional icebreakers, and call on Congress to work with the Administration to provide sufficient resources to fund these critical investments. These heavy icebreakers will ensure that the United States can meet our national interests, protect and manage our natural resources, and strengthen our international, state, local, and tribal relationships.

[101] "*USCG Polar Class Icebreaker Replacement Program*," accessed January 15, 2016, at https://www.fbo.gov/index?s=opportunity&mode=form&id=a778c49349c443d2658666e 19cc100e9&tab=core&tabmode=list&=.

[102] "Heavy Polar Icebreaker Industry Engagement Activities," accessed April 4, 2016, at http://www.uscg.mil/ACQUISITION/icebreaker/Industry_Day_031816.asp.

FY2017 Submission

The Coast Guard's proposed FY2017 budget requested $150 million in acquisition funding for a new polar icebreaker. The figure of $150 million included $147.6 million in the polar icebreaker line of the Coast Guard's Acquisition, Construction, and Improvements (AC&I) account, and $2.4 million that was embedded in the personnel and management line in the AC&I account.[103] The Coast Guard's FY2017-FY2021 five-year Capital Investment Plan (CIP) included a total of $780 million in acquisition funding for a new polar icebreaker. As shown in *Table C-1*, the $150 million requested for FY2017 was the first major increment of acquisition funding requested (not just projected for a future fiscal year) for a new polar icebreaker.

FY2018 Submission

The Coast Guard's proposed FY2018 budget requested $19 million in acquisition funding for a new polar icebreaker and includes a total of $949 million over the five-year period FY2018- FY2022. The Coast Guard states that

> This request supports activities to complete and release a Request for Proposal (RFP) for Detail Design and Construction in FY 2018. Specifically, this funding supports program-wide activities including open water and ice tank model testing; review of Industry Studies contract deliverables; Integrated Program Office (IPO) and Ship Design Team (SDT) support; logistics and integration development for government furnished information and equipment; and additional modeling efforts to inform the evaluation and source selection process for the Detail Design & Construction RFP....
>
> Currently, the Program is maturing the system specification, developing the RFP for Detail Design & Construction, and completing required documentation to transition to the "Obtain" phase - planned for early FY 2018. In July 2016, the Coast Guard established an Integrated Program Office with the Navy to continue efforts to accelerate the construction timeline and leverage the expertise and best practices from shipbuilding programs in both services. Based on this collaboration and lessons learned by the Navy, the Program was able to significantly mature the acquisition approach with the incorporation of Industry Studies to identify solutions to minimize cost, schedule, production and technology risks. Industry Studies are focusing on leveraging industry perspectives, existing vessel designs, and use of mature technology to inform the iterative development of the Heavy Polar Icebreaker system specification. Future "Obtain" phase activities include award of a contract for Detail Design & Construction for the heavy polar icebreaker.[104]

---

[103] Department of Homeland Security, United States Coast Guard, *Fiscal Year 2017 Congressional Justification*, pp. CG-AC&I-28 and CG-AC&I-47 (PDF pages 170 and 189 of 407).

[104] Department of Homeland Security, U.S. Coast Guard, *Fiscal Year 2018 Congressional Justification*, undated but released May 2017, pp. AC&I-50 and AC&I-51.

Actual Prior-Year Funding in FY2013-FY2018

In each line of *Table C-1*, the first figure shown (e.g., $8 million in the case of the FY2013 budget) is the amount of funding that was requested for that fiscal year. Actual funding figures for FY2013-FY2017 are as follows: $7.609 million in FY2013; $2.0 million in FY2014; zero in FY2015; $6.0 million in FY2016; $175 million in FY2017; and $169 million for FY2018, for a total of $359.609 million for the period FY2013-FY2018. (An additional $30 million in FY2016 funding was subsequently reprogrammed to other uses.)

APPENDIX D. FUNDING LEVEL IN PC&I ACCOUNT

This appendix presents additional discussion of the funding level of the Coast Guard's Procurement, Construction, and Improvements (PC&I) account.[105]

Overview

The Coast Guard has testified that funding the PC&I account at a level of about $1 billion to $1.2 billion per year—the approximate average annual funding level programmed in the FY2014, FY2015, and FY2016 budget submissions, as shown in *Table C-2*—would make it difficult to fund various Coast Guard acquisition projects, including a new polar icebreaker and improvements to Coast Guard shore installations. Coast Guard plans call for procuring Offshore Patrol Cutters (OPCs) at an eventual rate of two per year.[106] If each OPC costs roughly $400 million, procuring two OPCs per year in an PC&I account of about $1 billion to $1.2 billion per year would leave about $200 million to $400 million per year for all other PC&I-funded programs.

Since 2017, Coast Guard officials have been stating more regularly what they stated only infrequently in earlier years: that executing the Coast Guard's various acquisition programs fully and on a timely basis would require the PC&I account to be funded in coming years at a level of about $2 billion per year. Statements from Coast Guard officials on this issue in past years have sometimes put this figure as high as about $2.5 billion per year.

---

[105] Prior to FY2019, the PC&I account was called the Acquisition, Construction, and Improvements (AC&I) account.

[106] For more on the OPC program, see CRS Report R42567, *Coast Guard Cutter Procurement: Background and Issues for Congress*, by Ronald O'Rourke.

Using Past PC&I Funding Levels as a Guide for Future PC&I Funding Levels

In assessing future funding levels for executive branch agencies, a common practice is to assume or predict that the figure in coming years will likely be close to where it has been in previous years. While this method can be of analytical and planning value, for an agency like the Coast Guard, which goes through periods with less acquisition of major platforms and periods with more acquisition of major platforms, this approach might not always be the best approach, at least for the PC&I account.

More important, in relation to maintaining Congress's status as a co-equal branch of government, including the preservation and use of congressional powers and prerogatives, an analysis that assumes or predicts that future funding levels will resemble past funding levels can encourage an artificially narrow view of congressional options regarding future funding levels, depriving Congress of agency in the exercise of its constitutional power to set funding levels and determine the composition of federal spending.

Past Coast Guard Statements About Required PC&I Funding Level

At an October 4, 2011, hearing on the Coast Guard's major acquisition programs before the Coast Guard and Maritime Transportation subcommittee of the House Transportation and Infrastructure Committee, the following exchange occurred:

> REPRESENATIVE FRANK LOBIONDO:
>
> Can you give us your take on what percentage of value must be invested each year to maintain current levels of effort and to allow the Coast Guard to fully carry out its missions?
>
> ADMIRAL ROBERT J. PAPP, COMMANDANT OF THE COAST GUARD:
>
> I think I can, Mr. Chairman. Actually, in discussions and looking at our budget—and I'll give you rough numbers here, what we do now is we have to live within the constraints that we've been averaging about $1.4 billion in acquisition money each year.
>
> If you look at our complete portfolio, the things that we'd like to do, when you look at the shore infrastructure that needs to be taken care of, when you look at renovating our smaller icebreakers and other ships and aircraft that we have, we've done some rough estimates that it would really take close to about $2.5 billion a year, if we were to do all the things that we would like to do to sustain our capital plant.
>
> So I'm just like any other head of any other agency here, as that the end of the day, we're given a top line and we have to make choices and tradeoffs and basically, my tradeoffs boil down to sustaining frontline operations balancing that, we're trying to

recapitalize the Coast Guard and there's where the break is and where we have to define our spending.[107]

An April 18, 2012, blog entry stated the following:

> If the Coast Guard capital expenditure budget remains unchanged at less than $1.5 billion annually in the coming years, it will result in a service in possession of only 70 percent of the assets it possesses today, said Coast Guard Rear Adm. Mark Butt.
>
> Butt, who spoke April 17 [2012] at [a] panel [discussion] during the Navy League Sea Air Space conference in National Harbor, Md., echoed Coast Guard Commandant Robert Papp in stating that the service really needs around $2.5 billion annually for procurement.[108]

At a May 9, 2012, hearing on the Coast Guard's proposed FY2013 budget before the Homeland Security subcommittee of the Senate Appropriations Committee, Admiral Papp testified, "I've gone on record saying that I think the Coast Guard needs closer to $2 billion dollars a year [in acquisition funding] to recapitalize—[to] do proper recapitalization."[109]

At a May 14, 2013, hearing on the Coast Guard's proposed FY2014 budget before the Homeland Security Subcommittee of the Senate Appropriations Committee, Admiral Papp stated the following regarding the difference between having about $1.0 billion per year rather than about $1.5 billion per year in the PC&I account:

> Well, Madam Chairman, $500 million—a half a billion dollars—is real money for the Coast Guard. So, clearly, we had $1.5 billion in the [FY]13 budget. It doesn't get everything
>
> I would like, but it—it gave us a good start, and it sustained a number of projects that are very important to us.

---

[107] Source: Transcript of hearing.

[108] David Perera, "The Coast Guard Is Shrinking," *FierceHomelandSecurity.com*, April 18, 2012, accessed July 20, 2012, at http://www.fiercehomelandsecurity.com/story/coast-guard-shrinking/2012-04-18.

[109] Source: transcript of hearing. Papp may have been referring to remarks he made to the press before giving his annual state of the Coast Guard speech on February 23, 2012, in which reportedly stated that the Coast Guard would require about $2 billion per year in acquisition funding to fully replace its current assets. (See Adam Benson, "Coast Guard Cutbacks Will Cost 1,000 Jobs," *Norwich Bulletin*, February 23, 2012, accessed May 31, 2012, at http://www.norwichbulletin.com/x1138492141/Coast-Guard-cutbacks-will-cost-1-000-jobs. See also "Coast Guard Leader Calls For More Ships," *MilitaryFeed.com*, February 24, 2012, accessed May 31, 2012, at http://militaryfeed.com/coast-guard-leader-calls-for-more-ships-5/; Associated Press, "Coast Guard Commandant Calls for New Ships," *TheLog.com*, March 10, 2012, accessed May 31, 2012, at http://www.thelog.com/SNW/Article/CoastGuard-Commandant-Calls-for-New-Ships-to-Replace-Aging-Fleet; Mickey McCarter, "Congress Poised to Give Coast Guard More Money Than Requested for FY 2013," *HSToday.us*, May 10, 2012, accessed May 31, 2012, at http:// www.hstoday.us/ focused-topics/customs-immigration/single-article-page/congress-poised-to-give-coast-guardmore-money-than-requested-for-fy-2013.html.) See also "Interview, Adm. Robert Papp, US Coast Guard Commandant," *Defense News*, November 11, 2013: 30.

> When we go down to the $1 billion level this year, it gets my highest priorities in there, but we have to either terminate or reduce to minimum order quantities for all the other projects that we have going.
>
> If we're going to stay with our program of record, things that have been documented that we need for our service, we're going to have to just stretch everything out to the right. And when we do that, you cannot order in economic order quantities. It defers the purchase. Ship builders, aircraft companies—they have to figure in their costs, and it inevitably raises the cost when you're ordering them in smaller quantities and pushing it off to the right.
>
> Plus, it almost creates a death spiral for the Coast Guard because we are forced to sustain older assets—older ships and older aircraft—which ultimately cost us more money, so it eats into our operating funds, as well, as we try to sustain these older things.
>
> So, we'll do the best we can within the budget. And the president and the secretary have addressed my highest priorities, and we'll just continue to go on the—on an annual basis seeing what we can wedge into the budget to keep the other projects going.[110]

At a March 12, 2014, hearing on the Coast Guard's proposed FY2015 budget before the Homeland Security subcommittee of the House Appropriations Committee, Admiral Papp stated the following:

> Well, that's what we've been struggling with, as we deal with the five-year plan, the capital investment plan, is showing how we are able to do that. And it will be a challenge, particularly if it sticks at around $1 billion [per year]. As I've said publicly, and actually, I said we could probably—I've stated publicly before that we could probably construct comfortably at about 1.5 billion [dollars] a year. But if we were to take care of all the Coast Guard's projects that are out there, including shore infrastructure that that fleet that takes care of the Yemen [sic: inland] waters is approaching 50 years of age, as well, but I have no replacement plan in sight for them because we simply can't afford it. Plus, we need at some point to build a polar icebreaker. Darn tough to do all that stuff when you're pushing down closer to 1 billion [dollars per year], instead of 2 billion [dollars per year].
>
> As I said, we could fit most of that in at about the 1.5 billion [dollars per year] level, but the projections don't call for that. So we are scrubbing the numbers as best we can.[111]

At a March 24, 2015, hearing on the Coast Guard's proposed FY2016 budget before the Homeland Security subcommittee of the House Appropriations Committee, Admiral Paul Zukunft, Admiral Papp's successor as Commandant of the Coast Guard, stated the following:

> I look back to better years in our acquisition budget when we had a—an acquisition budget of—of $1.5 billion. That allows me to move these programs along at a much more

---

[110] Transcript of hearing. The remarks were made in response to a question from Sen. Mary Landrieu.

[111] Transcript of hearing.

rapid pace and, the quicker I can build these at full-rate production, the less cost it is in the long run as well. But there's an urgent need for me to be able to deliver these platforms in a timely and also in an affordable manner. But to at least have a reliable and a predictable acquisition budget would make our work in the Coast Guard much easier. But when we see variances of—of 30, 40% over a period of three or four years, and not knowing what the Budget Control Act may have in store for us going on, yes, we are treading water nowbut any further reductions, and now I am—I am beyond asking for help. We are taking on water.[112]

An April 13, 2017, press report states (emphasis added):

[Then-]Coast Guard Commandant Adm. Paul Zukunft on Wednesday [April 12] said that for the Coast Guard to sustain its recapitalization plans and operations the service needs a $2 billion annual acquisition budget that grows modestly overtime to keep pace with inflation.

The Coast Guard needs a "predictable, reliable" acquisition budget "and within that we need 5 percent annual growth to our operations and maintenance (O&M) accounts," Zukunft told reporters at a Defense Writers Group breakfast. Inflation will clip 2 to 3 percent from that, but "at 5 percent or so it puts you on a moderate but positive glide slope so you can execute, so you can build the force," he said.[113]

In an interview published on June 1, 2017, Zukunft said (emphasis added):

We cannot be more relevant than we are now. But what we need is predictable funding. We have been in over 16 continuing resolutions since 2010. I need stable and repeatable funding. *An acquisition budget with a floor of $2 billion.* Our operating expenses as I said, they've been funded below the Budget Control Act floor for the past five years. I need 5 percent annualized growth over the next five years and beyond to start growing some of this capability back.

But more importantly, we [need] more predictable, more reliable funding so we can execute what we need to do to carry out the business of the world's best Coast Guard.[114]

APPENDIX E. ACQUISITION VS. LEASING

In addition to the issues for Congress discussed earlier in this report, another potential issue for Congress that arises from time to time is whether future polar icebreakers should be acquired through a traditional acquisition (i.e., the government

---

[112] Transcript of hearing. The remarks were made in response to a question from Rep. John Culberson.

[113] Calvin Biesecker, "Zukunft Wants $2 Billion Baseline Acquisition Budget; Sustained Growth In O&M Funding," *Defense Daily*, April 13, 2017: 1.

[114] Jill Aitoro, "Interview: Adm. Paul Zukunft Demands Coast Guard Respect," *Defense News*, June 1, 2017.

procuring the ship and owning it throughout its service life) or through a leasing arrangement (under which the icebreakers would be privately built and privately owned, leased to the Coast Guard, and crewed by an all-Coast Guard crew or a mix of Coast Guard personnel and civilian mariners). This appendix provides background information on this issue.

Factors to consider in assessing whether future polar icebreakers should be acquired through a traditional acquisition or a leasing arrangement include the comparative costs of the two options and the potential differences between them in terms of factors such as average number of days of operation each year and capability for performing various missions. Comparing the potential costs of leasing versus purchasing a capital asset often involves, among other things, calculating the net present value of each option.

As mentioned earlier, U.S. polar ice operations support 9 of the Coast Guard's 11 statutory missions. Of these nine missions, the Coast Guard states that a leased ship would be able to execute four (search and rescue, aids to navigation, ice operations, and marine environmental protection) and would not be able to execute five (ports, waterways, and costal security; living marine resources; other law enforcement; marine safety; and defense readiness).[115]

At a December 1, 2011, hearing before the Coast Guard and Maritime Transportation subcommittee of the House Transportation and Infrastructure Committee that focused on the polar icebreaker fleet, Admiral Robert Papp, the Commandant of the Coast Guard at the time, stated the following:

> As far as we can determine, there are no icebreakers available—no heavy icebreakers available for leasing right now. They would have to be constructed [and then leased].
>
> If we were to lease an icebreaker, I'm sure that a company building an icebreaker outside of the government does not have to contend with the same federal acquisition rules that we have to if we were to construct an icebreaker. It could probably be done quicker.
>
> Personally, I'm ambivalent in terms of how we get an icebreaker for the Coast Guard. We've done the legal research. If we lease an icebreaker, we can put a Coast Guard crew on it and still have it as a U.S. vessel supporting U.S. sovereignty.
>
> But the—but they aren't available right now. And the other challenge that we face is the federal acquisition rules and [Office of Management and Budget Circular] A-11 requirements that [direct how to] score the money [in the budget] for leasing. We'd have to put up a significant amount of upfront money even with a lease that we don't have room for within our budget currently.[116]

---

[115] Government Accountability Office, *Coast Guard: Status of Polar Icebreaking Fleet Capability and Recapitalization Plan*, GAO-17-698R, September 25, 2017, p. 22 (briefing slide 13).

[116] Source: Transcript of hearing.

At another point in the hearing, Admiral Papp stated the following:

> We have looked at various business case scenarios, each and every time looking at, once again, from our normal perspective, the Coast Guard perspective, which has been owning ships forever. And generally, we keep ships 30-40 years or beyond. There is a point where leasing becomes more expensive, it's at or about the 20-25-year timeline.
>
> I just don't have the experience with leasing to be able to give you a good opinion on it. And once again, I'm ambivalent. We just need the icebreaking capability, I think it's for people who can do the analysis, the proper analysis of—but also have to take into account the capabilities required and we need to get about the business of determining the exact capabilities that we need which would take into account National Science Foundation requirements, Coast Guard requirements, requirements to break-in at McMurdo, to come up with a capable ship.[117]

At another point in the hearing, he stated the following:

> As I said, sir, I am truly ambivalent to this except from what I experienced. I do have now two points, yes the Navy leases some ships, but we've got a Navy that has well over 300 ships.
>
> So if they lose a leased vessel or something is pulled back or something happens, they have plenty of other ships they can fall back upon. Right now, all I am falling back on is the Coast Guard cutter Healy. And it feels good to know that we own that and that is our ship for 30 or 40 years and we can rely upon it.
>
> In terms of leasing, I don't know. My personal experience is I lease one of my two cars and I pay a lot of money leasing my car. But at the end of the lease period, I have no car and I've spent a lot of money. So I don't know if that's directly applicable to ships as well, but right now I got half my garage is empty because I just turned one in.[118]

At another point in the hearing, he stated the following:

> We've looked through the legal considerations on this, as long as we have a Coast Guard crew. In fact, you can even make a mixed crew of civilians and Coast Guard people. But as long as it's commanding by—commanded by [a] commissioned officer, you can assert sovereignty, you can take it into war zones and, in fact, the Navy does that as well.[119]

Another witness at the hearing—Mead Treadwell, the lieutenant governor of Alaska—stated the following:

---

[117] Source: Transcript of hearing.
[118] Source: Transcript of hearing.
[119] Source: Transcript of hearing.

> [Regarding] The issue of the ships, the company that is building these ships for Shell [Oil] has visited with me and other state officials, and that's why you heard us say in our testimony that we think the leasing option should be considered. We don't have a way to judge the relative cost. But if on the face of it, it seems like it may be a way to get us the capability that the admiral needs.[120]

Another witness at the hearing—Jeffrey Garrett, a retired Coast Guard admiral who spent much of his career on polar icebreakers—stated the following:

> The perspective I could offer was when I was a member of the Cameron [sic: Commandant's?] staff back in the last '80s here in Washington, we were directed to pursue exactly the same sort of lease versus buy analysis, and in fact, the Coast Guard had a two track procurement strategy to compare leasing a new Polar icebreaker or buying it.
>
> And after over a year of analysis, studies, discussion with other agencies looking around, what became clear was, number one, there was no off-the-shelf asset readily available. And secondly, that in the long run, if you—when you cost it all out and the value of the stream of payments, leasing would actually cost more.
>
> And when we did the recapitalization analysis recently, we also reviewed leasing again, and the I think the findings in that report indicate more expensive over the life of the vessel by about 12 percent.[121]

When asked why this was the finding, Garrett stated the following:

> A couple of technical things. First of all, whoever builds the ship—and again, this will have to be ship built for the Coast Guard since there's not something off-the-shelf out there that you could lease. Whoever builds it has to raise capital, and nobody can raise capital more inexpensively than the federal government.
>
> Secondly, whoever leases the ship is obviously going to make—want to make a profit on that lease. So just like as Admiral Papp referred to leasing your car, you know, there's going to be a profit involved. And so, if you take the net present value of all of those, of those payments, you got come out with the more expensive package for the same, if you're comparing the same vessel.
>
> The other, the other issue I think is more intangible and that's just the fact that we're really not talking about an auxiliary like the Naval, like the Navy leases a supply ship or something like that. We're talking about a frontline Coast Guard capital asset, if you will, capital ship that's going to be doing frontline government missions projecting U.S. sovereignty.

---

[120] Source: Transcript of hearing. The transcript reviewed by CRS attributes this quote to the GAO witness, Stephen Caldwell, but this appears to be a mistake, as the statement is made by a member of the first witness panel, which included the Commandant of the Coast Guard and the Lieutenant Governor. The GAO witness was a member of the second witness panel. The reference in the quote to "me and other state officials" indicates that the witness speaking was the Lieutenant Governor and not the Commandant.

[121] Source: Transcript of hearing.

And you know, the Navy doesn't lease those kinds of ships for its frontline fleet and the Coast Guard doesn't lease those kinds of ships for its mission capabilities, and that's what we're really talking about in terms of the ship we need here.

So while a lease may look attractive, I think there are several things that indicate it may not be the right way to go. And the—I think that's what we came down to. And again, this is all documented in the past and that late '80s analysis was re-summarizing the president's 1990 report to Congress which basically says leasing is more expensive and it's not the way to go for a new ship. That was the ship that actually became the Healy then.[122]

The prepared statement of Stephen Caldwell, the GAO witness at the hearing, states the following:

The three reports discussed earlier in this [GAO] statement all identify funding as a central issue in addressing the existing and anticipated challenges related to icebreakers. In addition to the Coast Guard budget analysis included in the Recapitalization report, all three reports reviewed alternative financing options, including the potential for leasing icebreakers, or funding icebreakers through the National Science Foundation (NSF) or the Department of Defense (DOD). Although DOD has used leases and charters in the past when procurement funding levels were insufficient to address mission requirements and capabilities, both the Recapitalization report and the High Latitude Study determined that the lack of existing domestic commercial vessels capable of meeting the Coast Guard's mission requirements reduces the availability of leasing options for the Coast Guard. Additionally, an initial cost-benefit analysis of one type of available leasing option included in the Recapitalization report and the High Latitude Study suggests that it may ultimately be more costly to the Coast Guard over the 30-year icebreaker lifespan.[123]

In July 2016, the Coast Guard stated that

NSF leased the icebreaker KRASIN from Russia from 2005-2006, ODEN from the Swedish government from 2007-2010, and VLADIMIR IGNATYUK from Russia in 2012 to support the McMurdo resupply mission. All leases were time charters, and crews were supplied with the leases. As a contingency measure, NSF obtained assurances of assistance from other vessels in the area, such as the Chinese flagged [icebreaking] vessel XUE LONG, in the event they encountered difficulty. They also hired icebreaker captains with previous McMurdo experience to supplement the crew. NSF acquired these leases through a RFP process, and had no assurances that icebreakers would be available to perform the mission, or what price would be quoted.

This process came with risks, as there was no way to gauge icebreaker availability until NSF received responses to their RFP. Additionally, a foreign-flagged commercial or

[122] Source: Transcript of hearing.

[123] Government Accountability Office, *Coast Guard[:] Observations on Arctic Requirements, Icebreakers, and Coordination with Stakeholders, Testimony Before the Subcommittee on Coast Guard and Maritime Transportation, Committee on Transportation and Infrastructure, House of Representatives, Statement of Stephen L. Caldwell, Director, Homeland Security and Justice*, GAO-12-254T, December 1, 2011, p. 24.

state vessel can become unavailable for a variety of environmental and political reasons. For example, the Swedish government abruptly terminated their contract during the spring/summer of 2011, and NSF was left without a platform to conduct its mission. NSF requested support from CGC HEALY, but it was employed in the Arctic. NSF ultimately leased the Russian icebreaker VLADIMIR IGNATYUK. After that incident, NSF decided to utilize CGC POLAR STAR to support the McMurdo mission, which it has been doing since 2013.[124]

At a June 14, 2016, hearing on the Coast Guard before the Coast Guard and Maritime Transportation subcommittee of the House Transportation and Infrastructure Committee, the following exchange occurred:

REPRESENTATIVE HUNTER (Chairman):

How do you plan on—on filling the capability gap until you get a heavy icebreaker, which is 10 years at the least based on the best projections of Congress and everybody working together? You still haven't answered that one.

ADMIRAL CHARLES MICHEL (Vice Commandant of the Coast Guard):

Well, right—the alternatives now, since we'll provide the answer to that, and it's probably going to be either a rolling recapitalization of the *Polar Star* or to try to bring—let *Polar Star* taper off and then try to bring *Polar Sea* back on and bridge out to the new icebreaker.

I do not know which one at this point, which path we would want to take. I'm not aware of any other—we've looked out there for vessels to lease for heavy icebreaking capabilities. There's nothing out there on planet earth that you can lease in the heavy icebreaking area. So that's kind of where we are, sir.

HUNTER:

Was it the—the Finns that came into my office?

(UNKNOWN)

Mm-hmm.

HUNTER:

Can't remember whether we had the Norwegians or the Finns. I mean, they—have you— you've obviously looked at that, right?

MICHEL:

Yes. As a matter of fact I—I traveled to Sweden and Finland...

HUNTER:

Yeah.

---

[124] Source: Email from Guard Office of Congressional Affairs to CRS, July 8, 2016.

MICHEL:

... and talked to them. And they do not have heavy icebreaking capability that will meet the needs as in the FedBizOpps. As a matter of fact, in—when I'm talking FedBizOpps [I mean] there's a technical package that the Coast Guard put out for our [new] heavy icebreaker [i.e., the one that the Obama Administration wanted to begin building in 2020].

It kind of lays out our basic requirements including the long pole in the tent which is the icebreaking requirement, which is six foot minimum at three knots, desirable eight-foot minimum at three knots and then 21 feet backing and ramming.

When I talked to the shipbuilders over there, they said there is not a vessel like that that currently exists that will meet those requirements in the—in the FedBizOpps technical package. So you'd have to build a vessel like that. And that's the type of vessel that we're looking for.[125]

The congressionally mandated July 2017 NASEM report on acquisition and operation of polar icebreakers states (emphasis as in original):

2. Recommendation: The United States Congress should fund the construction of four polar icebreakers of common design that would be owned and operated by the United States Coast Guard (USCG)....

Government ownership of new polar icebreakers would be less costly than the use of lease financing.... The government has a lower borrowing cost than any U.S.-based leasing firm or lessor. In addition, the lessor would use higher-cost equity (on which it would expect to make a profit) to cover a portion of the lease financing. The committee's analysis shows that direct purchase by the government would cost, at a minimum, 19 percent less than leasing on a net present value basis (after tax). There is also the risk of the lessor going bankrupt and compromising the availability of the polar icebreaker to USCG. For its analysis, the committee not only relied on its extensive experience with leveraged lease financing but also reviewed available Government Accountability Office reports and Office of Management and Budget rules, examined commercial leasing economics and current interest rates, and validated its analysis by consulting an outside expert on the issue....

Chartering (an operating lease) is not a viable option.... The availability of polar icebreakers on the open market is extremely limited. (The committee is aware of the sale of only one heavy icebreaker since 2010.) U.S. experience with chartering a polar icebreaker for the McMurdo resupply mission has been problematic on two prior charter attempts. Chartering is workable only if the need is short term and mission specific. The committee notes that chartering may preclude USCG from performing its multiple missions....[126]

[125] Transcript of hearing.

[126] National Academies of Sciences, Engineering, and Medicine, Division on Earth and Life Studies and Transportation Research Board, *Acquisition and Operation of Polar Icebreakers: Fulfilling the Nation's Needs*, Letter Report, with cover letter dated July 11, 2017, pp. 10, 12-13.

APPENDIX F. GREAT LAKES ICEBREAKERS

This appendix provides a brief discussion of the Coast Guard's Great Lakes icebreakers.[127] The Coast Guard's current Great Lakes icebreaker fleet consists of nine cutters:

- one heavy icebreaker—Mackinaw (WLBB-30), a 240-foot ship displacing 3,500 tons;
- six 140-foot Bay-class icebreaking tugs displacing 662 tons each; and
- two 225-foot Juniper-class seagoing buoy tenders displacing about 2,000 tons each that have a light icebreaking capability.[128]

Although *Mackinaw* is referred to as a heavy icebreaker, the word *heavy* in this instance is being used in the context of Great Lakes icebreaking—*Mackinaw* is much larger and has more icebreaking capability than the eight other ships listed above.[129] *Mackinaw* would not, however, qualify as a heavy polar icebreaker, as it is much smaller and has much less icebreaking capability than a heavy polar icebreaker.[130]

Coast Guard officials have stated that they do not view the procurement of additional Great Lakes icebreakers as an urgent near-term acquisition need. In support of this assessment, they cite the capabilities of the current Great Lakes icebreaking fleet, the relatively young age of *Mackinaw* (which entered service in 2006), service life extension work being done on the ice-breaking tugs that is designed to add 15 years to their service lives,[131] and Canada's own Great Lakes icebreaking capabilities. A 2016 Coast Guard report to Congress on the Great Lakes icebreaking mission stated:

---

[127] This appendix is adapted from the section entitled "Great Lakes Icebreakers" on pages 7-10 of CRS Testimony TE10030, *Icebreaker Acquisition and the Need for a National Maritime Strategy*, by Ronald O'Rourke, November 29, 2018.

[128] Source: U.S.. Coast Guard, "*Ninth Coast Guard District Units*," accessed November 19, 2018, at: https://www.atlanticarea.uscg.mil/Atlantic-Area/Units/District-9/Ninth-District-Units/. A total of 10 cutters are assigned to the Ninth District, which is responsible for the Great Lakes, the Saint Lawrence Seaway, and parts of the surrounding states. The tenth cutter assigned to the Ninth District is a 100-foot inland buoy tender whose primary missions do not include icebreaking.

[129] At continuous speeds of 3 knots, *Mackinaw* can break ice up to 32 inches thick, the 140-foot icebreaking tugs can break ice up to 22 inches thick, and the 225-foot seagoing buoy tenders can break ice up to 14 inches thick.

[130] As discussed earlier in this report, the Coast Guard's two heavy polar icebreakers—the operational *Polar Star* and the non-operational *Polar Sea*, are 399 feet long and displace about 13,200 tons each. *Polar Star* can break ice up to six feet (72 inches) thick at a continuous speed of 3 knots. The Coast Guard states that *Mackinaw* is equivalent to the Canadian Coast Guard ship *Samuel Risley*, a Great Lakes-homeported icebreaker and buoy tender that Canada classifies as a light icebreaker in a comparison conducted across its entire icebreaking fleet, including its Arctic icebreakers. (U.S. Coast Guard, *Great Lakes Icebreaking Mission Analysis, Fiscal Year 2016 Report to Congress*, August 30, 2016, p. 5.)

[131] For more on this service life extension work, see U.S. Coast Guard, "*In-Service Vessel Sustainment Program*," accessed November 19, 2018, at: https://www.dcms. uscg.mil/Our-Organization/Assistant-Commandant-for-Acquisitions-CG-9/Programs/Surface-Programs/In-Service-Vessel-Sustainment-Program/.

> The current mix of heavy and medium [Great Lakes] icebreakers is capable of managing priorities and requests for icebreaking in Tier 1 and 2 waterways. When a severe ice season stresses Coast Guard asset capabilities, the existing agreement and partnership with Canada fills the capability gap and brings in extra heavy-icebreaking resources to manage the ice.... [T]he 2014 and 2015 ice seasons were a 20-year anomaly, consuming almost twice as many cutter resource hours as in any other year since 2005.
>
> The Coast Guard cannot reliably predict the economic impact of maintaining a single heavy Great Lakes icebreaker. Additionally, given the extreme conditions when ice coverage exceeds 90 percent, it is not clear that shipping delays would be significantly mitigated by an increase in icebreaking capability. Delays can be associated with several factors such as slow transit speeds, availability of pilots, and simultaneous and competing demand signals for icebreaking services across the Great Lakes.[132]

The Coast Guard's position notwithstanding, some Members of Congress in recent years have expressed interest in the possibility of bolstering the Coast Guard's Great Lakes icebreaking fleet by procuring a second icebreaker with capabilities generally similar to those of *Mackinaw*. Interest in this option was reinforced by the winters of 2013-2014 and 2014-2015, which featured particularly high levels of ice coverage on the Great Lakes.[133] The committee report language requiring the above-quoted Coast Guard report to Congress is one example of this interest.[134] Another example is Section 820 of the Frank LoBiondo Coast Guard Authorization Act of 2018 (S. 140/P.L. 115-282 of December 4, 2018), which states:

---

[132] U.S. Coast Guard, *Great Lakes Icebreaking Mission Analysis, Fiscal Year 2016 Report to Congress*, August 30, 2016, p. 11. The report was required by S.Rept. 114-68 of June 18, 2015, the Senate Appropriations Committee's report on S. 1619, the Department of Homeland Security Appropriations Bill, 2016 (see page 75).

[133] Although interest in procuring a second heavy Great Lakes icebreaker was reinforced by high levels of ice coverage in the winters of 2013-2014 and 2014-2015, interest in Congress in procuring such a ship dates back further than 2013. See, for example, H.R. 1747 of the 111th Congress, the Great Lakes Icebreaker Replacement Act, which was introduced on March 26, 2009, reported by the Committee on Transportation and Infrastructure on April 21, 2009 (H.Rept. 111- 81), and agreed to by the House by voice vote on April 27, 2009. A similar bill, S. 1024, was introduced in the Senate on May 12, 2009.

[134] S.Rept. 114-68 stated:

GREAT LAKES ICEBREAKING CAPACITY

> The Coast Guard is required by law to maintain a heavy icebreaking capability on the Great Lakes to assist in keeping channels and harbors open to navigation in response to the reasonable demands of commerce to meet the winter shipping needs of industry. The Committee is concerned that the Coast Guard does not possess adequate capacity to meet its statutorily required icebreaking mission on the Great Lakes, with negative consequences to the regional and national economy as well as to the safety of local communities. While the Committee fully supports the Coast Guard's Service Life Extension Project for its nine-vessel 140-foot icebreaking tugs as part of the In-Service Vessel Sustainment Program, it notes that additional assets may be necessary to successfully operate in the heavy ice conditions often experienced by the Great Lakes. The Committee directs the Coast Guard to undertake an updated mission analysis study to determine the assets necessary to effectively carry out its icebreaking requirements on the Great Lakes, including consideration of a second heavy icebreaker for the Great Lakes, consistent with the capabilities of the Mackinaw. The updated mission analysis should factor in recent historically high levels of ice coverage and the economic costs of reduced Great Lakes shipping associated with maintaining only one heavy icebreaker. The updated mission analysis shall be submitted to the Committee not later than 180 days after the date of enactment of this act. (Page 75)

SEC. 820. Great Lakes icebreaker acquisition.

(a) Icebreaking on the Great Lakes.—For fiscal years 2018 and 2019, the Commandant of the Coast Guard may use funds made available pursuant to section 4902 of title 14, United States Code, as amended by this Act, for the construction of an icebreaker that is at least as capable as the Coast Guard Cutter Mackinaw to enhance icebreaking capacity on the Great Lakes.

(b)Acquisition plan.—Not later than 45 days after the date of enactment of this Act, the Commandant shall submit a plan to the Committee on Commerce, Science, and Transportation of the Senate and the Committee on Transportation and Infrastructure of the House of Representatives for acquiring an icebreaker described in subsections (a) and (b). Such plan shall include—

(1) the details and schedule of the acquisition activities to be completed; and

(2) a description of how the funding for Coast Guard acquisition, construction, and improvements that was appropriated under the Consolidated Appropriations Act, 2017 (Public Law 115–31) will be allocated to support the acquisition activities referred to in paragraph (1).[135]

An examination of procurement costs for *Mackinaw*, the National Science Foundation's ice-capable research ship *Sikuliaq*, new oceanographic research ships being procured for NOAA, and OPCs suggests that a new *Mackinaw*-sized heavy Great Lakes icebreaker built in a U.S. shipyard might have a design and construction cost between $175 million and $300 million, depending on its exact capabilities and the acquisition strategy employed.[136] The design portion of the ship's cost might be reduced if

---

[135] In addition, Section 819 of S. 140/P.L. 115-282 states:

SEC. 819. Acquisition plan for inland waterway and river tenders and bay-class icebreakers.

(a)Acquisition plan.—Not later than 270 days after the date of the enactment of this Act, the Commandant of the Coast Guard shall submit to the Committee on Commerce, Science, and Transportation of the Senate and the Committee on Transportation and Infrastructure of the House of Representatives a plan to replace or extend the life of the Coast Guard fleet of inland waterway and river tenders, and the Bay-class icebreakers.

(b)Contents.—The plan under subsection (a) shall include—

(1)an analysis of the work required to extend the life of vessels described in subsection (a);

(2)recommendations for which, if any, such vessels it is cost effective to undertake a ship-life extension or enhanced maintenance program;

(3)an analysis of the aids to navigation program to determine if advances in navigation technology may reduce the needs for physical aids to navigation;

(4)recommendations for changes to physical aids to navigation and the distribution of such aids that reduce the need for the acquisition of vessels to replace the vessels described in subsection (a);

(5)a schedule for the acquisition of vessels to replace the vessels described in subsection (a), including the date on which the first vessel will be delivered;

(6)the date such acquisition will be complete;

(7)a description of the order and location of replacement vessels;

(8)an estimate of the cost per vessel and of the total cost of the acquisition program of record; and

(9)an analysis of whether existing vessels can be used.

[136] Source: CRS analysis of cost per weight for *Mackinaw* (adjusted for inflation), *Sikuliaq*, new NOAA oceanographic research ships now being procured, and OPCs.

Some press reports in 2015 and 2016 cited a cost of about $200 million for a new heavy Great Lakes icebreaker. (See, for example, Todd Spangler, "A New Icebreaker for the Great Lakes? It's Far from Certain," *Detroit Free Press*, August 7, 2015; "Frozen Commerce: Great Lakes Businesses Need a New Icebreaker," *Pittsburgh Post-Gazette*, August 17, 2015; Todd Spangler, "Call for Arctic Icebreakers Could Hurt Great Lakes," *Detroit Free Press*, September 1, 2015; Bob Gross, "Congress Authorizes New Icebreaker for Great

Mackinaw's design or the design of some other existing icebreaker were to be used as the parent design. Depending on the capabilities and other work load of the shipyard selected to build the ship, the construction time for a new heavy Great Lakes icebreaker might be less than that of a new heavy polar icebreaker.

Lakes," *Times Herald (Port Huron, MI)*, February 3, 2016; "Task Force Calls Anew for More Great Lakes Icebreakers, Second Poe-Sized Lock," *Professional Mariner*, February 17, 2016 [the article states that it presents the text of a news release from the Great Lakes Maritime Task Force].) An opinion column in 2016 cited a figure of $240 million. (John Hageman, "Is Winter Great Lakes Shipping Necessary?" *Sandusky Register*, February 18, 2016.)

The Great Lakes Maritime Task Force, an organization that states that it "was founded in 1992 in Toledo, Ohio, to promote waterborne commerce and related industries on the Great Lakes" (see Great Lakes Maritime Task Force, "About Us," accessed November 26, 2018, at: http://www.glmtf.org/about), states in its annual report for 2017 that a second heavy Great Lakes icebreaker "is projected to cost $240 million." (*2017 Annual Report of Great Lakes Maritime Task Force*, PDF page 3 of 6, accessed November 26, 2018, at: http://www.glmtf.org/wpcontent/uploads/2018/05/2017-Annual-Report.pdf.) The same figure is cited in the organization's annual report for 2016. The organization's annual report for 2015 cited a figure of approximately $200 million.

In: Conventional Weapons Systems
Editor: Stewart Keller
ISBN: 978-1-53615-414-6

Chapter 4

COAST GUARD CUTTER PROCUREMENT: BACKGROUND AND ISSUES FOR CONGRESS (UPDATED)*

Ronald O'Rourke

SUMMARY

The Coast Guard's acquisition program of record (POR) calls for procuring 8 National Security Cutters (NSCs), 25 Offshore Patrol Cutters (OPCs), and 58 Fast Response Cutters (FRCs) as replacements for 90 aging Coast Guard high-endurance cutters, medium-endurance cutters, and patrol craft. The Coast Guard's proposed FY2019 budget requests a total of $705 million in acquisition funding for the NSC, OPC, and FRC programs.

NSCs are the Coast Guard's largest and most capable general-purpose cutters; they are intended to replace the Coast Guard's 12 aged Hamilton-class high-endurance cutters. NSCs have an estimated average procurement cost of about $682 million per ship. Although the Coast Guard's POR calls for procuring a total of 8 NSCs to replace the 12 Hamilton-class cutters, Congress through FY2018 has funded 11 NSCs, including two (the 10th and 11th) in FY2018. Six NSCs are now in service. The seventh is scheduled to be commissioned into service in January 2019, and the eighth and ninth are scheduled for delivery in 2019 and 2020, respectively. The Coast Guard's proposed FY2019 budget requests $65 million in acquisition funding for the NSC program; this request does not include additional funding for a 12th NSC.

OPCs are to be smaller, less expensive, and in some respects less capable than NSCs; they are intended to replace the Coast Guard's 29 aged medium-endurance cutters. Coast Guard officials describe the OPC program as the service's top acquisition priority. OPCs have an estimated average procurement cost of about $391 million per ship. On

* This is an edited, reformatted and augmented version of Congressional Research Service Publication No. R42567, Updated December 7, 2018.

September 15, 2016, the Coast Guard announced that it was awarding a contract with options for building up to nine ships in the class to Eastern Shipbuilding Group of Panama City, FL. The first OPC was funded in FY2018 and is to be delivered in 2021. The Coast Guard's proposed FY2019 budget requests $400 million in acquisition funding for the OPC program for the construction of the second OPC (which is scheduled for delivery in 2022) and procurement of long lead time materials (LLTM) for the third OPC (which is scheduled for delivery in 2023).

FRCs are considerably smaller and less expensive than OPCs; they are intended to replace the Coast Guard's 49 aging Island-class patrol boats. FRCs have an estimated average procurement cost of about $58 million per boat. A total of 50 have been funded through FY2018. The 29th was commissioned into service on November 8, 2018. The Coast Guard's proposed FY2019 budget requests $240 million in acquisition funding for the procurement of four more FRCs.

The NSC, OPC, and FRC programs pose several issues for Congress, including the following:

- whether to fully or partially fund the acquisition of a 12th NSC in FY2019;
- whether to fund the acquisition of four FRCs in FY2019, as requested, or some other number, such as six, which is the maximum number that has been acquired in some prior fiscal years;
- whether to use annual or multiyear contracting for procuring OPCs;
- the procurement rate for the OPC program;
- the impact of Hurricane Michael on Eastern Shipbuilding of Panama City, FL, the shipyard that is to build the first nine OPCs;
- planned procurement quantities for NSCs, OPCs, and FRCs; and
- initial testing of the NSC.

Congress's decisions on these programs could substantially affect Coast Guard capabilities and funding requirements, and the U.S. shipbuilding industrial base.

INTRODUCTION

This report provides background information and potential oversight issues for Congress on the Coast Guard's programs for procuring 8 National Security Cutters (NSCs), 25 Offshore Patrol Cutters (OPCs), and 58 Fast Response Cutters (FRCs). The Coast Guard's proposed FY2019 budget requests a total of $705 million in acquisition funding for the NSC, OPC, and FRC programs.

The issue for Congress is whether to approve, reject, or modify the Coast Guard's funding requests and acquisition strategies for the NSC, OPC, and FRC programs. Congress's decisions on these three programs could substantially affect Coast Guard capabilities and funding requirements, and the U.S. shipbuilding industrial base.

The NSC, OPC, and FRC programs have been subjects of congressional oversight for several years, and were previously covered in an earlier CRS report that is now archived.[1]

---

[1] The earlier report was, *Coast Guard Deepwater Acquisition Programs: Background, Oversight Issues, and Options for Congress*, by Ronald O'Rourke. From the late 1990s until 2007, the Coast Guard's efforts to

CRS testified on the Coast Guard's cutter acquisition programs most recently on July 25, 2017.[2] The Coast Guard's plans for modernizing its fleet of polar icebreakers are covered in a separate CRS report.[3]

BACKGROUND

Older Ships to be Replaced by NSCs, OPCs, and FRCs

The 91 planned NSCs, OPCs, and FRCs are intended to replace 90 older Coast Guard ships—12 high-endurance cutters (WHECs), 29 medium-endurance cutters (WMECs), and 49 110-foot patrol craft (WPBs).[4] The Coast Guard's 12 Hamilton (WHEC-715) class high-endurance cutters entered service between 1967 and 1972.[5] The Coast Guard's 29 medium-endurance cutters include 13 Famous (WMEC-901) class ships that entered service between 1983 and 1991,[6] 14 Reliance (WMEC-615) class ships that entered service between 1964 and 1969,[7] and 2 one-of-a kind cutters that originally entered service with the Navy in 1944 and 1971 and were later transferred to the Coast Guard.[8] The Coast Guard's 49 110-foot Island (WPB-1301) class patrol boats entered service between 1986 and 1992.[9]

Many of these 90 ships are manpower-intensive and increasingly expensive to maintain, and have features that in some cases are not optimal for performing their

acquire NSCs, OPCs, and FRCs were parts of a larger, integrated Coast Guard acquisition effort aimed at acquiring several new types of cutters and aircraft that was called the Integrated Deepwater System (IDS) program, or Deepwater for short. In 2007, the Coast Guard broke up the Deepwater effort into a series of individual cutter and aircraft acquisition programs, but continued to use the term Deepwater as a shorthand way of referring collectively to these now-separated programs. In its FY2012 budget submission, the Coast Guard stopped using the term Deepwater entirely as a way of referring to these programs. Congress, in acting on the Coast Guard's proposed FY2012 budget, did not object to ending the use of the term Deepwater. Reflecting this development, CRS Report RL33753, *Coast Guard Deepwater Acquisition Programs: Background, Oversight Issues, and Options for Congress*, by Ronald O'Rourke, was archived in early 2012, following final congressional action on the FY2012 budget, and remains available to congressional readers as a source of historical reference information on Deepwater acquisition efforts.

[2] See CRS Testimony TE10020, *Building a 21st Century Infrastructure for America: Coast Guard Sea, Air, and Land Capabilities: Part II*, by Ronald O'Rourke.

[3] CRS Report RL34391, *Coast Guard Polar Security Cutter (Polar Icebreaker) Program: Background and Issues for Congress*, by Ronald O'Rourke.

[4] In the designations WHEC, WMEC, and WPB, W means Coast Guard ship, HEC stands for high-endurance cutter, MEC stands for medium-endurance cutter, and PB stands for patrol boat.

[5] Hamilton-class cutters are 378 feet long and have a full load displacement of about 3,400 tons.

[6] Famous-class cutters are 270 feet long and have a full load displacement of about 1,800 tons.

[7] Reliance-class cutters are 210 feet long and have a full load displacement of about 1,100 tons.

[8] The two one-of-a-kind cutters are the *Acushnet* (WMEC-167), which originally entered service with the Navy in 1944, and the *Alex Haley* (WMEC-39), which originally entered service with the Navy in 1971. The *Acushnet* served in the Navy from until 1946, when it was transferred to the Coast Guard. The ship was about 214 feet long and had a displacement of about 1,700 tons. The *Alex Haley* served in the Navy until 1996. It was transferred to the Coast Guard in 1997, converted into a cutter, and reentered service with the Coast Guard in 1999. It is 282 feet long and has a full load displacement of about 2,900 tons.

[9] Island-class boats are 110 feet long and have a full load displacement of about 135 to 170 tons.

assigned missions. Some of them have already been removed from Coast Guard service: eight of the Island-class patrol boats were removed from service in 2007 following an unsuccessful effort to modernize and lengthen them to 123 feet; the one-of-a-kind cutter that originally entered service with the Navy in 1944 was decommissioned in 2011; and the Hamilton-class cutters are being decommissioned as new NSCs enter service. A July 2012 Government Accountability Office (GAO) report discusses the generally poor physical condition and declining operational capacity of the Coast Guard's older high-endurance cutters, medium-endurance cutters, and 110-foot patrol craft.[10]

Missions of NSCs, OPCs, and FRCs

NSCs, OPCs, and FRCs, like the ships they are intended to replace, are to be multimission ships for routinely performing 7 of the Coast Guard's 11 statutory missions, including

- search and rescue (SAR);
- drug interdiction;
- migrant interdiction;
- ports, waterways, and coastal security (PWCS);
- protection of living marine resources;
- other/general law enforcement; and
- defense readiness operations.[11]

Smaller Coast Guard patrol craft and boats contribute to the performance of some of these seven missions close to shore. NSCs, OPCs, and FRCs perform them both close to shore and in the deepwater environment, which generally refers to waters more than 50 miles from shore.

[10] Government Accountability Office, *Coast Guard[:]Legacy Vessels' Declining Conditions Reinforce Need for More Realistic Operational Targets*, GAO-12-741, July 2012, 71 pp.

[11] The four statutory Coast Guard missions that are not to be routinely performed by NSCs, OPCs, and FRCs are marine safety, aids to navigation, marine environmental protection, and ice operations. These missions are performed primarily by other Coast Guard ships. The Coast Guard states, however, that "while [NSCs, OPCs, and FRCs] will not routinely conduct [the] Aids to Navigation, Marine Safety, or Marine Environmental Protection missions, they may periodically be called upon to support these missions (i.e., validate the position of an Aid to Navigation, transport personnel or serve as a Command and Control platform for a Marine Safety or Marine Environmental Response mission, etc.)." (Source: Coast Guard information paper provided to CRS on June 1, 2012.)

NSC Program

National Security Cutters (*Figure 1*)—also known as Legend (WMSL-750)[12] class cutters because they are being named for legendary Coast Guard personnel[13]—are the Coast Guard's largest and most capable general-purpose cutters.[14] They are larger and technologically more advanced than Hamilton-class cutters, and are built by Huntington Ingalls Industries' Ingalls Shipbuilding of Pascagoula, MS (HII/Ingalls).

The Coast Guard's acquisition program of record (POR)—the service's list, established in 2004, of planned procurement quantities for various new types of ships and aircraft—calls for procuring 8 NSCs as replacements for the service's 12 Hamilton-class high-endurance cutters. The Coast Guard's November 2017 acquisition program baseline (APB) for the NSC program estimated the total acquisition cost of a nine-ship NSC program at $6.135 billion, or an average of about $682 million per ship.[15]

Source: U.S. Coast Guard photo accessed May 2, 2012, at http://www.flickr.com/photos/coast_guard/5617034780/sizes/l/in/set-72157629650794895/.

Figure 1. National Security Cutter.

---

[12] In the designation WMSL, W means Coast Guard ship and MSL stands for maritime security cutter, large.

[13] For a Coast Guard news release that mentions the naming rule for the class, see U.S. Coast Guard, "Acquisition Update: Keel Authenticated for the Fifth National Security Cutter," May 17, 2013.

[14] The NSC design is 418 feet long and has a full load displacement of about 4,500 tons. The displacement of the NSC design is about equal to that of Navy's Oliver Hazard Perry (FFG-7) class frigates, which are 453 feet long and have a full load displacement of about 4,200 tons. The Coast Guard's three polar icebreakers are much larger than NSCs, but are designed for a more specialized role of operations in polar waters. The Coast Guard states that

The largest and most technologically advanced of the Coast Guard's newest classes of cutters, the NSCs replace the aging 378-foot high endurance cutters, which have been in service since the 1960s. Compared to legacy cutters, the NSCs' design provides better sea-keeping and higher sustained transit speeds, greater endurance and range, and the ability to launch and recover small boats from astern, as well as aviation support facilities and a flight deck for helicopters and unmanned aerial vehicles.

("National Security Cutter," accessed April 19, 2018, at https://www.dcms.uscg.mil/Our-Organization/Assistant-Commandant-for-Acquisitions-CG-9/Programs/Surface-Programs/National-Security-Cutter/.)

[15] Government Accountability Office, *Home Security Acquisitions[:] Leveraging Programs' Results Could Further DHS's Progress to Improve Portfolio Management*, May 2018, GAO-18-339SP, p. 91.

Although the Coast Guard's POR calls for procuring a total of 8 NSCs to replace the 12 Hamilton-class cutters, Congress through FY2018 has funded 11 NSCs, including two (the 10th and 11th) in FY2018. Six NSCs are now in service (the sixth was commissioned into service on April 1, 2017). The seventh is scheduled to be commissioned into service in January 2019, and the eighth and ninth are scheduled for delivery in 2019 and 2020, respectively. The Coast Guard's proposed FY2019 budget requests $65 million in acquisition funding for the NSC program; this request does not include additional funding for a 12th NSC.

For additional information on the status and execution of the NSC program from May 2018 and March 2016 GAO reports, see *Appendix C*.

OPC Program

Offshore Patrol Cutters (*Figure 2*, *Figure 3*, and *Figure 4*)—also known as Heritage (WMSM915)[16] class cutters because they are being named for past cutters that played a significant role in the history of the Coast Guard and the Coast Guard's predecessor organizations[17]—are to be smaller, less expensive, and in some respects less capable than NSCs.[18] In terms of full load displacement, OPCs are to be about 80% as large as NSCs.[19] Coast Guard officials describe the OPC program as the service's top acquisition priority. OPCs are being built by Eastern Shipbuilding Group of Panama City, FL.

The Coast Guard's POR calls for procuring 25 OPCs as replacements for the service's 29 medium-endurance cutters. As of June 2017, the Coast Guard estimated the

---

[16] In the designation WMSM, W means Coast Guard ship and MSM stands for maritime security cutter, medium.

[17] For the naming rule for the class and a list of the names of the first 11 OPCs, see U.S. Coast Guard, "*The Offshore Patrol Cutter (OPC) Is The Coast Guard's Highest Investment Priority and Will Play A Critical Role in the Service's Future*," undated, accessed August 15, 2017, at http://www.dcms.uscg.mil/Our-Organization/Assistant-Commandantfor-Acquisitions-CG-9/Newsroom/OPC_Day/. See also Sam LaGrone, "Coast Guard Celebrates Birthday by Naming 11 Planned Offshore Patrol Cutters," *USNI News*, August 4, 2017 (updated August 5, 2017).

[18] The service states that OPCs

> The OPCs will provide the majority of offshore presence for the Coast Guard's cutter fleet, bridging the capabilities of the 418-foot national security cutters, which patrol the open ocean, and the 154-foot fast response cutters, which serve closer to shore. The OPCs will conduct missions including law enforcement, drug and migrant interdiction, search and rescue, and other homeland security and defense operations. Each OPC will be capable of deploying independently or as part of task groups and serving as a mobile command and control platform for surge operations such as hurricane response, mass migration incidents and other events. The cutters will also support Arctic objectives by helping regulate and protect emerging commerce and energy exploration in Alaska.
>
> ("*Offshore Patrol Cutter*," accessed April 20, 2018, https://www.dcms.uscg.mil/Our-Organization/Assistant-Commandant-for-Acquisitions-CG-9/ Programs/ Surface-Programs/ Offshore-PatrolCutter/ Offshore-Patrol-Cutter-Program-Profile/.)

[19] As of May 26, 2017, the OPC's light ship displacement (i.e., its "empty" displacement, without fuel, water, ballast, stores, and crew) was preliminarily estimated at about 2,640 to 2,800 tons, and its full load displacement was preliminarily estimated at about 3,500 to 3,730 tons. (Source: Figures provided to CRS by Cost Guard liaison office, May 26, 2017.) In terms of full load displacement, this would make OPCs roughly 80% as large as NSCs.

total acquisition cost of the 25 ships at $10.449 billion, or an average of about $391 million per ship.[20] The first OPC was funded in FY2018 and is to be delivered in 2021. The Coast Guard's proposed FY2019 budget requests $400 million in acquisition funding for the OPC program for the construction of the second OPC (which is scheduled for delivery in 2022) and for procurement of long leadtime materials (LLTM) for the third OPC (which is scheduled for delivery in 2023).

Source: "Offshore Patrol Cutter Notional Design Characteristics and Performance," accessed September 16, 2016, at https://www.uscg.mil/acquisition/opc/pdf/OPC% 20Placemat%203 6x24.pdf.

Figure 2. Offshore Patrol Cutter. (Artist's rendering)

Source: "Offshore Patrol Cutter Notional Design Characteristics and Performance," accessed September 16, 2016, at https://www.uscg.mil/acquisition/opc/pdf/OPC%20 Placemat%2036x 24.pdf.

Figure 3. Offshore Patrol Cutter. (Artist's rendering)

---

[20] Government Accountability Office, *Home Security Acquisitions[:] Leveraging Programs' results Could Further DHS's Progress to Improve Portfolio Management*, May 2018, GAO-18-339SP, p. 93.

Source: Image received from Coast Guard liaison office, May 25, 2017.

Figure 4. Offshore Patrol Cutter. (Artist's rendering)

The Coast Guard's Request for Proposal (RFP) for the OPC program, released on September 25, 2012, established an affordability requirement for the program of an average unit price of $310 million per ship, or less, in then-year dollars (i.e., dollars that are not adjusted for inflation) for ships 4 through 9 in the program.[21] This figure represents the shipbuilder's portion of the total cost of the ship; it does not include the cost of government-furnished equipment (GFE) on the ship,[22] or other program costs—such as those for program management, system integration, and logistics—that contribute to the above-cited figure of $391 million per ship.[23]

At least eight shipyards expressed interest in the OPC program.[24] On February 11, 2014, the Coast Guard announced that it had awarded Preliminary and Contract Design (P&CD) contracts to three of those eight firms—Bollinger Shipyards of Lockport, LA;

---

[21] Source: *Section C.5 of the RFP*, accessed October 31, 2012, at http://www.uscg.mil/ ACQUISITION/newsroom/ updates/opc092512.asp.

[22] GFE is equipment that the government procures and then delivers to the shipyard for installation on the ship.

[23] Source: Coast Guard emails to CRS dated June 25, 2013.

[24] The firms were the following: Bollinger Shipyards of Lockport, LA; Eastern Shipbuilding Group of Panama City, FL; General Dynamics Bath Iron Works (GD/BIW) of Bath, ME; Huntington Ingalls Industries (HII) of Pascagoula, MS; Marinette Marine Corporation of Marinette, WS; General Dynamics National Steel and Shipbuilding Company (GD/NASSCO) of San Diego, CA; Vigor Shipyards of Seattle, WA; and VT Halter Marine of Pascagoula, MS. (Source: U. S. Coast Guard Offshore Patrol Cutter (OPC) List of Interested Contractors Updated July 2012, accessed online October 23, 2012, at http://www.uscg.mil/ ACQUISITION/opc/pdf/companiesinterested.pdf; and Kevin Brancato and Anne Laurent, *Coast Guard's $12 Billion Cutter Competition Spurs Eight Shipyards to Dive In*, Bloomberg Government Study, November 8, 2012, 6 pp. The Coast Guard document states that these firms "expressed interest in the Offshore Patrol Cutter acquisition and have agreed to their names provided on the Coast Guard website." See also Stew Magnuson, "New Coast Guard Cutter Sparks Fierce Competition among Shipbuilders," *National Defense* (*www.national defensemagazine.org*), April 2013, accessed March 26, 2013, at http://www.nationaldefensemagazine.org/ articles/2013/4/1/2013april-new-coast-guard-cutter-sparks-fiercecompetition-among-shipbuilders.)

Eastern Shipbuilding Group of Panama City, FL; and General Dynamics' Bath Iron Works (GD/BIW) of Bath, ME.[25] On September 15, 2016, the Coast Guard announced that it had awarded the detail design and construction (DD&C) contract to Eastern Shipbuilding. The contract covers detail design and production of up to 9 OPCs and has a potential value of $2.38 billion if all options are exercised.[26]

Section 223 of the Howard Coble Coast Guard and Maritime Transportation Act of 2014 (S. 2444/P.L. 113-281 of December 18, 2014) states the following:

SEC. 223. Multiyear Procurement Authority For Offshore Patrol Cutters.

In fiscal year 2015 and each fiscal year thereafter, the Secretary of the department in which the Coast Guard is operating may enter into, in accordance with section 2306b of title 10, United States Code, multiyear contracts for the procurement of Offshore Patrol Cutters and associated equipment.

For additional information on the status and execution of the OPC program from May 2018 and March 2016 GAO reports, see *Appendix C.*

---

[25] "*Acquisition Update: U.S. Coast Guard Awards Three Contracts for Offshore Patrol Cutter Preliminary and Contract Design*," February 11, 2014, accessed February 14, 2014, at http://www.uscg.mil/hq/cg9/newsroom/updates/opc021114.asp. HII and VT Halter Marine reportedly filed protests of the Coast Guard's award decision on February 24 and 25, respectively. The Coast Guard issued stop work orders to Bollinger, Eastern, and GD/BIW pending GAO's rulings on the protests. (Calvin Biesecker, "Coast Guard Issues Stop Work Orders On OPC Following Protests," *Defense Daily*, February 28, 2014: 2-3. See also Christopher P. Cavas, "Ingalls Protesting US Coast Guard Cutter Contract," *DefenseNews.com*, February 26, 2014.) On June 5, 2014, it was reported that GAO had rejected the protests, and that the Coast Guard had directed Bollinger, Eastern, and GD/BIW to resume their work. (Calvin Biesecker, "Coast Guard Directs Design Work Continue On OPC After GAO Denies Protests," *Defense Daily*, June 5, 2014: 1; Christopher P. Cavas, "US Coast Guard Cutter Award Upheld," *Defense News* (http://www.defensenews.com), June 5, 2014. For the text of the decision, see Government Accountability Office, Decision in the Matter of Huntington Ingalls Industries, Inc.; VT Halter Marine, Inc., June 2, 2014.)

[26] "*Acquisition Update: Coast Guard Selects Offshore Patrol Cutter Design*," September 15, 2016, accessed September 16, 2016, at https://www.uscg.mil/acquisition/newsroom/ updates/OPC091516.asp. An October 7, 2016, press report states that "after no protests were filed by the losing bidders to build the Coast Guard's new class of medium-endurance cutters, the service this week directed Eastern Shipbuilding Group to proceed with detail design and construction of the Offshore Patrol Cutter (OPC).... The period for the losing bidders to file a protest ended at close of business on Monday [October 3]." (Calvin Biesecker, "Coast Guard Directs Eastern Shipbuilding to Move Forward With Offshore Patrol Cutter," *Defense Daily*, October 7, 2016, p. 3. See also "*Acquisition Update: Coast Guard Moves Forward to Next Phase of OPC Acquisition*," October 5, 2016, accessed March 20, 2016, at https://www.uscg.mil/acquisition/newsroom/updates/OPC100516.asp.)
On September 7, 2017, the Coast Guard exercised a fixed-price option to its contract with Eastern Shipbuilding to procure long lead time materials (LLTM) for the first OPC; the total value of the option is $41.68 million. ("*Coast Guard Exercises Long Lead Time Materials Option for First Offshore Patrol Cutter*," September 7, 2017, accessed October 25, 2017, at http://www.dcms.uscg.mil/Our-Organization/Assistant-Commandant-for-Acquisitions-CG-9/ Newsroom/OPC090717/.)

FRC Program

Fast Response Cutters (*Figure 5*)—also called Sentinel (WPC-1101)[27] class patrol boats because they are being named for enlisted leaders, trailblazers, and heroes of the Coast Guard and its predecessor services of the U.S. Revenue Cutter Service, U.S. Lifesaving Service, and U.S. Lighthouse Service[28]—are considerably smaller and less expensive than OPCs, but are larger than the Coast Guard's older patrol boats.[29] FRCs are built by Bollinger Shipyards of Lockport, LA.

Source: U.S. Coast Guard photo accessed May 4, 2012, at http://www.flickr.com/photos/coast_guard/6871815460/sizes/l/in/set-72157629286167596/.

Figure 5. Fast Response Cutter. (With an older Island-class patrol boat behind)

The Coast Guard's POR calls for procuring 58 FRCs as replacements for the service's 49 Island-class patrol boats.[30] The Coast Guard as of June 2017 estimated the total acquisition cost of the 58 cutters at $3.342 billion, or an average of about $58 million per cutter.[31] A total of 50 FRCs have been funded through FY2018. The 29th was

[27] In the designation WPC, W means Coast Guard ship and PC stands for patrol craft.

[28] Source for class naming rule: U.S. Coast Guard bulletin, "*ALCOAST 349/17 - Nov 2017 New Fast Response Cutters Named for Coast Guard Heroes*," November 22, 2017, accessed November 20, 2017, at https://content.govdelivery.com/accounts/USDHSCG/bulletins/ 1c6c844.

[29] FRCs are 154 feet long and have a full load displacement of 353 tons.

[30] The Coast Guard states that

> The planned fleet of FRCs will conduct primarily the same missions as the 110' patrol boats being replaced. In addition, the FRC will have several increased capabilities enhancing overall mission execution. The FRC is designed for rapid response, with approximately a 28 knot speed capability, and will typically operate in the coastal zones. Examples of missions that FRCs will complete include SAR, Migrant Interdiction, Drug Interdiction and Ports Waterways and Coastal Security.

[31] Government Accountability Office, *Home Security Acquisitions[:] Leveraging Programs' results Could Further DHS's Progress to Improve Portfolio Management*, May 2018, GAO-18-339SP, p. 81.

commissioned into service on November 8, 2018. The Coast Guard's proposed FY2019 budget requests $240 million in acquisition funding for the procurement of four more FRCs.

For additional information on the status and execution of the FRC program from May 2018 and March 2016 GAO reports, see *Appendix C.*

Funding in FY2013-FY2019 Budget Submissions

Table 1 shows annual requested and programmed acquisition funding for the NSC, OPC, and FRC programs in the Coast Guard's FY2013-FY2019 budget submissions. Actual appropriated figures differ from these requested and projected amounts.

Issues for Congress

Whether to Fund a 12th NSC in FY2019

One issue for Congress is whether to fully or partially fund the acquisition of a 12th NSC in FY2019. Based on funding provided by Congress for the procurement of the 11th NSC in FY2018, fully funding the procurement of a 12th in FY2019 might require about $635 million.

Supporters of procuring a 12th NSC in FY2019 could argue that a total of 12 NSCs would provide one-for-one replacements for the 12 Hamilton-class cutters, that the Coast Guard's POR record includes only about 61% as many new cutters as the Coast Guard has calculated would be required to fully perform the Coast Guard's anticipated missions in coming years (see "Planned NSC, OPC, and FRC Procurement Quantities" below, as well as *Appendix A*); and that funding the 12th NSC in FY2019 would allow the Coast Guard and HII/Ingalls, in identifying an efficient production profile for the ship, to consider options for building the 10th, 11th, and 12th NSCs at intervals of less than 12 months.

Skeptics or opponents of procuring a 12th NSC in FY2019 could argue that the Coast Guard's POR includes only 8 NSCs, that the Coast Guard's fleet mix analyses (see "Planned NSC, OPC, and FRC Procurement Quantities" below, as well as *Appendix A*) have not shown a potential need for more than 9 NSCs, and that in a situation of finite Coast Guard budgets, funding a 12th NSC in FY2018 might require reducing funding for other Coast Guard programs. They might also argue that deferring procurement of a 12th NSC to FY2020 would permit an efficient oneper-year production profile for the 10th, 11th, and 12th NSCs.

Table 1. NSC, OPC, and FRC Funding in FY2013-FY2019 Budget Submissions (Figures in millions of then-year dollars)

Budget	FY13	FY14	FY15	FY16	FY17	FY18	FY19	FY20	FY21	FY22	FY23
NSC program											
FY13	683	0	0	0 38 75	0 0 130	45 30 95					
FY14		616	710	91.4	132 127	95 54					
FY15			638								
FY16											
FY17									21		
FY18							47 30 65	15 65 65	21	6.6	
FY19							65 65	n/a	n/a	n/a	n/a
OPC program											
FY13	30	50	40	200	530						
FY14		25	65	200	530	430					
FY15			20	90	100	530	430				
FY16				18.5	100	530	430	430			
FY17					100	530	430	530	770		
FY18						500	400	457	716	700	
FY19							400	n/a	n/a	n/a	n/a
FRC program											
FY13	139	360	360	360	360						
FY14		75	110	110	110	110					
FY15			110	340	220	220	315				
FY16				340	325	240	240	325			
FY17					240	240	325	325	18		
FY18						240	335	335	26	18	
FY19							240	n/a	n/a	n/a	n/a
Total											
FY13	852	410	400	560	890						
FY14		716	885	348	640	585					
FY15			768	505	450	780	792				
FY16				449.9	557	865	700	370			
FY17					467	865	820	920	809		
FY18						794	800	857	763	724.6	
FY19							705	n/a	n/a	n/a	n/a

Source: Table prepared by CRS based on FY2013-FY2019 budget submissions.

Note: n/a means not available.

Number of FRCs to Fund in FY2019

Another issue for Congress is whether to fund the acquisition of four FRCs in FY2019, as requested, or some other number, such as six, which is the maximum number that has been acquired in some prior fiscal years. As shown in *Table 1*, the Coast Guard's FY2019 budget requests \$240 million for the procurement of four FRCs. Based on prior-year appropriations for FRCs, procuring a total of six FRCs in FY2019 would require \$340 million, or \$100 million more than the requested amount.

Supporters of funding the acquisition of six FRCs in FY2019 could argue that it would increase production economies of scale and thus reduce the unit acquisition cost of the ships, and help the Coast Guard to close more quickly a gap in patrol boat capacity that is limiting the Coast Guard's ability to interdict illegal drugs and carry out other missions. Supporters of funding the acquisition of four FRCs in FY2019 could argue that in a situation of finite Coast Guard budgets, funding two additional FRCs at an additional cost of $100 million might require reducing funding for other Coast Guard programs.

Annual or Multiyear Contracting for OPCs

Another issue for Congress is whether to acquire OPCs using annual contracting or multiyear contracting in the form of multiyear procurement (MYP) or block buy contracting. The Coast Guard currently plans to use a contract with options for procuring the first nine OPCs. Although a contract with options may look like a form of multiyear contracting, it operates more like a series of annual contracts. Contracts with options do not achieve the reductions in acquisition costs that are possible with MYP and block buy contracting. Using MYP or block buy contracting involves accepting certain tradeoffs.[32]

As mentioned earlier, Section 223 of the Howard Coble Coast Guard and Maritime Transportation Act of 2014 (S. 2444/P.L. 113-281 of December 18, 2014) grants authority to use MYP in the OPC program. MYP typically cannot be used on the first several ships in a shipbuilding program because the law that regulates MYP (10 U.S.C. 2306b) requires a stable design for an acquisition program to qualify for MYP. In a shipbuilding program, a stable design is typically demonstrated by completing the construction of the first ship in the class, by which time the first several ships in the class typically have been funded and put under contract.

Block buy contracting, by comparison, can be used at the start of a shipbuilding program, beginning with the first ship. (Indeed, this was a principal reason why block buy contracting was in effect invented in FY1998, as the contracting method for procuring the Navy's first four Virginia-class attack submarines.) As with MYP, authority for using block buy contracting must be granted by Congress. Since Section 223 of P.L. 113-281 grants authority to use MYP but not block buy contracting, Congress

---

[32] These tradeoffs include the following:

- reduced congressional control over year-to-year spending, and tying the hands of future Congresses;
- reduced flexibility for making changes in Coast Guard acquisition programs in response to unforeseen changes in strategic or budgetary circumstances (which can cause any needed funding reductions to fall more heavily on acquisition programs not covered by multiyear contracts);
- a potential need to shift funding from later fiscal years to earlier fiscal years to fund economic order quantity (EOQ) purchases (i.e., up-front batch purchases) of components;
- the risk of having to make penalty payments to shipbuilders if multiyear contracts need to be terminated due to unavailability of funds needed for the continuation of the contracts; and
- the risk that materials and components purchased for ships to be procured in future years might go to waste if those ships are not eventually procured.

would need to grant authority to the Coast Guard to use block buy contracting in the OPC program.

CRS estimates that if the Coast Guard were to use block buy contracting with economic order quantity (EOQ) purchases (i.e., up-front batch purchases) of components for acquiring the first several OPCs, and either MYP or block buy contracting with EOQ purchases for acquiring the remaining ships in the program, the savings on the total acquisition cost of the 25 OPCs (compared to costs under contracts with options) could amount to roughly $1 billion. CRS also estimates that acquiring the first nine ships in the OPC program under the current contract with options could forego roughly $350 million of the $1 billion in potential savings.

One potential option for the subcommittee would be to look into the possibility of having the Coast Guard either convert the current OPC contract at an early juncture into a block buy contract with EOQ authority, or, if conversion is not possible, replace the current contract at an early juncture with a block buy contract with EOQ authority.[33] Replacing the current contract with a block buy contract might require recompeting the program, which would require effort on the Coast Guard's part and could create business risk for Eastern Shipbuilding Group, the shipbuilder that holds the current contract. On the other hand, the cost to the Coast Guard of recompeting the program would arguably be small relative to a potential additional savings of perhaps $300 million, and Eastern arguably would have a learning curve advantage in any new competition by virtue of its experience in building the first OPC.

OPC Procurement Rate

The current procurement profile for the OPC, which reaches a maximum projected rate of two ships per year, would deliver OPCs many years after the end of the originally planned service lives of the medium-endurance cutters that they are to replace. Coast Guard officials have testified that the service plans to extend the service lives of the medium-endurance cutters until they are replaced by OPCs. There will be maintenance and repair expenses associated with extending the service lives of medium-endurance cutters, and if the Coast Guard does not also make investments to increase the capabilities of these ships, the ships may have less capability in certain regards than OPCs.[34]

One possible option for addressing this situation would be to increase the maximum annual OPC procurement rate from the currently planned two ships per year to three or four ships per year. Doing this could result in the 25th OPC being delivered about four

---

[33] As part of the replacement scenario, the Coast Guard could end the implementation of the current contract with options by not exercising an option.

[34] For further discussion, see Government Accountability Office, *Coast Guard Acquisitions[:] Actions Needed to Address Longstanding Portfolio Management Challenges*, GAO 18-454, July 2018, pp. 32-36.

years or six years sooner, respectively, than under the currently planned maximum rate. Increasing the OPC procurement rate to three or four ships per year would require a substantial increase to the Coast Guard's Procurement, Construction, and Improvements (PC&I) account,[35] an issue discussed in *Appendix B*.

Increasing the maximum procurement rate for the OPC program could, depending on the exact approach taken, reduce OPC unit acquisition costs due to improved production economies of scale. Doubling the rate to four ships per year, for example, could reduce unit procurement costs by as much as 10%, which could result in hundreds of millions of dollars in additional savings in acquisition costs for the program. Increasing the maximum procurement rate could also create new opportunities for using competition in the OPC program. Notional alternative approaches for increasing the OPC procurement rate to three or four ships per year include but are not necessarily limited to the following:

- increasing the production rate to three or four ships per year at Eastern Shipbuilding—an option that would depend on Eastern Shipbuilding's production capacity;
- introducing a second shipyard to build Eastern's design for the OPC;
- introducing a second shipyard (such as one of the other two OPC program finalists) to build its own design for the OPC—an option that would result in two OPC classes; or
- building additional NSCs in the place of some of the OPCs—an option that might include descoping equipment on those NSCs where possible to reduce their acquisition cost and make their capabilities more like that of the OPC. Such an approach would be broadly similar to how the Navy is planning to use a descoped version of the San Antonio (LPD-17) class amphibious ship as the basis for its LPD-17 Flight II (LPD-30) class amphibious ships.[36]

Impact of Hurricane Michael on Eastern Shipbuilding

Another potential issue for Congress concerns the impact of Hurricane Michael on Eastern Shipbuilding of Panama City, FL, the shipyard that is to build the first nine OPCs. An October 22, 2018, press report states:

---

[35] Prior to FY2019, the PC&I account was called the Acquisition, Construction, and Improvements (AC&I) account.

[36] For additional discussion, see CRS Report R43543, *Navy LPD-17 Flight II (LX[R]) Amphibious Ship Program: Background and Issues for Congress*, by Ronald O'Rourke.

U.S. Coast Guard officials and Eastern Shipbuilding Group are still assessing the damage caused by deadly category 4 Hurricane Michael to the Panama City, Fla.-based yard contracted to build the new class of Offshore Patrol Cutters.

On September 28, the Coast Guard awarded Eastern Shipbuilding a contract to build the future USCGC *Argus* (WMSM-915), the first offshore patrol cutter (OPC). The yard was also set to build a second OPC, the future USCGC *Chase* (WMSM-916). Eastern Shipbuilding's contract is for nine OPCs, with options for two additional cutters. Ultimately, the Coast Guard plans to buy 25 OPCs.

However, just as the yard was preparing to build *Argus*, Hurricane Michael struck the Florida Panhandle near Panama City on October 10. Workers from the shipyard and Coast Guard project managers evacuated and are just now returning to assess damage to the yard facilities, Brian Olexy, communications manager for the Coast Guard's Acquisitions Directorate, told USNI News.

"Right now we haven't made any decisions yet on shifts in schedule," Olexy said....

Since the yard was just the beginning stages of building *Argus*, Olexy said the hull wasn't damaged. "No steel had been cut," he said.

Eastern Shipbuilding is still in the process of assessing damage to the yard and trying to reach its workforce. Many employees evacuated the area and have not returned, or are in the area but lost their homes, Eastern Shipbuilding spokesman Justin Smith told USNI News.

At first, about 200 workers returned to work, but by week's end about 500 were at the yard, Smith said. The company is providing meals, water, and ice for its workforce.

"Although we were significantly impacted by this catastrophic weather event, we are making great strides each day thanks to the strength and resiliency of our employees," Joey D'Isernia, president of Eastern Shipbuilding, said in a statement.[37]

A November 1, 2018, statement from Eastern Shipbuilding states that the firm resumed operations at both of its two main shipbuilding facilities just two weeks after Hurricane Michael devastated Panama City Florida and the surrounding communities....

... the majority of ESG's [Eastern Shipbuilding Group's] workforce has returned to work very quickly despite the damage caused by the storm. "Our employees are a resourceful and resilient group of individuals with the drive to succeed in the face of adversity. This has certainly been proven by their ability to bounce back over the two weeks following the storm. Our employees have returned to work much faster than anticipated and brought with them an unbreakable spirit, that I believe sets this shipyard and our community apart" said [Eastern Shipbuilding] President Joey D'Isernia. "Today, our staffing levels exceed 80% of our pre-Hurricane Michael levels and is rising daily."

---

[37] Ben Werner, "Coast Guard, Shipbuilder Assessing Hurricane Damage to Yard Building Offshore Patrol Cutter," *USNI News*, October 22, 2018. See also Paul McLeary, "Hurricane Michael Hits Coast Guard's Largest Program, Leaving Devastation," *Breaking Defense*, October 18, 2018; Marex, "Despite Hurricane Michael, Eastern Shipbuilding Keeps Working," *Maritime Executive*, October 16, 2018; Samuel Hill, "Eastern Shipbuilding Hit Hard by Hurricane Michael," *Workboat*, October 16, 2018.

Immediately following the storm, ESG set out on an aggressive initiative to locate all of its employees and help get them back on the job as soon as practical after they took necessary time to secure the safety and security of their family and home. Together with its network of friends, partners, and customers in the maritime community, ESG organized daily distribution of meals and goods to employees in need. Additionally, ESG created an interest free deferred payback loan program for those employees in need and has organized Go Fund Me account to help those employees hardest hit by the storm. ESG also knew temporary housing was going to be a necessity in the short term and immediately built a small community located on greenfield space near its facilities for those employees with temporary housing needs.

ESG has worked closely with its federal, state and commercial partners over the past two weeks to provide updates on the shipyard as well as on projects currently under construction. Power was restored to ESG's Nelson Facility on 10-21-18 and at ESG's Allanton Facility on 10-24-18 and production of vessels under contract is ramping back up. Additionally, all of the ESG personnel currently working on the US Coast Guard's Offshore Patrol Cutter contract have returned to work....

"We are grateful to our partners and the maritime business community as a whole for their support and confidence during the aftermath of this historic storm. Seeing our incredible employees get back to building ships last week was an inspiration," said D'Isernia. "While there is no doubt that the effects of Hurricane Michael will linger with our community for years to come, I can say without reservation that we are open for business and excited about delivering quality vessels to our loyal customers."[38]

Planned NSC, OPC, and FRC Procurement Quantities

Another issue for Congress concerns the Coast Guard's planned NSC, OPC, and FRC procurement quantities. The POR's planned force of 91 NSCs, OPCs, and FRCs is about equal in number to the Coast Guard's legacy force of 90 high-endurance cutters, medium-endurance cutters, and 110-foot patrol craft. NSCs, OPCs, and FRCs, moreover, are to be individually more capable than the older ships they are to replace. Even so, Coast Guard studies have concluded that the planned total of 91 NSCs, OPCs, and FRCs would provide 61% of the cutters that would be needed to fully perform the service's statutory missions in coming years, in part because Coast Guard mission demands are expected to be greater in coming years than they were in the past. For further discussion of this issue, about which CRS has testified and reported on since 2005,[39] see *Appendix A*.

---

[38] Eastern Shipbuilding news release, November 1, 2018, entitled "Eastern Shipbuilding Group, Inc. Resumes Operations."

[39] See Statement of Ronald O'Rourke, Specialist in National Defense, Congressional Research Service, Before the Senate Commerce, Science, and Transportation Committee, Subcommittee on Fisheries and the Coast Guard, Hearing on The Coast Guard's Revised Deepwater Implementation Plan, June 21, 2005, pp. 1-5.

NSC Program: Initial Testing

Another potential oversight issue for Congress concerns the results of initial testing of the NSC. A January 2016 GAO report stated the following:

> The U.S. Navy, the Coast Guard's independent test agent, completed initial testing for the National Security Cutter (NSC) in April 2014 and rated the NSC as operationally effective and suitable. Still, testing revealed 10 major deficiencies.... Initial testing is an event designed to verify performance of critical systems to ensure assets are capable of meeting mission requirements. The event tests critical operational issues and key performance parameters. The NSC fully met 12 of 19 key performance parameters. Tests of one key performance parameter, as well as other critical systems, were deferred to follow-on testing. The Coast Guard and the U.S. Navy disagree on the NSC's requirements for cutter boat operations. Without clear requirements the Navy and Coast Guard will not have a basis for determining actions to resolve any performance issues. Coast Guard officials acknowledged that clarifying these requirements would be beneficial.
>
> The Coast Guard plans to begin follow-on testing in fall 2016. It must submit corrective action plans to the U.S. Navy to close any deficiencies. According to Coast Guard documentation, it may choose not to correct all deficiencies due to the cost of changes. Department of Homeland Security (DHS) acquisition guidance does not specify the timing of follow-on testing or the actions to be taken in response to the findings. Without a definite time frame DHS risks encountering the same problems as the NSC program experienced with future acquisitions and fielding assets without knowing the full capabilities. During operations, the NSC has experienced performance issues that were not identified during initial testing, and the Coast Guard has planned design changes to some of the cutters' equipment.... However, the Coast Guard has not yet found the causes for problems affecting the NSC's propulsion systems. As a result of these and other equipment failures, the NSC has been operating in a degraded condition in some mission areas. DHS has no plans for additional acquisition review boards for the NSC, which would provide oversight going forward. Continued management-level oversight by DHS would help ensure that problems identified during testing and operations are addressed.[40]

Legislative Activity in 2018

Summary of Appropriations Action on FY2019 Acquisition Funding Request

Table 2 summarizes appropriations action on the Coast Guard's request for FY2019 acquisition funding for the NSC, OPC, and FRC programs.

---

[40] Government Accountability Office, *National Security Cutter[:]Enhanced Oversight Needed to Ensure Problems Discovered during Testing and Operations Are Addressed*, GAO-16-148, January 2016, summary page.

Table 2. Summary of Appropriations Action on FY2019 Acquisition Funding Request (Figures in millions of dollars, rounded to nearest tenth)

Request	Request	HAC	SAC	Final
NSC program	65	140	72.6	
OPC program	400	400	400	
FRC program	240	340	240	
TOTAL	705	880	712.6	

Source: Table prepared by CRS based on Coast Guard's FY2019 budget submission, HAC committee report, and SAC chairman's mark and explanatory statement on FY2019 DHS Appropriations Act. HAC is House Appropriations Committee; SAC is Senate Appropriations Committee.

FY2019 DHS Appropriations Act (H.R. XXXX/S. 3109)

House

The House Appropriations Committee marked up the FY2019 DHS Appropriations Act (referred to here as H.R. XXXX) on July 25, 2018. The text of the bill as marked up, and the committee's report reflecting the markup, were not available as of August 3, 2018. The figures shown in the HAC column of *Table 2* and the discussion below are based on the bill text and draft committee report (referred to here as H.Rept. 115-XXX) going into July 25, 2018, markup meeting, combined with a summary of the amendments adopted at the markup meeting, which were posted on the committee's website in conjunction with the pre-markup bill text and draft committee report.[41]

H.Rept. 115-XXX recommends the funding levels shown in the HAC column of *Table 2*. In addition, the committee states that at the July 25, 2018, markup, the committee adopted by voice vote an amendment that "allows certain existing, unobligated funds to be used to purchase long-lead time materials for the 12th National Security Cutter."

H.Rept. 115-XXX states:

> *National Security Cutter (NSC).* The Consolidated Appropriations Act, 2018 (Public Law 115–141) provided $1,241,000,000 for the NSC program, which included funds for construction of the tenth and eleventh NSC, a contrast from the historic approach of funding construction for one NSC per fiscal year. The Committee's fiscal year 2019 recommendation includes $140,000,000 for the NSC program, $75,000,000 more than requested. Included in this amount is an additional $75,000,000 above the request to continue support of Post Delivery Activities (PDA) for the seventh through ninth hulls and other program-wide activities.

[41] Houser Appropriations Committee, "*Appropriations Committee Approves Fiscal Year 2019 Homeland Security Funding Bill," July 25, 2019*, accessed August 3, 2018, at https://appropriations.house.gov/news/document single.aspx?DocumentID=395388.

Offshore Patrol Cutter (OPC). The recommendation includes $400,000,000 for the OPC program, as requested, to fund construction of the second OPC, long lead time materials for the third, and program management costs.

Fast Response Cutter (FRC). The recommendation provides $340,000,000 for six FRCs, four for the current program of record, as requested, and two to continue replacement of the 110-foot Island Class Cutters supporting U.S. Central Command in Southwest Asia. The Committee strongly encourages the Coast Guard to transition the 110-foot patrol boats supporting U.S. Central Command in Southwest Asia to FRCs in the most expedient manner possible, and to update the Committee of any changes to its FRC deployment strategy. The Committee understands the current patrol boats are well past their service life and wants to ensure the Coast Guard men and women serving in this challenging area of operations have the right equipment necessary to meet these evolving threats. (Page 39)

H.Rept. 115-XXX also states:

Similar to the other Armed Services, the Coast Guard must maintain military readiness in order to meet its mission requirements. Within 180 days of enactment of this Act, the Coast Guard is directed to report to the Committee on any lost operational time due to unplannedmaintenance or supply shortfalls for cutters, aircraft, and boats, as well as the current operations and support (O&S) maintenance backlog for cutters, aircraft, shore facilities, and information technology systems, including the operational impact of this backlog.

he Committee recommends $7,620,209,000 for [the Coast Guard] O&S [operation and support account], $27,071,000 above the request to fund additional full-time equivalents, increase child care subsidy benefits, fund an independent analysis of the current and projected air and sea fleet requirements, and address rising costs for fuel and rent. Also included in this amount is $1,000,000 to equip the Fast Response Cutter fleet with hailing and acoustic laser light tactical systems. (Page 36)

H.Rept. 115-XXX also states:

The Commandant of the Coast Guard is directed to provide to the Committee not later than one year after the date of enactment of this Act, a report that examines the number and type of Coast Guard assets required to meet the Service's current and foreseeable needs in accordance with the Service's statutory missions. The report shall include, but not be limited to, an assessment of the required number and types of cutters and aircraft for current and planned asset acquisitions. The report shall specifically address regional mission requirements in the Western Hemisphere, including the Polar regions, support provided to Combatant Commanders, and trends in illicit activity and illegal migration. In order to provide an impartial assessment, the recommendation includes an increase of $3,300,000 for the report to be prepared by a Federally Funded Research and Development Center experienced in similar examinations. (Page 37)

Senate

The Senate Appropriations Committee, in its report (S.Rept. 115-283 of June 21, 2018) on S. 3109, recommends the funding levels shown in the SAC column of *Table 2*. S.Rept. 115-283 states the following:

> *National Security Cutter*—Legend Class National Security Cutters [NSCs] are replacing the legacy High Endurance Cutters, built between 1967 and 1972. In fiscal year 2017, the Coast Guard interdicted 2,512 illegal migrants and removed 224 metric tons of cocaine with an estimated street value of over $6,600,000,000, which surpassed the previous record amount of cocaine removal set the previous fiscal year. Of the 224 metric tons of cocaine removed, four NSCs interdicted 72.6 metric tons of cocaine with an estimated street value of $2,144,000,000. In a single deployment, the USCGC JAMES (WMSL–754) removed 16.8 metric tons of uncut cocaine with a street value in excess of $496,000,000. The Committee recommends $72,600,000 for the NSC program including $7,600,000 for advance purchase of several systems for the tenth and eleventh NSCs, including wind indicating and measurement systems, homing beacons for aircraft, and navigation and sensor data distribution systems. The Committee worked diligently to ensure resources were provided by the Consolidated Appropriations Act, 2018 to construct a tenth and eleventh NSC. The Committee is pleased that the Coast Guard has indicated that contracts for these NSCs are on track to be awarded on time. The Committee continues to believe that the Coast Guard's fleet of twelve High Endurance Cutters should be replaced with twelve NSCs. The Committee intends to continue to work with the Coast Guard to understand the costs, operational benefits of, and recommended schedule for acquisition of a twelfth NSC.
>
> *Offshore Patrol Cutter*—The OPC will replace the fleet of Medium Endurance Cutters and further enhance the Coast Guard's layered security strategy. The recommendation includes $400,000,000 for the OPC program. This funding will provide for production of a second OPC, LLTM for a third OPC, program activities, test and evaluation, government furnished equipment, and training aids. The Committee encourages the Coast Guard to evaluate the requirements for a sensitive compartmented information facility and amultimodal radar system onboard the OPC and determine whether and when these requirements should be incorporated into revised design and construction.
>
> *Fast Response Cutter*—The Committee recommends $240,000,000 to acquire four FRCs, as requested. (Page 68)

S.Rept. 115-283 also states the following:

> Due in large part to the Committee's efforts, the Coast Guard's surface and air fleets are in the midst of unprecedented modernization. With the expansion of the National Security Cutter fleet, continuation of Fast Response Cutter production, beginning of the Offshore Patrol Cutter acquisition, and initiation of the first Heavy Polar Icebreaker

acquisition in more than four decades, the Coast Guard's cutter fleet is growing into a state-of-the-art force adaptable to any mission. (Page 6)

S.Rept. 115-283 also states the following:

Coast Guard Yard—The Committee has urged the Coast Guard to expedite planning for facility and equipment upgrades necessary for service life extensions of the Fast Response Cutter [FRC] and other vessels at the Coast Guard Yard at Curtis Bay in Baltimore, Maryland. The nearest travel lift of sufficient size and capacity to service the FRC is in Hampton Roads, Virginia. Transporting the travel lift between Hampton Roads and Baltimore is a costly and time consuming procedure that removes the lift from service during transport. The recommendation includes funding within the PC&I appropriation to acquire necessary equipment and make physical modifications to wharves or other parts of the Coast Guard Yard facility to accommodate FRCs and other vessels. (Page 63)

S.Rept. 115-283 also states the following:

Full-Funding Policy—The Committee again directs an exception to the administration's current acquisition policy that requires the Coast Guard to attain the total acquisition cost for a vessel, including long lead time materials [LLTM], production costs, and postproduction costs, before a production contract can be awarded. This policy has the potential to make shipbuilding less efficient, to force delayed obligation of production funds, and to require post-production funds far in advance of when they will be used. The Department should position itself to acquire vessels in the most efficient manner within the guidelines of strict governance measures. The Committee expects the administration to adopt a similar policy for the acquisition of the Offshore Patrol Cutter [OPC] and heavy polar icebreaker. (Page 67)

S.Rept. 115-283 also states the following:

Homeport of New Vessels—The Committee supports the current and planned homeport locations for National Security Cutters. However, the Committee recognizes the challenges that replacing the 378-foot Hamilton-class cutter with 418-foot NSCs has imposed on Coast Guard facilities where waterfront space is limited and reiterates the importance of pier availability for Coast Guard cutters and other surface vessels to minimize operational delays or other unnecessary costs that would undermine the Coast Guard's ability to conduct its missions. Therefore, not later than 180 days after enactment of this act, the Coast Guard shall report to the Committee on infrastructure requirements associated with the homeporting of new vessels. At a minimum, the Coast Guard shall assess if major acquisition system infrastructure is required, identify associated funding needs, and provide a plan to address these requirements to the Committee. (Pages 70-71)

Frank LoBiondo Coast Guard Authorization Act of 2018 (S. 140/P.L. 115-282)

S. 140 was introduced in the Senate on January 12, 2017, as a bill to amend the White Mountain Apache Tribe [WMAT] Water Rights Quantification Act of 2010 to clarify the use of amounts in the WMAT Settlement Fund. The bill retained that purpose and was acted on by Congress into 2018. Later in 2018, S. 140 became a bill to authorize appropriations for the Coast Guard, and for other purposes. On November 14, 2018, the Senate concurred, 94-6, in the House amendment to S. 140 with an amendment (S.Amdt. 4054 as modified). On November 27, 2018, the House agreed to by voice vote a motion that the House suspend the rules and agree to the Senate amendment to the House amendment. S. 140 was signed into law as P.L. 115-282 on December 4, 2018.

Section 204 of S. 140/P.L. 115-282 states:

SEC. 204. Authorization of amounts for Fast Response Cutters.

(a) In general.—Of the amounts authorized under section 4902 of title 14, United States Code, as amended by this Act, for each of fiscal years 2018 and 2019 up to $167,500,000 is authorized for the acquisition of 3 Fast Response Cutters.

(b) Treatment of acquired cutters.—Any cutters acquired pursuant to subsection (a) shall be in addition to the 58 cutters approved under the existing acquisition baseline.

Section 304 of S. 140/P.L. 115-282 states: SEC. 304. Unmanned aircraft.

(a) Land-based unmanned aircraft system program.—Chapter 3 of title 14, United States Code, is amended by adding at the end the following:

"§ 319. Land-based unmanned aircraft system program

(a) In general.—Subject to the availability of appropriations, the Secretary shall establish a land-based unmanned aircraft system program under the control of the Commandant.

(b) Unmanned aircraft system defined.—In this section, the term 'unmanned aircraft system' has the meaning given that term in section 331 of the FAA Modernization and Reform Act of 2012 (49 U.S.C. 40101 note).".

(b) Limitation on unmanned aircraft systems.—Chapter 11 of title 14, United States Code, is amended by inserting after section 1155 the following:

"§ 1156. Limitation on unmanned aircraft systems

"(a) In general.—During any fiscal year for which funds are appropriated for the design or construction of an Offshore Patrol Cutter, the Commandant—

(1) may not award a contract for design of an unmanned aircraft system for use by the Coast Guard; and

(2) may lease, acquire, or acquire the services of an unmanned aircraft system only if such system—

"(A) has been part of a program of record of, procured by, or used by a Federal entity (or funds for research, development, test, and evaluation have been received from a Federal entity with regard to such system) before the date on which the Commandant leases, acquires, or acquires the services of the system; and

"(B) is leased, acquired, or utilized by the Commandant through an agreement with a Federal entity, unless such an agreement is not practicable or would be less cost-effective than an independent contract action by the Coast Guard.

(b) Small unmanned aircraft exemption.—Subsection (a)(2) does not apply to small unmanned aircraft.

(c) Definitions.—In this section, the terms 'small unmanned aircraft' and 'unmanned aircraft system' have the meanings given those terms in section 331 of the FAA Modernization and Reform Act of 2012 (49 U.S.C. 40101 note).".

(c) Clerical amendments.—

CHAPTER 3.—The analysis for chapter 3 of title 14, United States Code, is amended by adding at the end the following:

"319. Land-based unmanned aircraft system program.".

CHAPTER 11.—The analysis for chapter 11 of title 14, United States Code, is amended by inserting after the item relating to section 1155 the following:

"1156. Limitation on unmanned aircraft systems.".

(c) Conforming amendment.—Subsection (c) of section 1105 of title 14, United States Code, is repealed.

Section 311 of S. 140/P.L. 115-282 states:

SEC. 311. Contracting for major acquisitions programs.

(a) General acquisition authority.—Section 501(d) of title 14, United States Code, is amended by inserting "aircraft, and systems," after "vessels,".

(b) Contracting authority.—Chapter 11 of title 14, United States Code, as amended by this Act, is further amended by inserting after section 1136 the following:

"§ 1137. Contracting for major acquisitions programs

"(a) In general.—In carrying out authorities provided to the Secretary to design, construct, accept, or otherwise acquire assets and systems under section 501(d), the Secretary, acting through the Commandant or the head of an integrated program office established for a major acquisition program, may enter into contracts for a major acquisition program.

"(b) Authorized methods.—Contracts entered into under subsection (a)—

"(1) may be block buy contracts;

"(2) may be incrementally funded;

"(3) may include combined purchases, also known as economic order quantity purchases, of—

materials and components; and

long lead time materials; and

"(4) as provided in section 2306b of title 10, may be multiyear contracts.

"(c) Subject to appropriations.—Any contract entered into under subsection (a) shall provide that any obligation of the United States to make a payment under the contract is subject to the availability of amounts specifically provided in advance for that purpose in subsequent appropriations Acts.".

(c) Clerical amendment.—The analysis for chapter 11 of title 14, United States Code, as amended by this Act, is further amended by inserting after the item relating to section 1136 the following:

"1137. Contracting for major acquisitions programs."

(d) Conforming amendments.—The following provisions are repealed:

(1) Section 223 of the Howard Coble Coast Guard and Maritime Transportation Act of 2014 (14 U.S.C. 1152 note), and the item relating to that section in the table of contents in section 2 of such Act.

(2) Section 221(a) of the Coast Guard and Maritime Transportation Act of 2012 (14 U.S.C. 1133 note).

Section 207(a) of the Coast Guard Authorization Act of 2016 (14 U.S.C. 561 note).

(e) Internal regulations and policy.—Not later than 180 days after the date of enactment of this Act, the Secretary of the department in which the Coast Guard is operating shall establish the internal regulations and policies necessary to exercise the authorities provided under this section, including the amendments made in this section.

(f) Multiyear contracts.—The Secretary of the department in which the Coast Guard is operating is authorized to enter into a multiyear contract for the procurement of a tenth, eleventh, and twelfth National Security Cutter and associated government-furnished equipment.

Section 817 of S. 140/P.L. 115-282 states:

SEC. 817. Fleet requirements assessment and strategy.

(a) Report.—Not later than 1 year after the date of enactment of this Act, the Secretary of the department in which the Coast Guard is operating, in consultation with interested Federal and non-Federal stakeholders, shall submit to the Committee on Commerce, Science, and Transportation of the Senate and the Committee on Transportation and Infrastructure of the House of Representatives a report including—

(1) an assessment of Coast Guard at-sea operational fleet requirements to support its statutory missions established in the Homeland Security Act of 2002 (6 U.S.C. 101 et seq.); and

(2) a strategic plan for meeting the requirements identified under paragraph (1).

(b) Contents.—The report under subsection (a) shall include—

(1) an assessment of—

(A) the extent to which the Coast Guard at-sea operational fleet requirements referred to in subsection (a)(1) are currently being met;

(B) the Coast Guard's current fleet, its operational lifespan, and how the anticipated changes in the age and distribution of vessels in the fleet will impact the ability to meet at-sea operational requirements;

(C) fleet operations and recommended improvements to minimize costs and extend operational vessel life spans; and

(D) the number of Fast Response Cutters, Offshore Patrol Cutters, and National Security Cutters needed to meet at-sea operational requirements as compared to planned acquisitions under the current programs of record;

(2) an analysis of—

(A) how the Coast Guard at-sea operational fleet requirements are currently met, including the use of the Coast Guard's current cutter fleet, agreements with partners, chartered vessels, and unmanned vehicle technology; and

(B) whether existing and planned cutter programs of record (including the Fast Response Cutter, Offshore Patrol Cutter, and National Security Cutter) will enable the Coast Guard to meet at-sea operational requirements; and

(3) a description of—

planned manned and unmanned vessel acquisition; and

how such acquisitions will change the extent to which the Coast Guard at-sea operational requirements are met.

(c) Consultation and transparency.—

(1) CONSULTATION.—In consulting with the Federal and non-Federal stakeholders under subsection (a), the Secretary of the department in which the Coast Guard is operating shall—

(A) provide the stakeholders with opportunities for input—

(i) prior to initially drafting the report, including the assessment and strategic plan; and

(ii) not later than 3 months prior to finalizing the report, including the assessment and strategic plan, for submission; and

(B) document the input and its disposition in the report.

(2) TRANSPARENCY.—All input provided under paragraph (1) shall be made available to the public.

(d) Ensuring maritime coverage.—In order to meet Coast Guard mission requirements for search and rescue, ports, waterways, and coastal security, and maritime environmental response during recapitalization of Coast Guard vessels, the Coast Guard shall ensure continuity of the coverage, to the maximum extent practicable, in the locations that may lose assets.

Section 818 of S. 140/P.L. 115-282 states:

SEC. 818. National Security Cutter.

(a) Standard method for tracking.—The Commandant of the Coast Guard may not certify an eighth National Security Cutter as Ready for Operations before the date on which the Commandant provides to the Committee on Transportation and Infrastructure of the House of Representatives and the Committee on Commerce, Science, and Transportation of the Senate—

(1) a notification of a new standard method for tracking operational employment of Coast Guard major cutters that does not include time during which such a cutter is away from its homeport for maintenance or repair; and

(2) a report analyzing cost and performance for different approaches to achieving varied levels of operational employment using the standard method required by paragraph (1) that, at a minimum—

(A) compares over a 30-year period the average annualized baseline cost and performances for a certified National Security Cutter that operated for 185 days away from homeport or an equivalent alternative measure of operational tempo—

(i) against the cost of a 15 percent increase in days away from homeport or an equivalent alternative measure of operational tempo for a National Security Cutter; and

(ii) against the cost of the acquisition and operation of an additional National Security Cutter; and

(B) examines the optimal level of operational employment of National Security Cutters to balance National Security Cutter cost and mission performance.

(b) Conforming amendments.—

(1) Section 221(b) of the Coast Guard and Maritime Transportation Act of 2012 (126 Stat. 1560) is repealed.

(2) Section 204(c)(1) of the Coast Guard Authorization Act of 2016 (130 Stat. 35) is repealed.

Section 822 of S. 140/P.L. 115-282 states:

SEC. 822. Strategic assets in the Arctic.

(a) Definition of arctic.—In this section, the term "Arctic" has the meaning given the term in section 112 of the Arctic Research and Policy Act of 1984 (15 U.S.C. 4111).

(b) Sense of congress.—It is the sense of Congress that—

(1) the Arctic continues to grow in significance to both the national security interests and the economic prosperity of the United States; and

(2) the Coast Guard must ensure it is positioned to respond to any accident, incident, or threat with appropriate assets.

(c) Report.—Not later than 1 year after the date of enactment of this Act, the Commandant of the Coast Guard, in consultation with the Secretary of Defense and taking into consideration the Department of Defense 2016 Arctic Strategy, shall submit to the Committee on Commerce, Science, and Transportation of the Senate and the Committee on Transportation and Infrastructure of the House of Representatives a report on the progress toward implementing the strategic objectives described in the United States Coast Guard Arctic Strategy dated May 2013.

(d) Contents.—The report under subsection (c) shall include—

(1) a description of the Coast Guard's progress toward each strategic objective identified in the United States Coast Guard Arctic Strategy dated May 2013;

(2) an assessment of the assets and infrastructure necessary to meet the strategic objectives identified in the United States Coast Guard Arctic Strategy dated May 2013 based on factors such as—

(A) response time;

(B) coverage area;

(C) endurance on scene;

(D) presence; and

(E)deterrence;

(3) an analysis of the sufficiency of the distribution of National Security Cutters, Offshore Patrol Cutters, and Fast Response Cutters both stationed in various Alaskan ports and in other locations to meet the strategic objectives identified in the United States Coast Guard Arctic Strategy, dated May 2013;

(4) plans to provide communications throughout the entire Coastal Western Alaska Captain of the Port zone to improve waterway safety and mitigate close calls, collisions, and other dangerous interactions between the shipping industry and subsistence hunters;

(5) plans to prevent marine casualties, when possible, by ensuring vessels avoid environmentally sensitive areas and permanent security zones;

(6) an explanation of—

(A) whether it is feasible to establish a vessel traffic service, using existing resources or otherwise; and

(B) whether an Arctic Response Center of Expertise is necessary to address the gaps in experience, skills, equipment, resources, training, and doctrine to prepare, respond to, and recover spilled oil in the Arctic; and

(7) an assessment of whether sufficient agreements are in place to ensure the Coast Guard is receiving the information it needs to carry out its responsibilities.

Coast Guard Authorization Act of 2017 (Division D of National Defense Authorization Act for Fiscal Year 2019 [H.R. 5515/P.L. 115- 232])

House Floor Action

On May 22, 2018, as part of its consideration of the National Defense Authorization Act for Fiscal Year 2019 (H.R. 5515), the House agreed to by voice vote H.Amdt. 641, an *en bloc* amendment that included, *inter alia*, amendment number 52 as printed in H.Rept. 115-702 of May 22, 2018, on H.Res. 908, providing for the further consideration of H.R. 5515. Amendment number 52 added to H.R. 5515, as a new Division D, the *Coast Guard Authorization Act of 2017.*

Section 4204 within Division D states the following:

SEC. 4204. Authorization of amounts for Fast Response Cutters.

(a) In general.—Of the amounts authorized under section 4902 of title 14, United States Code, as amended by this division, for each of fiscal years 2018 and 2019 up to $167,500,000 is authorized for the acquisition of 3 Fast Response Cutters.

(b) Treatment of acquired cutters.—Any cutters acquired pursuant to subsection (a) shall be in addition to the 58 cutters approved under the existing acquisition baseline.

Section 4304 within Division D states the following (emphasis added):

SEC. 4304. Unmanned aircraft.

(a) Land-based unmanned aircraft system program.—Chapter 3 of title 14, United States Code, is amended by adding at the end the following:

"§ 319. Land-based unmanned aircraft system program

(a) In general.—Subject to the availability of appropriations, the Secretary shall establish a land-based unmanned aircraft system program under the control of the Commandant.

(b) Unmanned aircraft system defined.—In this section, the term 'unmanned aircraft system' has the meaning given that term in section 331 of the FAA Modernization and Reform Act of 2012 (49 U.S.C. 40101 note).".

(b) Limitation on unmanned aircraft systems.—Chapter 11 of title 14, United States Code, is amended by inserting after section 1154 the following:

"§ 1155. Limitation on unmanned aircraft systems

"(a) In general.—During any fiscal year for which funds are appropriated for the design or construction of an Offshore Patrol Cutter, the Commandant—

"(1) may not award a contract for design of an unmanned aircraft system for use by the Coast Guard; and"(2) may lease, acquire, or acquire the services of an unmanned aircraft system only if such system—

(A) has been part of a program of record of, procured by, or used by a Federal entity (or funds for research, development, test, and evaluation have been received from a Federal entity with regard to such system) before the date on which the Commandant leases, acquires, or acquires the services of the system; and

(B) is leased, acquired, or utilized by the Commandant through an agreement with a Federal entity, unless such an agreement is not practicable or would be less cost-effective than an independent contract action by the Coast Guard.

(b) Small unmanned aircraft exemption.—Subsection (a)(2) does not apply to small unmanned aircraft.

(c) Definitions.—In this section, the terms 'small unmanned aircraft' and 'unmanned aircraft system' have the meanings given those terms in section 331 of the FAA Modernization and Reform Act of 2012 (49 U.S.C. 40101 note).".

(c) Clerical amendments.—

CHAPTER 3.—The analysis for chapter 3 of title 14, United States Code, is amended by adding at the end the following:

"319. Land-based unmanned aircraft system program.".

CHAPTER 11.—The analysis for chapter 11 of title 14, United States Code, is amended by inserting after the item relating to section 1154 the following:

"1155. Limitation on unmanned aircraft systems."

(d) Conforming amendment.—Subsection (c) of section 1105 of title 14, United States Code, is repealed.

Section 4311 within Division D states the following:

SEC. 4311. Contracting for major acquisitions programs.

(a) General acquisition authority.—Section 501(d) of title 14, United States Code, is amended by inserting "aircraft, and systems," after "vessels,".

(b) Contracting authority.—Chapter 11 of title 14, United States Code, as amended by this division, is further amended by inserting after section 1136 the following:

"§ 1137. Contracting for major acquisitions programs

(a) In general.—In carrying out authorities provided to the Secretary to design, construct, accept, or otherwise acquire assets and systems under section 501(d), the Secretary, acting through the Commandant or the head of an integrated program office established for a major acquisition program, may enter into contracts for a major acquisition program.

(b) Authorized methods.—Contracts entered into under subsection (a)—

"(1) may be block buy contracts;

"(2) may be incrementally funded;

"(3) may include combined purchases, also known as economic order quantity purchases, of—

(A) materials and components; and

(B) long lead time materials; and

"(4) as provided in section 2306b of title 10, may be multiyear contracts.

"(c) Subject to appropriations.—Any contract entered into under subsection (a) shall provide that any obligation of the United States to make a payment under the contract is subject to the availability of amounts specifically provided in advance for that purpose in subsequent appropriations Acts.".

(c) Clerical amendment.—The analysis for chapter 11 of title 14, United States Code, as amended by this division, is further amended by inserting after the item relating to section 1136 the following:

"1137. Contracting for major acquisitions programs.".

(d) Conforming amendments.—The following provisions are repealed:

(1) Section 223 of the Howard Coble Coast Guard and Maritime Transportation Act of 2014 (14 U.S.C. 1152 note), and the item relating to that section in the table of contents in section 2 of such Act.

(2) Section 221(a) of the Coast Guard and Maritime Transportation Act of 2012 (14 U.S.C. 1133 note).

Section 207(a) of the Coast Guard Authorization Act of 2016 (14 U.S.C. 561 note).

(e) Internal regulations and policy.—Not later than 180 days after the date of enactment of this Act, the Secretary of the department in which the Coast Guard is operating shall establish the internal regulations and policies necessary to exercise the authorities provided under this section, including the amendments made in this section.

(f) Multiyear contracts.—The Secretary of the department in which the Coast Guard is operating is authorized to enter into a multiyear contract for the procurement of a tenth, eleventh, and twelfth National Security Cutter and associated government-furnished equipment.

Section 4817 within Division D states the following:

SEC. 4817. Fleet requirements assessment and strategy.

(a) Report.—Not later than 1 year after the date of enactment of this Act, the Secretary of the department in which the Coast Guard is operating, in consultation with interested Federal and non-Federal stakeholders, shall submit to the Committee on Commerce, Science, and Transportation of the Senate and the Committee on Transportation and Infrastructure of the House of Representatives a report including—

(1) an assessment of Coast Guard at-sea operational fleet requirements to support its statutory missions established in the Homeland Security Act of 2002 (6 U.S.C. 101 et seq.); and

(2) a strategic plan for meeting the requirements identified under paragraph (1).

(b) Contents.—The report under subsection (a) shall include— (1) an assessment of—

(A) the extent to which the Coast Guard at-sea operational fleet requirements referred to in subsection (a)(1) are currently being met;

(B) the Coast Guard's current fleet, its operational lifespan, and how the anticipated changes in the age and distribution of vessels in the fleet will impact the ability to meet at-sea operational requirements;

fleet operations and recommended improvements to minimize costs and extend operational vessel life spans; and

(D) the number of Fast Response Cutters, Offshore Patrol Cutters, and National Security Cutters needed to meet at-sea operational requirements as compared to planned acquisitions under the current programs of record;

(2) an analysis of—

(A) how the Coast Guard at-sea operational fleet requirements are currently met, including the use of the Coast Guard's current cutter fleet, agreements with partners, chartered vessels, and unmanned vehicle technology; and

(B) whether existing and planned cutter programs of record (including the Fast Response Cutter, Offshore Patrol Cutter, and National Security Cutter) will enable the Coast Guard to meet at-sea operational requirements; and

(3) a description of—

(A) planned manned and unmanned vessel acquisition; and

(B) how such acquisitions will change the extent to which the Coast Guard at-sea operational requirements are met.

(c) Consultation and transparency.—

(1) CONSULTATION.—In consulting with the Federal and non-Federal stakeholders under subsection (a), the Secretary of the department in which the Coast Guard is operating shall—

(A) provide the stakeholders with opportunities for input—

prior to initially drafting the report, including the assessment and strategic plan; and

not later than 3 months prior to finalizing the report, including the assessment and strategic plan, for submission; and

(B) document the input and its disposition in the report.

(2) TRANSPARENCY.—All input provided under paragraph (1) shall be made available to the public.

(d) Ensuring maritime coverage.—In order to meet Coast Guard mission requirements for search and rescue, ports, waterways, and coastal security, and maritime environmental response during recapitalization of Coast Guard vessels, the Coast Guard shall ensure continuity of the coverage, to the maximum extent practicable, in the locations that may lose assets.

Section 4818 within Division D states the following:

SEC. 4818. National Security Cutter.

(a) Standard method for tracking.—The Commandant of the Coast Guard may not certify an eighth National Security Cutter as Ready for Operations before the date on which the Commandant provides to the Committee on Transportation and Infrastructure of the House of Representatives and the Committee on Commerce, Science, and Transportation of the Senate—

(1) a notification of a new standard method for tracking operational employment of Coast Guard major cutters that does not include time during which such a cutter is away from its homeport for maintenance or repair; and a report analyzing cost and performance for different approaches to achieving varied levels of operational employment using the standard method required by paragraph (1) that, at a minimum—

(A) compares over a 30-year period the average annualized baseline cost and performances for a certified National Security Cutter that operated for 185 days away from homeport or an equivalent alternative measure of operational tempo—

(i) against the cost of a 15 percent increase in days away from homeport or an equivalent alternative measure of operational tempo for a National Security Cutter; and

(ii) against the cost of the acquisition and operation of an additional National Security Cutter; and

(B) examines the optimal level of operational employment of National Security Cutters to balance National Security Cutter cost and mission performance.

(b) Conforming amendments.—

(1) Section 221(b) of the Coast Guard and Maritime Transportation Act of 2012 (126 Stat. 1560) is repealed.

(2) Section 204 (c)(1) of the Coast Guard Authorization Act of 2016 (130 Stat. 35) is repealed.

Section 4822 within Division D states (emphasis added):

SEC. 4822. Strategic assets in the Arctic.

(a) Definition of arctic.—In this section, the term "Arctic" has the meaning given the term in section 112 of the Arctic Research and Policy Act of 1984 (15 U.S.C. 4111).

(b) Sense of congress.—It is the sense of Congress that—

(1) the Arctic continues to grow in significance to both the national security interests and the economic prosperity of the United States; and

(2) the Coast Guard must ensure it is positioned to respond to any accident, incident, or threat with appropriate assets.

(c) Report.—Not later than 1 year after the date of enactment of this Act, the Commandant of the Coast Guard, in consultation with the Secretary of Defense and taking into consideration the Department of Defense 2016 Arctic Strategy, shall submit to the Committee on Commerce, Science, and Transportation of the Senate and the Committee on Transportation and Infrastructure of the House of Representatives a report on the progress toward implementing the strategic objectives described in the United States Coast Guard Arctic Strategy dated May 2013.

(d) Contents.—The report under subsection (c) shall include—

(1) a description of the Coast Guard's progress toward each strategic objective identified in the United States Coast Guard Arctic Strategy dated May 2013;

(2) an assessment of the assets and infrastructure necessary to meet the strategic objectives identified in the United States Coast Guard Arctic Strategy dated May 2013 based on factors such as—

(A) response time;

(B) coverage area;

(c) endurance on scene;

(D) presence; and

(E) deterrence;

(3) an analysis of the sufficiency of the distribution of National Security Cutters, Offshore Patrol Cutters, and Fast Response Cutters both stationed in various Alaskan ports and in other locations to meet the strategic objectives identified in the United States Coast Guard Arctic Strategy, dated May 2013;

(4) plans to provide communications throughout the entire Coastal Western Alaska Captain of the Port zone to improve waterway safety and mitigate close calls, collisions, and other dangerous interactions between the shipping industry and subsistence hunters;

(5) plans to prevent marine casualties, when possible, by ensuring vessels avoid environmentally sensitive areas and permanent security zones;

(6) an explanation of—

(A) whether it is feasible to establish a vessel traffic service, using existing resources or otherwise; and

(B) whether an Arctic Response Center of Expertise is necessary to address the gaps in experience, skills, equipment, resources, training, and doctrine to prepare, respond to, and recover spilled oil in the Arctic; and

(7) an assessment of whether sufficient agreements are in place to ensure the Coast Guard is receiving the information it needs to carry out its responsibilities.

Conference

The conference report (H.Rept. 115-874 of July 25, 2018) on H.R. 5515/P.L. 115-232 of August 13, 2018, states:

> Coast Guard Authorization Act of 2018
>
> The House bill contained a division (Division D) that would authorize certain aspects of the Coast Guard.
>
> The Senate amendment contained no similar provisions. The House recedes. (Page 1137)

APPENDIX A. PLANNED NSC, OPC, AND FRC PROCUREMENT QUANTITIES

This appendix provides further discussion on the issue of the Coast Guard's planned NSC, OPC, and FRC procurement quantities.

Overview

The Coast Guard's program of record for NSCs, OPCs, and FRCs includes only about 61% as many cutters as the Coast Guard calculated in 2011 would be needed to fully perform its projected future missions. The Coast Guard's planned force levels for NSCs, OPCs, and FRCs have remained unchanged since 2004. In contrast, the Navy since 2004 has adjusted its ship force-level goals eight times in response to changing strategic and budgetary circumstances.[42]

Although the Coast Guard's strategic situation and resulting mission demands may not have changed as much as the Navy's have since 2004, the Coast Guard's budgetary circumstances may have changed since 2004. The 2004 program of record was heavily conditioned by Coast Guard expectations in 2004 about future funding levels in the PC&I account. Those expectations may now be different, as suggested by the willingness of Coast Guard officials in 2017 to begin regularly mentioning the need for an PC&I funding level of $2 billion per year (see *Appendix B*).

---

[42] See Table 1 and Table B-1 of CRS Report RL32665, *Navy Force Structure and Shipbuilding Plans: Background and Issues for Congress*, by Ronald O'Rourke. As shown in those tables, the Navy's force-level goal of 2002-2004 was followed by new force- level goals in early 2005, February 2006, mid-2011, September 2011, March 2012, January 2013, March 2015, and December 2016.

It can also be noted that continuing to, in effect, use the Coast Guard's 2004 expectations of future funding levels for the PC&I account as an implicit constraint on planned force levels for NSCs, OPCs, and FRCs can encourage an artificially narrow view of Congress's options regarding future Coast Guard force levels and associated funding levels, depriving Congress of agency in the exercise of its constitutional power to provide for the common defense and general welfare of the United States, and to set funding levels and determine the composition of federal spending.

2009 Coast Guard Fleet Mix Analysis

The Coast Guard estimated in 2009 that with the POR's planned force of 91 NSCs, OPCs, and FRCs, the service would have capability or capacity gaps[43] in 6 of its 11 statutory missions— search and rescue (SAR); defense readiness; counterdrug operations; ports, waterways, and coastal security (PWCS); protection of living marine resources (LMR); and alien migrant interdiction operations (AMIO). The Coast Guard judges that some of these gaps would be "high risk" or "very high risk."

Public discussions of the POR frequently mention the substantial improvement that the POR force would represent over the legacy force. Only rarely, however, have these discussions explicitly acknowledged the extent to which the POR force would nevertheless be smaller in number than the force that would be required, by Coast Guard estimate, to fully perform the Coast Guard's statutory missions in coming years. Discussions that focus on the POR's improvement over the legacy force while omitting mention of the considerably larger number of cutters that would be required, by Coast Guard estimate, to fully perform the Coast Guard's statutory missions in coming years could encourage audiences to conclude, contrary to Coast Guard estimates, that the POR's planned force of 91 cutters would be capable of fully performing the Coast Guard's statutory missions in coming years.

In a study completed in December 2009 called the Fleet Mix Analysis (FMA) Phase 1, the Coast Guard calculated the size of the force that in its view would be needed to fully perform the service's statutory missions in coming years. The study refers to this larger force as the objective fleet mix. *Table A-1* compares planned numbers of NSCs, OPCs, and FRCs in the POR to those in the objective fleet mix.

---

[43] The Coast Guard uses *capability* as a qualitative term, to refer to the kinds of missions that can be performed, and *capacity* as a quantitative term, to refer to how much (i.e., to what scale or volume) a mission can be performed.

Table A-1. Program of Record Compared to Objective Fleet Mix (From Fleet Mix Analysis Phase 1 (2009))

		Objective Fleet Mix	**Objective Fleet Mix compared to POR**	
	Program of	From FMA		
Ship type	Record (POR)	Phase 1	Number	%
NSC	8	9	+1	+13%
OPC	25	57	+32	+128%
FRC	58	91	+33	+57%
Total	91	157	+66	+73%

Source: Fleet Mix Analysis Phase 1, Executive Summary, Table ES-8 on page ES-13.

As can be seen in *Table A-1*, the objective fleet mix includes 66 additional cutters, or about 73% more cutters than in the POR. Stated the other way around, the POR includes about 58% as many cutters as the 2009 FMA Phase I objective fleet mix.

As intermediate steps between the POR force and the objective fleet mix, FMA Phase 1 calculated three additional forces, called FMA-1, FMA-2, and FMA-3. (The objective fleet mix was then relabeled FMA-4.) *Table A-2* compares the POR to FMAs 1 through 4.

Table A-2. POR Compared to FMAs 1 Through 4 (From Fleet Mix Analysis Phase 1 (2009))

	Program of Record				**FMA-4 (Objective**
Ship type	(POR)	FMA-1	FMA-2	FMA-3	Fleet Mix)
NSC	8	9	9	9	9
OPC	25	32	43	50	57
FRC	58	63	75	80	91
Total	91	104	127	139	157

Source: Fleet Mix Analysis Phase 1, Executive Summary, Table ES-8 on page ES-13.

FMA-1 was calculated to address the mission gaps that the Coast Guard judged to be "very high risk." FMA-2 was calculated to address both those gaps and additional gaps that the Coast Guard judged to be "high risk." FMA-3 was calculated to address all those gaps, plus gaps that the Coast Guard judged to be "medium risk." FMA-4—the objective fleet mix—was calculated to address all the foregoing gaps, plus the remaining gaps, which the Coast Guard judge to be "low risk" or "very low risk." *Table A-3* shows the POR and FMAs 1 through 4 in terms of their mission performance gaps.

Figure A-1, taken from FMA Phase 1, depicts the overall mission capability/ performance gap situation in graphic form. It appears to be conceptual rather than drawn to precise scale. The black line descending toward 0 by the year 2027 shows the declining capability and performance of the Coast Guard's legacy assets as they gradually age out of the force. The purple line branching up from the black line shows the added capability from ships and aircraft to be procured under the POR, including the 91 planned NSCs, OPCs, and FRCs.

Table A-3. Force Mixes and Mission Performance Gaps (From Fleet Mix Analysis Phase 1 (2009)—an X mark indicates a mission performance gap)

Missions with performance gaps	Risk levels of these Performance gaps	Program of Record (POR)	FMA-1	FMA-2	FMA-3	FMA-4 (Objective Fleet Mix)
Search and Rescue (SAR) capability	Very high	X				
Defense Readiness capacity	Very high	X				
Counter Drug capacity	Very high	X				
Ports, Waterways, and Coastal	High	X	X			
Security (PWCS) capacity[a]						
Living Marine Resources (LMR) capability and capacity[a]	High	X	X			[all gaps addressed]
PWCS capacity[b]	Medium	X	X	X		
LMR capacity[c]	Medium	X	X	X		
Alien Migrant Interdiction	Low/very low	X	X	X	X	
Operations (AMIO) capacity[d]						
PWCS capacity[e]	Low/very low	X	X	X	X	

Source: Fleet Mix Analysis Phase 1, Executive Summary, page ES-11 through ES-13.

Notes: In the first column, The Coast Guard uses capability as a qualitative term, to refer to the kinds of missions that can be performed, and capacity as a quantitative term, to refer to how much (i.e., to what scale or volume) a mission can be performed.

[a] This gap occurs in the Southeast operating area (Coast Guard Districts 7 and 8) and the Western operating area (Districts 11, 13, and 14).

[b] This gap occurs in Alaska.

[c] This gap occurs in Alaska and in the Northeast operating area (Districts 1 and 5).

[d] This gap occurs in the Southeast and Western operating areas.

[e] This gap occurs in the Northeast operating area.

The level of capability to be provided when the POR force is fully in place is the green line, labeled "2005 Mission Needs Statement." As can be seen in the graph, this level of capability is substantially below a projection of Coast Guard mission demands made after the terrorist attacks of September 11, 2001 (the red line, labeled "Post-9/11 CG Mission Demands"), and even further below a Coast Guard projection of future mission demands (the top dashed line, labeled "Future Mission Demands"). The dashed blue lines show future capability levels that would result from reducing planned procurement quantities in the POR or executing the POR over a longer time period than originally planned.

FMA Phase 1 was a fiscally unconstrained study, meaning that the larger force mixes shown in *Table A-2* were calculated primarily on the basis of their capability for performing missions, rather than their potential acquisition or life-cycle operation and support (O&S) costs.

Although the FMA Phase 1 was completed in December 2009, the figures shown in *Table A-2* were generally not included in public discussions of the Coast Guard's future force structure needs until April 2011, when GAO presented them in testimony.[44] GAO again presented them in a July 2011 report.[45]

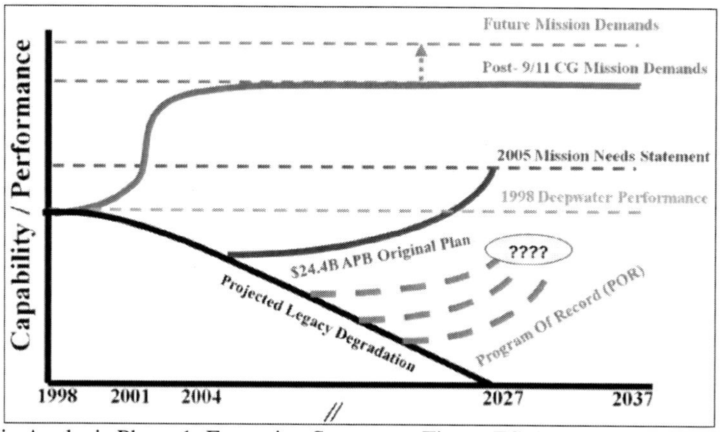

Source: Fleet Mix Analysis Phase 1, Executive Summary, Figure ES-1 on p. ES-2.

Figure A-1. Projected Mission Demands vs. Projected Capability/Performance. (From Fleet Mix Analysis Phase 1, Executive Summary)

The Coast Guard completed a follow-on study, called Fleet Mix Analysis (FMA) Phase 2, in May 2011. Among other things, FMA Phase 2 includes a revised and updated objective fleet mix called the refined objective mix. *Table A-4* compares the POR to the objective fleet mix from FMA Phase 1 and the refined objective mix from FMA Phase 2.

As can be seen in *Table A-4*, compared to the objective fleet mix from FMA Phase 1, the refined objective mix from FMA Phase 2 includes 49 OPCs rather than 57. The refined objective mix includes 58 additional cutters, or about 64% more cutters than in the POR. Stated the other way around, the POR includes about 61% as many cutters as the refined objective mix.

---

[44] Government Accountability Office, *Coast Guard[:]Observations on Acquisition Management and Efforts to Reassess the Deepwater Program, Testimony Before the Subcommittee on Coast Guard and Maritime Transportation, Committee on Transportation and Infrastructure, House of Representatives, Statement of John P. Hutton, Director Acquisition and Sourcing Management*, GAO-11-535T, April 13, 2011, p. 10.

[45] Government Accountability Office, *Coast Guard[:]Action Needed As Approved Deepwater Program Remains Unachievable*, GAO-11-743, July 2011, p. 46.

Table A-4. POR Compared to Objective Mixes in FMA Phases 1 and 2 (From Fleet Mix Analysis Phase 1 (2009) and Phase 2 (2011))

Ship type	Program of Record (POR)	Objective Fleet Mix from FMA Phase 1	Refined Objective Mix from FMA Phase 2
NSC	8	9	9
OPC	25	57	49
FRC	58	91	91
Total	91	157	149

Source: Fleet Mix Analysis Phase 1, Executive Summary, Table ES-8 on page ES-13, and Fleet Mix Analysis Phase 2, Table ES-2 on p. iv.

Compared to the POR, the larger force mixes shown in *Table A-2* and *Table A-4* would be more expensive to procure, operate, and support than the POR force. Using the average NSC, OPC, and FRC procurement cost figures presented earlier (see "Background"), procuring the 58 additional cutters in the Refined Objective Mix from FMA Phase 2 might cost an additional $10.7 billion, of which most (about $7.8 billion) would be for the 24 additional FRCs. (The actual cost would depend on numerous factors, such as annual procurement rates.) O&S costs for these 58 additional cutters over their life cycles (including crew costs and periodic ship maintenance costs) would require billions of additional dollars.[46]

The larger force mixes in the FMA Phase 1 and 2 studies, moreover, include not only increased numbers of cutters, but also increased numbers of Coast Guard aircraft. In the FMA Phase 1 study, for example, the objective fleet mix included 479 aircraft—93% more than the 248 aircraft in the POR mix. Stated the other way around, the POR includes about 52% as many aircraft as the objective fleet mix. A decision to procure larger numbers of cutters like those shown in *Table A-2* and *Table A-4* might thus also imply a decision to procure, operate, and support larger numbers of Coast Guard aircraft, which would require billions of additional dollars. The FMA Phase 1 study estimated the procurement cost of the objective fleet mix of 157 cutters and 479 aircraft at $61 billion to $67 billion in constant FY2009 dollars, or about 66% more than the procurement cost of $37 billion to $40 billion in constant FY2009 dollars estimated for the POR mix of 91 cutters and 248 aircraft. The study estimated the total ownership cost (i.e., procurement plus life-cycle O&S cost) of the objective fleet mix of cutters and aircraft at $201 billion to $208 billion in constant FY2009 dollars, or about 53% more than the total ownership cost of $132 billion to $136 billion in constant FY2009 dollars estimated for POR mix of cutters and aircraft.[47]

---

[46] The FMA Phase 1 and Phase 2 studies present acquisition and life-cycle ownership cost calculations for force mixes that include not only larger numbers of NSC, OPCs, and FRCs, but corresponding larger numbers of Coast Guard aircraft.

[47] Fleet Mix Analysis Phase 1, Executive Summary, Table ES-11 on page ES-19, and Table ES-10 on page ES-18. The life-cycle O&S cost was calculated through 2050.

A December 7, 2015, press report states the following:

> The Coast Guard's No. 2 officer said the small size and advanced age of its fleet is limiting the service's ability to carry out crucial missions in the Arctic and drug transit zones or to meet rising calls for presence in the volatile South China Sea.
>
> "The lack of surface vessels every day just breaks my heart," VADM Charles Michel, the Coast Guard's vice commandant, said Dec. 7.
>
> Addressing a forum on American Sea Power sponsored by the U.S. Naval Institute at the Newseum, Michel detailed the problems the Coast Guard faces in trying to carry out its missions of national security, law enforcement and maritime safety because of a lack of resources.
>
> "That's why you hear me clamoring for recapitalization," he said.
>
> Michel noted that China's coast guard has a lot more ships than the U.S. Coast Guard has, including many that are larger than the biggest U.S. cutter, the 1,800-ton [sic:4,800-ton] National Security Cutter. China is using those white-painted vessels rather than "gray-hull navy" ships to enforce its claims to vast areas of the South China Sea, including reefs and shoals claimed by other nations, he said.
>
> That is a statement that the disputed areas are "so much our territory, we don't need the navy. That's an absolutely masterful use of the coast guard," he said.
>
> The superior numbers of Chinese coast guard vessels and its plans to build more is something, "we have to consider when looking at what we can do in the South China Sea," Michel said.
>
> Although they have received requests from the U.S. commanders in the region for U.S. Coast Guard cutters in the South China Sea, "the commandant had to say 'no'. There's not enough to go around," he said.[48]

Potential oversight questions for Congress include the following:

- Under the POR force mix, how large a performance gap, precisely, would there be in each of the missions shown in *Table A-3*? What impact would these performance gaps have on public safety, national security, and protection of living marine resources?
- How sensitive are these performance gaps to the way in which the Coast Guard translates its statutory missions into more precise statements of required mission performance?
- Given the performance gaps shown in *Table A-3*, should planned numbers of Coast Guard cutters and aircraft be increased, or should the Coast Guard's statutory missions be reduced, or both?
- How much larger would the performance gaps in *Table A-3* be if planned numbers of Coast Guard cutters and aircraft are reduced below the POR figures?

---

[48] Otto Kreisher, "'Not Enough' USCG Vessels to Meet Demand for Presence in South China Sea, Arctic," *Seapower*, December 7, 2015.

- Has the executive branch made sufficiently clear to Congress the difference between the number of ships and aircraft in the POR force and the number that would be needed to fully perform the Coast Guard's statutory missions in coming years? Why has public discussion of the POR focused mostly on the capability improvement it would produce over the legacy force and rarely on the performance gaps it would have in the missions shown in *Table A-3*?

APPENDIX B. FUNDING LEVELS IN PC&I ACCOUNT

This appendix provides background information on funding levels in the Coast Guard's Procurement, Construction, and Improvements (PC&I) account.[49]

Overview

As shown in *Table B-1*, the FY2013 budget submission programmed an average of about $1.5 billion per year in the PC&I account. As also shown in the table, the FY2014-FY2016 budget submissions reduced that figure to between $1 billion and $1.2 billion per year.

The Coast Guard has testified that funding the PC&I account at a level of about $1 billion to $1.2 billion per year would make it difficult to fund various Coast Guard acquisition projects, including a new polar icebreaker and improvements to Coast Guard shore installations. Coast Guard plans call for procuring OPCs at an eventual rate of two per year. If each OPC costs roughly $400 million, procuring two OPCs per year in an PC&I account of about $1 billion to $1.2 billion per year, as programmed under the FY2014-FY2016 budget submissions, would leave about $200 million to $400 million per year for all other PC&I-funded programs.

Since 2017, Coast Guard officials have been stating more regularly what they stated only infrequently in earlier years: that executing the Coast Guard's various acquisition programs fully and on a timely basis would require the PC&I account to be funded in coming years at a level of about $2 billion per year. Statements from Coast Guard officials on this issue in past years have sometimes put this figure as high as about $2.5 billion per year.

---

[49] Prior to FY2019, the PC&I account was called the Acquisition, Construction, and Improvements (AC&I) account.

Table B-1. Funding in PC&I Account in FY2013-FY2019 Budgets (Figures in millions of dollars, rounded to nearest tenth)

Budget	FY13	FY14	FY15	FY16	FY17	FY18	FY19	FY20	FY21	FY22	FY23	Avg.	% change compared to avg. for FY13 budget
FY13	1,217.3	1,429.5	1,619.9	1,643.8	1,722.0							1,526.5	—
FY14		951.1	1,195.7	901.0	1,024.8	1,030.3						1,020.6	-33.1%
FY15			1,084.2	1,103.0	1,128.9	1,180.4	1,228.7					1,145.0	-25.0%
FY16				1,017.3	1,125.3	1,255.7	1,201.0	1,294.6				1,178.8	-22.8%
FY17					1,136.8	1,259.6	1,339.9	1,560.5	1,840.8			1,427.5	-6.5%
FY18						1,203.7	1,360.9	1,602.7	1,810.6	1,687.5		1,533.1	+0.4%
FY19							1,886.8	n/a	n/a	n/a	n/a	n/a	n/a

Source: Table prepared by CRS based on Coast Guard FY2013-FY2019 budget submissions.

Using Past PC&I Funding Levels as a Guide for Future PC&I Funding Levels

In assessing future funding levels for executive branch agencies, a common practice is to assume or predict that the figure in coming years will likely be close to where it has been in previous years. While this method can be of analytical and planning value, for an agency like the Coast Guard, which goes through periods with less acquisition of major platforms and periods with more acquisition of major platforms, this approach might not always be the best approach, at least for the PC&I account.

More important, in relation to maintaining Congress's status as a co-equal branch of government, including the preservation and use of congressional powers and prerogatives, an analysis that assumes or predicts that future funding levels will resemble past funding levels can encourage an artificially narrow view of congressional options regarding future funding levels, depriving Congress of agency in the exercise of its constitutional power to set funding levels and determine the composition of federal spending.

Past Coast Guard Statements About Required PC&I Funding Level

At an October 4, 2011, hearing on the Coast Guard's major acquisition programs before the Coast Guard and Maritime Transportation subcommittee of the House Transportation and Infrastructure Committee, the following exchange occurred:

> REPRESENATIVE FRANK LOBIONDO:
>
> Can you give us your take on what percentage of value must be invested each year to maintain current levels of effort and to allow the Coast Guard to fully carry out its missions?
>
> ADMIRAL ROBERT J. PAPP, COMMANDANT OF THE COAST GUARD:
>
> I think I can, Mr. Chairman. Actually, in discussions and looking at our budget—and I'll give you rough numbers here, what we do now is we have to live within the constraints that we've been averaging about $1.4 billion in acquisition money each year.
>
> If you look at our complete portfolio, the things that we'd like to do, when you look at the shore infrastructure that needs to be taken care of, when you look at renovating our smaller icebreakers and other ships and aircraft that we have, we've done some rough estimates that it would really take close to about $2.5 billion a year, if we were to do all the things that we would like to do to sustain our capital plant.
>
> So I'm just like any other head of any other agency here, as that the end of the day, we're given a top line and we have to make choices and tradeoffs and basically, my tradeoffs boil down to sustaining frontline operations balancing that, we're trying to recapitalize the Coast Guard and there's where the break is and where we have to define our spending.[1]

[1] Source: Transcript of hearing.

An April 18, 2012, blog entry stated the following:

If the Coast Guard capital expenditure budget remains unchanged at less than $1.5 billion annually in the coming years, it will result in a service in possession of only 70 percent of the assets it possesses today, said Coast Guard Rear Adm. Mark Butt.

Butt, who spoke April 17 [2012] at [a] panel [discussion] during the Navy League Sea Air Space conference in National Harbor, Md., echoed Coast Guard Commandant Robert Papp in stating that the service really needs around $2.5 billion annually for procurement.[2]

At a May 9, 2012, hearing on the Coast Guard's proposed FY2013 budget before the Homeland Security subcommittee of the Senate Appropriations Committee, Admiral Papp testified, "I've gone on record saying that I think the Coast Guard needs closer to $2 billion dollars a year [in acquisition funding] to recapitalize—[to] do proper recapitalization."[3]

At a May 14, 2013, hearing on the Coast Guard's proposed FY2014 budget before the Homeland Security Subcommittee of the Senate Appropriations Committee, Admiral Papp stated the following regarding the difference between having about $1.0 billion per year rather than about $1.5 billion per year in the PC&I account:

Well, Madam Chairman, $500 million—a half a billion dollars—is real money for the Coast Guard. So, clearly, we had $1.5 billion in the [FY]13 budget. It doesn't get everything

I would like, but it—it gave us a good start, and it sustained a number of projects that are very important to us.

When we go down to the $1 billion level this year, it gets my highest priorities in there, but we have to either terminate or reduce to minimum order quantities for all the other projects that we have going.

If we're going to stay with our program of record, things that have been documented that we need for our service, we're going to have to just stretch everything out to the right. And when we do that, you cannot order in economic order quantities. It defers the

[2] David Perera, "The Coast Guard Is Shrinking," FierceHomelandSecurity.com, April 18, 2012, accessed July 20, 2012, at http://www.fiercehomelandsecurity.com/story/coast-guard-shrinking/2012-04-18.

[3] Source: transcript of hearing. Papp may have been referring to remarks he made to the press before giving his annual state of the Coast Guard speech on February 23, 2012, in which reportedly stated that the Coast Guard would require about $2 billion per year in acquisition funding to fully replace its current assets. (See Adam Benson, "Coast Guard Cutbacks Will Cost 1,000 Jobs," *Norwich Bulletin*, February 23, 2012, accessed May 31, 2012, at http://www.norwichbulletin.com/x1138492141/Coast-Guard-cutbacks-will-cost-1-000-jobs. See also "Coast Guard Leader Calls For More Ships," *MilitaryFeed.com*, February 24, 2012, accessed May 31, 2012, at http://militaryfeed.com/coast-guard-leader-calls-for-more-ships-5/; Associated Press, "Coast Guard Commandant Calls for New Ships," *TheLog.com*, March 10, 2012, accessed May 31, 2012, at http://www.thelog.com/SNW/Article/CoastGuard-Commandant-Calls-for-New-Ships-to-Replace-Aging-Fleet; Mickey McCarter, "Congress Poised to Give Coast Guard More Money Than Requested for FY 2013," *HSToday.us*, May 10, 2012, accessed May 31, 2012, at http://www.hstoday.us/focused-topics/customs-immigration/single-article-page/congress-poised-to-give-coast-guardmore-money-than-requested-for-fy-2013.html.) See also "Interview, Adm. Robert Papp, US Coast Guard Commandant," *Defense News*, November 11, 2013: 30.

purchase. Ship builders, aircraft companies—they have to figure in their costs, and it inevitably raises the cost when you're ordering them in smaller quantities and pushing it off to the right.

Plus, it almost creates a death spiral for the Coast Guard because we are forced to sustain older assets—older ships and older aircraft—which ultimately cost us more money, so it eats into our operating funds, as well, as we try to sustain these older things.

So, we'll do the best we can within the budget. And the president and the secretary have addressed my highest priorities, and we'll just continue to go on the—on an annual basis seeing what we can wedge into the budget to keep the other projects going.[4]

At a March 12, 2014, hearing on the Coast Guard's proposed FY2015 budget before the Homeland Security subcommittee of the House Appropriations Committee, Admiral Papp stated the following:

Well, that's what we've been struggling with, as we deal with the five-year plan, the capital investment plan, is showing how we are able to do that. And it will be a challenge, particularly if it sticks at around $1 billion [per year]. As I've said publicly, and actually, I said we could probably—I've stated publicly before that we could probably construct comfortably at about 1.5 billion [dollars] a year. But if we were to take care of all the Coast Guard's projects that are out there, including shore infrastructure that that fleet that takes care of the Yemen [sic: inland] waters is approaching 50 years of age, as well, but I have no replacement plan in sight for them because we simply can't afford it. Plus, we need at some point to build a polar icebreaker. Darn tough to do all that stuff when you're pushing down closer to 1 billion [dollars per year], instead of 2 billion [dollars per year].

As I said, we could fit most of that in at about the 1.5 billion [dollars per year] level, but the projections don't call for that. So we are scrubbing the numbers as best we can.[5]

At a March 24, 2015, hearing on the Coast Guard's proposed FY2016 budget before the Homeland Security subcommittee of the House Appropriations Committee, Admiral Paul Zukunft, Admiral Papp's successor as Commandant of the Coast Guard, stated the following:

I look back to better years in our acquisition budget when we had a—an acquisition budget of—of $1.5 billion. That allows me to move these programs along at a much more rapid pace and, the quicker I can build these at full-rate production, the less cost it is in the long run as well. But there's an urgent need for me to be able to deliver these platforms in a timely and also in an affordable manner. But to at least have a reliable and a predictable acquisition budget would make our work in the Coast Guard much easier. But when we see variances of—of 30, 40% over a period of three or four years, and not knowing what the Budget Control Act may have in store for us going on, yes, we are

[4] Transcript of hearing. The remarks were made in response to a question from Sen. Mary Landrieu.

[5] Transcript of hearing.

treading water now but any further reductions, and now I am—I am beyond asking for help. We are taking on water.[6]

An April 13, 2017, press report states the following (emphasis added):

> Coast Guard Commandant Adm. Paul Zukunft on Wednesday [April 12] said that for the Coast Guard to sustain its recapitalization plans and operations the service needs a $2 billion annual acquisition budget that grows modestly overtime to keep pace with inflation.
>
> The Coast Guard needs a "predictable, reliable" acquisition budget "and within that we need 5 percent annual growth to our operations and maintenance (O&M) accounts," Zukunft told reporters at a Defense Writers Group breakfast. Inflation will clip 2 to 3 percent from that, but "at 5 percent or so it puts you on a moderate but positive glide slope so you can execute, so you can build the force," he said.[7]

In an interview published on June 1, 2017, Zukunft said the following (emphasis added):

> We cannot be more relevant than we are now. But what we need is predictable funding. We have been in over 16 continuing resolutions since 2010. I need stable and repeatable funding. An acquisition budget with a floor of $2 billion. Our operating expenses as I said, they've been funded below the Budget Control Act floor for the past five years. I need 5 percent annualized growth over the next five years and beyond to start growing some of this capability back.
>
> But more importantly, we [need] more predictable, more reliable funding so we can execute what we need to do to carry out the business of the world's best Coast Guard.[8]

Appendix C. Additional Information on Status and Execution of NSC, OPC, and FRC Programs from May 2018 and March 2016 GAO Reports

This appendix presents additional information on the status and execution of the NSC, OPC, and FRC programs from May 2018 and March 2016 GAO reports reviewing DHS acquisition programs.[9]

---

[6] Transcript of hearing. The remarks were made in response to a question from Rep. John Culberson.

[7] Calvin Biesecker, "Zukunft Wants $2 Billion Baseline Acquisition Budget; Sustained Growth In O&M Funding," *Defense Daily*, April 13, 2017: 1.

[8] Jill Aitoro, "Interview: Adm. Paul Zukunft Demands Coast Guard Respect," *Defense News*, June 1, 2017.

[9] Government Accountability Office, *Home Security Acquisitions[:] Leveraging Programs' results Could Further DHS's Progress to Improve Portfolio Management*, May 2018, GAO-18-339SP, 109 pp., and Government Accountability Office, *Homeland Security Acquisitions[:]DHS Has Strengthened Management, but Execution and Affordability Concerns Endure*, GAO-16-338SP, March 2016, 110 pp. Hereinafter GAO-18-339SP and GAO-16- 338SP, respectively.

NSC Program

May 2018 GAO Report

Regarding the NSC program, the May 2018 GAO report states the following:

DHS's Under Secretary for Management (USM) directed the USCG to complete follow-on operational test and evaluation (OT&E) by March 2019. According to USCG officials, the program's OTA began follow-on OT&E in October 2017, which will test unmet key performance parameters (KPP) and address deficiencies found during prior testing. The NSC completed initial operational testing in 2014, but did not fully demonstrate 7 of its 19 KPPs, including those related to unmanned aircraft and cutter-boat deployment in rough seas. According to USCG officials, operators have since demonstrated these KPPs during USCG operations. For example, USCG officials stated that they successfully demonstrated operations of a prototype unmanned aircraft on an NSC. However, the USCG will not evaluate the NSC's unmanned aircraft KPP until the unmanned aircraft undergoes initial OT&E, currently planned for June 2019. In addition, the NSC will be the first USCG asset to undergo cybersecurity testing. However, this test has been delayed over a year with the final cyber test event scheduled for August 2018 because of a change in NSC operational schedules, among other things.

The DHS USM also directed the USCG to complete a study to determine the root cause of the NSC's propulsion system issues by December 2017; however, as of January 2018, the study was not yet complete. GAO previously reported on these issues—including high engine temperatures, cracked cylinder heads, and overheating generator bearings that were impacting missions—in January 2016....

The USCG initially planned to implement a crew rotational concept in which crews would rotate while NSCs were underway to achieve a goal of 230 days away from the cutter's homeport. In February 2018, USCG officials told GAO they abandoned the crew rotational concept because the concept did not provide the USCG with the expected return on investment. Instead, USCG officials said a new plan has been implemented that does not rotate crew and is anticipated to increase the days away from home port from the current capability of 185 days to 200 days.[10]

March 2016 GAO Report

The March 2016 GAO report states the following:

Notably, officials from six [DHS acquisition] programs explained that their current KPPs [key performance parameters] are still poorly defined and may require revisions going forward. For example, USCG officials identified that the NSC's cutter boat requirements should have been written more clearly, and, in January 2016, we recommended the NSC program office clarify them....[11]

---

[10] GAO-18-339SP, p. 92.

[11] GAO-16-338SP, p. 31.

... DHS has not identified specific actions to improve the affordability of one of the programs that department leadership reviewed—USCG NSC—and this program continues to face a funding gap exceeding 10 percent. In this case, the USCG did not provide DHS leadership critical information necessary for addressing affordability issues [12]

... the USCG NSC program—one of the department's largest investments—continues to face a funding gap exceeding 10 percent even though it was reviewed in September 2014. We found that the funding certification memo that the USCG provided to the DHS ARB [Acquisition Review Board] did not include as much detail as the others we reviewed across DHS components. Specifically, the NSC funding certification, signed by the USCG CFO [Chief Financial Officer], consisted of only a high-level narrative discussion, stating that adjustments would be made, as necessary, to sustain and operate the NSC. Unlike the other funding certifications we reviewed, it did not include detailed tables that quantified cost estimates, funding streams, and the monetary value of proposed tradeoffs. We also found that DHS leadership did not document any tradeoffs to improve the program's affordability after the September 2014 ARB. While the DHS CFO's June 2014 memorandum identifies that the success of the ARB reviews is dependent on the quality of the information presented to the ARB, it does not specify what information the components should include in the memos. It does not specifically require detailed information, such as quantifying cost estimates, funding streams, and the monetary value of proposed tradeoffs. We have previously established that information should be communicated to management in a form that enables them to carry out their responsibilities. Without detailed information, the ARB will be unable to hold fully informed discussions about tradeoffs needed to improve program affordability.... [13]

The [Coast Guard's C4ISR acquisition] program is continuing to work to replace the C4ISR [command and control, communications, computers, intelligence, surveillance, and reconnaissance] system on the NSC because it relies on contractor-proprietary software that is becoming obsolete and is costly to maintain. This transition has been delayed by more than 7 years largely due to funding shortfalls and, according to program officials, difficulties scheduling system installations when the NSCs are in port. Future funding shortfalls would likely delay the transition further, and it appears that the program's cost estimate exceeds its funding plan significantly from fiscal year 2016 to 2020. However, the gap may not be as great as it appears. The C4ISR funding plans DHS has presented to Congress do not identify all of the funding the USCG plans to allocate to C4ISR operations. GAO has reported on USCG affordability issues since 2011 (GAO-11-743). In April 2015, GAO recommended DHS account for all of the operations and maintenance funding the USCG is allocating to its major acquisition programs in an annual report to Congress. DHS concurred with this recommendation, but USCG officials told GAO they have made no progress in accounting for these funding allocations. This issue obscures the size of future funding gaps, and the actual amount allocated through fiscal year 2015 may be greater than $797 million.

---

[12] GAO-16-338SP, p. 32.

[13] GAO-16-338SP, p. 40.

Test Activities [for the C4ISR program]

The USCG initially planned to test the C4ISR system separately from its planes and vessels, including the NSC, but officials subsequently decided to test the C4ISR system in conjunction with the planes and vessels to save money and avoid duplication. However, the C4ISR system's KPPs were not specifically evaluated during the NSC's operational test because the necessary testing activities were not fully integrated into the NSC test plan. The USCG deferred a significant portion of the C4ISR testing on the NSC to later dates including the testing of cybersecurity capabilities and real-time tactical communications with the Navy. In June 2014, GAO recommended the USCG fully integrate C4ISR assessments into other assets' operational test plans or test the C4ISR program independently in order to assess the operational effectiveness and suitability of the C4ISR system. The USCG concurred with GAO's recommendation and stated that it planned to test the C4ISR system's KPPs during follow-on testing for the NSC. The NSC's follow-on operational test and evaluation was scheduled for fiscal year 2015, but slipped to the end of fiscal year 2016 when the USCG refined the NSC's testing schedule. In the meantime, the USCG is using the C4ISR system on deployed NSCs....[14]

Performance

The USCG has been operating the NSC since 2010, and it initiated production of the eighth NSC in 2015, but it has not yet demonstrated the NSC can fully meet 7 of its 19 key performance parameters (KPP). In September 2015, USCG officials indicated they were in the process of validating data that would demonstrate the NSC could meet the KPP that establishes the NSC's transit range requirement. The NSC's other unmet KPPs include those related to unmanned aircraft, cutter-boat deployment, and interoperability requirements.

Acquisition Strategy

The USCG awarded delivery and task orders to produce the first three NSCs to Integrated Coast Guard Systems—a joint venture between Northrop Grumman and Lockheed Martin—as part of the now-defunct acquisition effort designated Deepwater. In 2006, the USCG revised its Deepwater acquisition strategy, citing cost increases, and took over the role of lead systems integrator, acknowledging that it had relied too heavily on contractors. In 2010, the USCG awarded the production contract for the fourth NSC to Northrop Grumman. In 2011, Northrop Grumman spun off its shipbuilding sector as an independent company named Huntington Ingalls Industries (HII). HII is producing the sixth, seventh, and eighth NSCs for the USCG, and plans to deliver the eighth NSC in 2019.

[14] GAO-16-338SP, p. 86.

Program Execution

From 2008 to 2014, the program's schedule for completing developmental testing slipped nearly 5 years, and its schedule for completing initial operational testing slipped nearly 3 years. In July 2011, GAO reported on a number of issues the USCG identified during developmental testing that the USCG needed to address before initiating operational testing, including performance and safety issues (GAO-11-743). The program's full operational capability date also slipped, from fiscal year 2016 to fiscal year 2020, although program officials anticipate it will occur sooner. USCG officials attributed the schedule slips to, among other things, funding shortfalls. The NSC has adhered to the revised schedule since January 2014, but going forward, the NSC is projected to face additional funding shortfalls. From fiscal year 2016 to fiscal year 2020, the NSC's costs are projected to exceed its funding by $401 million. However, the funding gap may not be as large as it appears. The NSC funding plan DHS has presented to Congress does not identify all of the funding the USCG plans to allocate to the NSC's operations. GAO has reported on USCG affordability issues since 2011. In April 2015, GAO recommended DHS account for all of the operations and maintenance funding the USCG is allocating to its major acquisition programs in an annual report to Congress. DHS concurred with this recommendation, but USCG officials told GAO they have made no progress in accounting for these funding allocations. This issue obscures the size of the future funding gaps, and the amount allocated through fiscal year 2015 may be greater than $5.7 billion.

From 2008 to 2014, the program's acquisition cost estimate increased from $4.7 billion to $5.7 billion. The USCG primarily attributed this increase to the lingering impacts of Hurricane Katrina, which struck the region where the NSCs are being built in 2005. USCG officials explained that the hurricane created labor shortages, which increased rates and decreased productivity. Alternatively, from 2008 to 2014, the program's life-cycle cost estimate decreased from $24.3 billion to $21.9 billion. USCG officials attributed this decrease to increasingly accurate cost estimates for personnel, materials, and maintenance. The program's approved cost thresholds remained stable from January 2014 to January 2016.

Test Activities

The NSC completed its initial operational testing in 2014, and the Department of Homeland Security's (DHS) Director of Operational Test and Evaluation (DOT&E) subsequently found the NSC operationally effective and suitable. However, testing identified several major deficiencies, and the USCG did not demonstrate the NSC could fully meet 7 of its 19 KPPs. For example, the USCG has not yet procured an unmanned aircraft system for the NSC, and has not yet demonstrated the NSC can meet the related KPP. Three of the NSC's unmet KPPs are related to cutter-boat deployment in rough seas. USCG officials indicated that challenges remain in determining a path forward to resolve these KPPs because the USCG and its operational test agent have different interpretations of the cutter boat requirements. In January 2016, GAO recommended the NSC program office clarify the KPPs for the cutter boats.

USCG officials have indicated that all deficiencies and unmet KPPs will be tested as part of follow-on operational test and evaluation (FOT&E), but it is unclear when the USCG will complete the NSC's FOT&E. The USCG has planned test activities through the end of fiscal year 2017 and USCG officials indicated that DOT&E will independently assess the FOT&E results. However, it is unclear when the USCG will actually demonstrate the NSC can meet its unmanned aircraft and intelligence requirements. In January 2016, GAO recommended DHS specify when the USCG must complete the NSC's FOT&E and any further actions the NSC program should take following FOT&E.

Other Issues

In May 2015, DHS reported the program office had 55 full time equivalents (FTE) but needed 62 FTEs. USCG officials have told GAO this has made it difficult to obligate funds in a timely manner. However, according to USCG officials, as of September 2015 the program office was in the process of hiring staff to fill several vacancies.

Program Office Comments [from the Coast Guard]

Cost estimates cited herein are threshold values taken from the approved NSC baseline. They do not reflect current estimates to complete based on updated data, which includes actual production contract award amounts for NSCs 7 and 8. The NSC program completed IOT&E [initial operational test and evaluation] in 2014 and continues to work with DHS to complete remaining testing and resolve pending discrepancies. Delaying IOT&E was a deliberate decision to ensure maximum benefits from the testing and resulted in the Navy evaluator's assessment that the NSC is "Operationally Effective and Suitable." Despite not fully completing all aspects of IOT&E, recent NSC operations have resulted in rarely seen magnitudes of law enforcement success. USCGC BERTHOLF recently seized nearly 29,000 pounds of cocaine, part of a remarkable 2015 interagency/partner nation effort which included more than 110 interdictions, the arrest of 700 suspected smugglers, and the seizure of 709,888 pounds of cocaine worth roughly $9.4 billion.

GAO Response

Across all 25 program assessments [in this GAO report], GAO has reported threshold cost estimates because they are the maximum costs authorized by DHS leadership. DHS leadership approved an updated NSC cost estimate in September 2014, but it has not changed the program's maximum authorized cost.[15]

[15] GAO-16-338SP, pp. 95-96.

OPC Program

May 2018 GAO Report

Regarding the OPC program, the May 2018 GAO report states the following:

DHS approved six key performance parameters (KPP) for the OPC related to the ship's operating range and duration, crew size, interoperability and maneuverability, and ability to support operations in moderate to rough seas. The first OPC has not yet been constructed, so the USCG has not yet demonstrated whether it can meet these KPPs. The USCG plans to use engineering reviews, and developmental and operational tests throughout the acquisition to measure the OPC's performance.

USCG officials told GAO that the program completed an early operational assessment on the basic ship design in August 2017, which entailed a review of the current design plans. The program plans to refine the ship's design as needed based on preliminary test results. However, as of December 2017, USCG officials had not received the results of this assessment.

The USCG plans to conduct initial operational test and evaluation (OT&E) on the first OPC in fiscal year 2023. However, the test results from initial OT&E will not be available to inform key decisions. For example, the results will not be available to inform the decision to build 2 OPCs per year—which USCG officials said is scheduled to begin in fiscal year 2021. Without test results to inform these key decisions, the USCG must make substantial commitments prior to knowing how well the ship will meet its requirements....

The USCG is in the process of completing the design of the OPC before starting construction, which is in-line with GAO shipbuilding best practices. In addition, USCG officials stated that the program is using state-of-the-market technology that has been proven on other ships as opposed to state-of-the-art technology, which lowers the risk of the program.[16]

March 2016 GAO Report

The March 2016 GAO report states the following:

Performance

Department of Homeland Security (DHS) leadership has approved six key performance parameters (KPP) for the OPC, establishing goals for the ship's operating range and duration, crew size, interoperability and maneuverability, and ability to conduct operations in moderate to rough seas. The first OPC has not yet been constructed, so the USCG has not yet demonstrated whether it can meet these KPPs. The USCG plans to use engineering reviews, and developmental and operational tests throughout the acquisition to measure the OPC's performance.

---

[16] GAO-18-339SP, p. 94.

Acquisition Strategy

The USCG is using a two-phased down-select strategy to select a contractor to deliver the OPC. First, the USCG conducted a full and open competition to select three contractors to perform preliminary and contract design work, and in February 2014, the USCG awarded fixed-price contracts to Eastern Shipbuilding Group, Bollinger Shipyards, and Bath Iron Works for phase 1. Second, in late fiscal year 2016, for phase 2, the USCG plans to select one of these three contractors to develop a detailed design of the OPC, and construct the first 9 to 11 ships.

Program Execution

From 2012 to 2016, the program's initial operational test and evaluation (IOT&E) date slipped 12 months, and its initial and full operational capability dates both slipped 15 months. Additionally, the program's preliminary design review date slipped 13 months, including 4 months during 2015. USCG officials said they completed the contract design review in March 2015, but they did not expect to complete the preliminary design review until January 2016. USCG officials attributed these schedule slips to delays in awarding the three preliminary and contract design contracts, and a subsequent bid protest that was filed with GAO. GAO denied the protest in June 2014.

In June 2014, GAO identified that the OPC's schedule had slipped 14 years between 2007 and 2014. Going forward, USCG officials have stated that additional OPC delays will decrease the USCG's operational capacity because the aging Medium Endurance Cutters will require increased downtime for maintenance and other issues, reducing their availability.

The OPC's acquisition and life-cycle cost estimates did not change from 2012 to 2015. However, in June 2014, GAO reported that the OPC program's acquisition cost estimate had increased by $4 billion from 2007 to 2012. USCG officials said this increase was largely due to invalid assumptions in the earlier cost estimate, along with schedule delays and inflation.

Test Activities

DHS's Director of Operational Test and Evaluation approved the OPC Test and Evaluation Master Plan (TEMP) in October 2011, but the USCG has issued an interim TEMP to reflect schedule changes resulting from the bid protest. The USCG now plans to conduct IOT&E on the first OPC in fiscal year 2023. USCG officials told GAO that they have been working closely with DHS's Office of Test and Evaluation and U.S. Navy test officials since 2010 to incorporate testing into the program.

Other Issues

The program is currently projected to have a $1.2 billion funding shortfall from fiscal years 2016 to 2020. Program officials said this is because the OPC's current cost estimate does not reflect its schedule delays, and that they are working to update the cost estimate. Nonetheless, in 2012, DHS's Chief Financial Officer raised concerns that the OPC's costs could grow as other shipbuilding programs' costs have grown in the past, and could ultimately affect the affordability of other USCG acquisition programs. In June 2014, GAO reported that the OPC will absorb about two-thirds of the USCG's acquisition funding from 2018 to 2032, and recommended that the USCG develop a 20-year fleet modernization plan that identifies all acquisitions needed to maintain the current level of service, along with tradeoffs if the funding needed to execute the plan is not consistent with annual budgets. The USCG concurred with this recommendation but did not identify an estimated date for completing the plan, and USCG officials told GAO they had not identified what tradeoffs they would make to address affordability issues.

In May 2015, DHS headquarters identified that the program office needed 26 full time equivalents (FTE) and actually had 20 FTEs. However, in December 2015, program officials told GAO the program now only needs 20 FTEs, but is still 3 short. Program officials also said that these shortfalls did not significantly affect the program.

Program Office Comments [from the Coast Guard]. The USCG provided technical comments that GAO addressed as appropriate.[17]

FRC Program

May 2018 GAO Report

Regarding the FRC program, the May 2018 GAO report states the following:

> In February 2017, DHS's Director, Office of Test and Evaluation (DOT&E) assessed the results from the program's July 2016 follow-on operational test and evaluation (OT&E) and determined that
>
> - the program met its six key performance parameters, and
> - the FRC was operationally effective and suitable.
>
> During follow-on OT&E, the OTA found that several deficiencies from the program's initial OT&E had been corrected. For example, the OTA closed a severe deficiency related to the engines based on modifications to the FRC's main diesel engines. However, five major deficiencies remain. According to USCG officials, the remaining deficiencies are related to ergonomics (e.g., improving the working environment for operators) and issues with stowage space. USCG officials stated that they plan to resolve the remaining deficiencies by fiscal year 2020.
>
> DOT&E noted that these deficiencies do not prevent mission completion or present a danger to personnel, but recommended that they be resolved as soon as possible. USCG

[17] GAO-16-338SP, pp. 97-98.

officials indicated that they plan to resolve the remaining deficiencies through engineering or other changes....

The USCG continues to work with the contractor—Bollinger Shipyards, LLC—to address issues covered by the warranty and acceptance clauses for each ship. For example, 18 engines—9 operational engines and 9 spare engines—have been replaced under the program's warranty. According to USCG documentation, 65 percent of the current issues with the engines have been resolved through retrofits; however, additional problems with the engines have been identified since our April 2017 review. For example, issues with water pump shafts are currently being examined through a root cause analysis and will be redesigned and are scheduled to undergo retrofits starting in December 2018. We previously found that the FRC's warranty resulted in improved cost and quality by requiring the shipbuilder to pay for the repair of defects. As of September 2017, USCG officials said the replacements and retrofits completed under the program's warranty allowed the USCG to avoid an estimated $104 million in potential unplanned costs—of which $63 million is related to the engines.[18]

For a discussion of some considerations relating to warranties in shipbuilding and other acquisition programs, see *Appendix D*.

March 2016 GAO Report

The March 2016 GAO report states the following:

Officials from 8 of the 11 programs that remained on track during 2015 said their programs were at risk of future schedule slips, cost growth, or both due to anticipated funding constraints, workforce challenges, expanded development efforts, and other reasons. These 8 programs include 7 that previously experienced schedule slips, cost growth, or both. For example, officials from the USCG Fast Response Cutter (FRC) program said that funding shortfalls could affect the number of cutters they are able to procure each year, which could increase costs. Currently, the USCG plans to award a contract by the end of June 2016 that will allow the USCG to purchase 4 to 6 cutters per year, depending on available funding levels. In June 2014, we reported that the USCG estimated a decision to order fewer ships per year would likely increase the program's costs by $600 million to $800 million beyond its current estimates.... [19]

Performance

The FRC partially met one of its six key performance parameters (KPP) during initial operational test and evaluation (IOT&E) in fiscal year 2013. The other five KPPs were not met or not tested. In September 2015, USCG officials told GAO the FRC had since demonstrated it could meet all six of its KPPs, but the Department of Homeland Security's (DHS) Director of Operational Test and Evaluation (DOT&E) has not

[18] GAO-18-339SP, p. 82. For additional discussions of warranties in acquisition programs, see XXX.

[19] GAO-16-338SP, p. 17.

validated the FRC's performance since IOT&E. The FRC's follow-on operational test and evaluation (FOT&E) is scheduled for June 2016.

Acquisition Strategy

In September 2008, USCG officials awarded Bollinger Shipyards Lockport a contract for 1 FRC with options to build up to 33 more. GAO subsequently received a bid protest, which was denied, and upheld the USCG's contract award in January 2009. In May 2014, the USCG established that it would only procure 32 FRCs through this contract. In June 2014, GAO reported that the USCG purchased the technical specifications and licenses from Bollinger that are necessary to build the FRC, and planned to use this information to conduct a full and open competition for the remaining 26 vessels. The USCG has designated this effort Phase 2 of the program.

The USCG began Phase 2 with a request for proposals (RFP), all of which were to be received by July 2015. According to program officials, they plan to award the Phase 2 contract by the end of June 2016. According to USCG officials, the Phase 2 RFP allowed the bidders to make certain changes to the design of the ship, though the key performance parameters remain the same as for Phase 1. In addition, the design for several critical systems—such as the propulsion system, generators, hull structure, and bridge layout—remained consistent with the Phase 1 design.

Program Execution

Previously, the program's initial operational capability date slipped from December 2012 to August 2013 because of the bid protest and the need for structural modifications. Additionally, the program's full operational capability date slipped from September 2022 to March 2027 because, according to USCG officials, the procurement quantities for the FRC changed under the Phase 1 contract. In fiscal years 2010 and 2011, the quantities decreased from six FRCs per year to four. Under the Phase 2 contract, program officials said the USCG will be able to purchase four to six FRCs per year. The USCG has established that the annual procurement quantity will be dictated by funding levels, and a $143 million gap appears to remain between the program's projected funding levels and estimated costs through fiscal year 2020. Program officials told GAO that funding shortfalls could cause further delays going forward, but maintained that the program is still on track to meet its cost goals. Nonetheless, in June 2014, GAO reported that the USCG estimated the decision to order fewer ships per year will likely increase the program's costs by $600 million to $800 million beyond its current estimates.

The FRC's projected funding gap may not actually be $143 million from fiscal year 2016 to fiscal year 2020. The FRC funding plan DHS has presented to Congress does not identify all of the funding the USCG plans to allocate to FRC operations. GAO has reported on USCG affordability issues since 2011 (GAO-11-743). In April 2015, GAO recommended DHS account for all of the operations and maintenance funding the USCG is allocating to its major acquisition programs in an annual report to Congress. DHS concurred with the recommendation, but USCG officials told GAO they have made no progress in accounting for these funding allocations. This issue obscures the size of future funding gaps, and the actual amount allocated through fiscal year 2015 may be greater than $2.1 billion.

Test Activities

In 2009, DOT&E approved the FRC program's Test and Evaluation Master Plan (TEMP). In 2012, USCG officials updated the TEMP in preparation for IOT&E, which was conducted in fiscal year 2013 and assessed three of the program's six KPPs. At that time, the FRC did not fully meet any of them. IOT&E also revealed several major deficiencies, the most significant of which involved the FRC's cutter boat, which exhibited problems operating in moderate sea conditions, and the FRC's main diesel engines, which had multiple equipment failures during testing. Subsequently, independent testers from the U.S. Navy concluded the FRC was operationally effective, but not operationally suitable.

USCG officials told GAO they have improved the FRC's performance since the 2013 IOT&E. For example, they replaced and successfully tested the FRC's cutter boat, worked with the engine manufacturer to determine the root cause of equipment failures, and have begun retrofitting the engines. USCG officials stated the FRC has demonstrated it can meet all six of its KPPs, but DOT&E will not validate the FRC's performance until the USCG completes its FOT&E, which is scheduled for June 2016.

In January 2015, USCG officials told GAO that they were updating the TEMP again in preparation for FOT&E, and that they expected DOT&E would approve the updated TEMP by June 2015. However, it has taken the USCG longer than anticipated to update the TEMP, and in September 2015, USCG officials stated that they were still working on the updates.

Other Issues

In May 2015, DHS headquarters reported that the program required five additional staff, but in September 2015, program officials told GAO that number was down to one. The open staff position was for a naval architect, and the officials said that they were in the process of filling the position.

Program Office Comments [from the Coast Guard]

The FRC program has delivered 15 FRCs and 14 of those have been commissioned. In the past year, FRCs have rescued 117 undocumented migrants. In a Joint Operation, an FRC intercepted a smuggling vessel carrying 212 kg of cocaine, worth an estimated value of over $7 million. The FRC program looks forward to demonstrating the capabilities of the FRC during FOT&E.[20]

---

[20] GAO-16-338SP, pp. 87-88.

Appendix D. Some Considerations Relating to Warranties in Shipbuilding and Other Acquisition Programs

This appendix presents some considerations relating to warranties in shipbuilding and other defense acquisition.[21]

In discussions of Navy and Coast Guard shipbuilding, one question that sometimes arises is whether including a warranty in a shipbuilding contract is preferable to not including one.

Including a warranty in a shipbuilding contract (or a contract for building some other kind of military end item), while potentially valuable, might not always be preferable to not including one—it depends on the circumstances of the acquisition, and it is not necessarily a valid criticism of an acquisition program to state that it is using a contract that does not include a warranty (or a weaker form of a warranty rather than a stronger one).

Including a warranty generally shifts to the contractor the risk of having to pay for fixing problems with earlier work. Although that in itself could be deemed desirable from the government's standpoint, a contractor negotiating a contract that will have a warranty will incorporate that risk into its price, and depending on how much the contractor might charge for doing that, it is possible that the government could wind up paying more in total for acquiring the item (including fixing problems with earlier work on that item) than it would have under a contract without a warranty.

When a warranty is not included in the contract and the government pays later on to fix problems with earlier work, those payments can be very visible, which can invite critical comments from observers. But that does not mean that including a warranty in the contract somehow frees the government from paying to fix problems with earlier work. In a contract that includes a warranty, the government will indeed pay something to fix problems with earlier work—but it will make the payment in the less-visible (but still very real) form of the up-front charge for including the warranty, and that charge might be more than what it would have cost the government, under a contract without a warranty, to pay later on for fixing those problems.

From a cost standpoint, including a warranty in the contract might or might not be preferable, depending on the risk that there will be problems with earlier work that need fixing, the potential cost of fixing such problems, and the cost of including the warranty in the contract. The point is that the goal of avoiding highly visible payments for fixing problems with earlier work and the goal of minimizing the cost to the government of

---

[21] This appendix is adapted from Appendix C of CRS Testimony TE10019, *Options and Considerations for Achieving a 355-Ship Navy*, by Ronald O'Rourke.

fixing problems with earlier work are separate and different goals, and that pursuing the first goal can sometimes work against achieving the second goal.[22]

The Department of Defense's guide on the use of warranties states the following:

> Federal Acquisition Regulation (FAR) 46.7 states that "the use of warranties is not mandatory." However, if the benefits to be derived from the warranty are commensurate with the cost of the warranty, the CO [contracting officer] should consider placing it in the contract. In determining whether a warranty is appropriate for a specific acquisition, FAR Subpart 46.703 requires the CO to consider the nature and use of the supplies and services, the cost, the administration and enforcement, trade practices, and reduced requirements. The rationale for using a warranty should be documented in the contract file....
>
> In determining the value of a warranty, a CBA [cost-benefit analysis] is used to measure the life cycle costs of the system with and without the warranty. A CBA is required to determine if the warranty will be cost beneficial. CBA is an economic analysis, which basically compares the Life Cycle Costs (LCC) of the system with and without the warranty to determine if warranty coverage will improve the LCCs. In general, five key factors will drive the results of the CBA: cost of the warranty + cost of warranty administration + compatibility with total program efforts + cost of overlap with Contractor support + intangible savings. Effective warranties integrate reliability, maintainability, supportability, availability, and life-cycle costs. Decision factors that must be evaluated include the state of the weapon system technology, the size of the warranted population, the likelihood that field performance requirements can be achieved, and the warranty period of performance.[23]

---

[22] It can also be noted that the country's two largest builders of Navy ships—General Dynamics (GD) and Huntington Ingalls Industries (HII)—derive about 60% and 96%, respectively, of their revenues from U.S. government work. (See General Dynamics, *2016 Annual Report*, page 9 of Form 10-K [PDF page 15 of 88]) and Huntington Ingalls Industries, *2016 Annual Report*, page 5 of Form 10-K [PDF page 19 of 134]). These two shipbuilders operate the only U.S. shipyards currently capable of building several major types of Navy ships, including submarines, aircraft carriers, large surface combatants, and amphibious ships. Thus, even if a warranty in a shipbuilding contract with one of these firms were to somehow mean that the government did not have pay under the terms of that contract—either up front or later on—for fixing problems with earlier work done under that contract, there would still be a question as to whether the government would nevertheless wind up eventually paying much of that cost as part of the price of one or more future contracts the government may have that firm.

[23] Department of Defense, *Department of Defense Warranty Guide*, Version 1.0, September 2009, accessed July 13, 2017, at https://www.acq.osd.mil/dpap/pdi/uid/docs/departmentof defensewarrantyguide[1].doc.

In: Conventional Weapons Systems
Editor: Stewart Keller

ISBN: 978-1-53615-414-6

Chapter 5

ARMS SALES: CONGRESSIONAL REVIEW PROCESS (UPDATED)*

Paul K. Kerr

SUMMARY

This report reviews the process and procedures that currently apply to congressional consideration of foreign arms sales proposed by the President. This includes consideration of proposals to sell major defense equipment, defense articles and services, or the retransfer to third-party states of such military items. Under Section 36(b) of the Arms Export Control Act (AECA), Congress must be formally notified *30* calendar days before the Administration can take the final steps to conclude a government-to-government foreign military sale of major defense equipment valued at $14 million or more, defense articles or services valued at $50 million or more, or design and construction services valued at $200 million or more. In the case of such sales to NATO member states, NATO, Japan, Australia, South Korea, Israel, or New Zealand, Congress must be formally notified *15* calendar days before the Administration can proceed with the sale. However, the prior notice threshold values are higher for sales to NATO members, Japan, Australia, South Korea, Israel, or New Zealand. Commercially licensed arms sales also must be formally notified to Congress *30* calendar days before the export license is issued if they involve the sale of major defense equipment valued at $14 million or more, or defense articles or services valued at $50 million or more (Section 36(c) AECA). In the case of such sales to NATO member states, NATO, Japan, Australia, South Korea, Israel, or New Zealand, Congress must be formally notified *15* calendar days before the Administration is authorized to proceed with a given sale. As with government-to-government sales, the prior notice threshold values are higher for sales to NATO members, Japan, Australia, South Korea, Israel, or New Zealand.

---

* This is an edited, reformatted and augmented version of Congressional Research Service, Publication No. RL31675, dated October 22, 2018.

Furthermore, commercially licensed arms sales cases involving defense articles that are firearms-controlled under category I of the United States Munitions List and valued at $1 million or more must also be formally notified to Congress for review *30* days prior to the license for export being approved. In the case of proposed licenses for such sales to NATO members, Japan, Australia, South Korea, Israel, or New Zealand, *15* days prior notification is required.

In general, the executive branch, after complying with the terms of applicable U.S. law, principally contained in the AECA, is free to proceed with an arms sales proposal unless Congress passes legislation prohibiting or modifying the proposed sale. Under current law Congress faces two fundamental obstacles to block or modify a presidential sale of military equipment: it must pass legislation expressing its will on the sale, and it must be capable of overriding a presumptive presidential veto of such legislation. Congress, however, is free to pass legislation to block or modify an arms sale *at any time* up to the point of delivery of the items involved. This chapter will be updated, if notable changes in these review procedures or applicable law occur.

Congressional Notification Requirements

This report reviews the process and procedures that currently apply to congressional consideration of foreign arms sales proposed by the President. This includes consideration of proposals to sell major defense equipment, defense articles and services, or the retransfer to other states of such military items. In general, the executive branch, after complying with the terms of applicable U.S. law, principally contained in the Arms Export Control Act[1] (AECA) (P.L. 90-629, 82 Stat. 1320), is free to proceed with an arms sales proposal unless Congress passes legislation prohibiting or modifying the proposed sale. The President has the obligation under the law to submit the arms sale proposal to Congress, but only after he has determined that he is prepared to proceed with any such notifiable arms sales transaction.

The Department of State (on behalf of the President) submits a preliminary or informal notification of a prospective major arms sale *20* calendar days before the executive branch takes further formal action. This informal notification is provided to the committees of primary jurisdiction for arms sales issues. In the Senate, this is the Senate Foreign Relations Committee; in the House, it is the Foreign Affairs Committee. It has been the practice for such informal notifications to be made for arms sales cases that would have to be formally notified to Congress under the provisions of Section 36(b) of the AECA.[2] The informal notification practice stemmed from a February 18, 1976, letter from the Department of Defense making a *nonstatutory* commitment to give Congress these preliminary classified notifications.[3] Beginning in 2012, the State Department implemented a new informal notification process, which the department calls a "tiered

---

[1] Originally titled The Foreign Military Sales Act.

[2] 22 U.S.C. 2776(b).

[3] Letter of February 18, 1976 from Lt. Gen. H.M. Fish, USAF, Director, Defense Security Assistance Agency to Senator Hubert H. Humphrey, Senate Committee on Foreign Relations.

review," in which the relevant committees are notified between 20 and 40 calendar days before receiving formal notification, depending on the system and destination in question.[4]

Under Section 36(b) of the AECA, Congress must be formally notified *30* calendar days before the Administration can take the final steps to conclude a government-to-government foreign military sale of major defense equipment valued at $14 million or more, defense articles or services valued at $50 million or more, or design and construction services valued at $200 million or more.[5] In the case of such sales to NATO member states, NATO, Japan, Australia, South Korea, Israel, or New Zealand, Congress must be formally notified *15* calendar days before the Administration can proceed with the sale. However, the prior notice threshold values are higher for NATO members, Japan, Australia, South Korea, Israel, or New Zealand. These higher thresholds are $25 million for the sale, enhancement, or upgrading of major defense equipment; $100 million for the sale, enhancement, or upgrading of defense articles and defense services; and $300 million for the sale, enhancement, or upgrading of design and construction services, so long as such sales to these countries do not include or involve sales to a country outside of this group of states.[6] Section 36(i) requires the President to notify both the Senate Foreign Relations Committee and House Foreign Affairs Committee at least 30 days in advance of a pending shipment of defense articles subject to the 36(b) requirements if the chairman and ranking member of either committee request such notification. Certain articles or services listed on the Missile Technology Control Regime are subject to a variety of additional reporting requirements.

Commercially licensed arms sales also must be formally notified to Congress *30* calendar days before the export license is issued if they involve the sale of major defense equipment valued at $14 million or more, or defense articles or services valued at $50 million or more (Section 36(c) AECA).[7] In the case of such sales to NATO member states, NATO, Japan, Australia, South Korea, Israel, or New Zealand, Congress must be formally notified *15* calendar days before the Administration can proceed with such a sale. However, the prior notice threshold values are higher for sales to NATO members, Japan, Australia, South Korea, Israel, or New Zealand, specifically: $25 million for the sale, enhancement, or upgrading of major defense equipment; $100 million for the sale,

[4] Prior to giving such notice, the State Department transmits to the committees any license applications for commercially licensed arms sales as soon as the department receives them. The State Department does not provide the same notice regarding government-to-government foreign military sales.

[5] The Foreign Assistance Act of 1974 (P.L. 93-559, 88 Stat. 1795) amended the Foreign Military Sales Act to add a $25 million dollar threshold for defense articles and services. The International Security Assistance and Arms Export Control Act of 1976 (P.L. 94-329, 90 Stat. 729) added major defense equipment with a $7 million threshold. The International Security and Development Cooperation Act of 1981 (P.L. 97-113, 95 Stat. 1519) changed the $7 million threshold to $14 million and the $25 million threshold to $50 million. The International Security and Development Cooperation Act of 1980 (P.L. 96-533, 94 Stat. 3131) added design and construction services with a $200 million threshold.

[6] The Foreign Relations Authorization Act, Fiscal Year 2003 (P.L. 107-228, 116 Stat. 1350), added these thresholds.

[7] 22 U.S.C. 2776(c).

enhancement, or upgrading of defense articles and defense services; and $300 million for the sale, enhancement, or upgrading of design and construction services, so long as such sales to these countries do not include or involve sales to a country outside of this group of states. Furthermore, commercially licensed arms sales cases involving defense articles that are firearms controlled under category I of the United States Munitions List and valued at $1 million or more must also be formally notified to Congress for review 30 days prior to the license for export being approved (15 days prior notice is required for proposed licenses for sales to NATO members, Japan, Australia, South Korea, Israel, or New Zealand).[8,9]

Section 36(b)(5)(A) contains a reporting requirement for defense articles or equipment items whose technology or capability has, prior to delivery, been "enhanced or upgraded from the level of sensitivity or capability described" in the original congressional notification. For such exports, the President must submit a report to the relevant committees at least 45 days before the exports' delivery that describes the enhancement or upgrade and provides "a detailed justification for such enhancement or upgrade." This requirement applies for 10 years after the Administration has notified Congress of the export.[10] According to Section 36(b)(5)(C), the Administration must, in the case of upgrades or enhancements meeting certain value thresholds, submit a new notification to Congress and the export will be considered "as if it were a separate letter of offer ... subject to all of the requirements, restrictions, and conditions set forth in this subsection." The threshold values are higher for sales to NATO members, Japan, Australia, South Korea, Israel, or New Zealand.

A congressional recess or adjournment does not stop the 30 calendar-day statutory review period. It should be emphasized that after Congress receives a statutory notification required under Sections 36(b) or 36(c) of the AECA, for example, and 30 calendar days elapse without Congress having blocked the sale, the executive branch is free to proceed with the sales process. This fact does not mean necessarily that the executive branch and the prospective arms purchaser will sign a sales contract and that the items will be transferred on the 31st day after the statutory notification of the proposal has been made. It would, however, be legal to do so at that time.

---

[8] Added by the Foreign Relations Authorization Act, Fiscal Year 2003.

[9] These notification requirements and reporting thresholds also apply to prospective retransfers of United States-origin major defense equipment, defense articles, or defense services as stipulated in Section 3(d) of the Arms Export Control Act (AECA); and leases or loans of defense articles from U.S. Defense Department stocks (see Sections 62 and 63 AECA). Section 36(d) contains similar notification requirements, though not reporting thresholds, for commercial technical assistance or manufacturing licensing agreements. As with arms sales, Congress can block any of these reportable transactions by enacting a joint resolution of disapproval as stipulated in the Arms Export Control Act (AECA) (see 22 U.S.C. 2753, 2776, 2796).

[10] This provision also applies to defense services and design and construction services.

CONGRESSIONAL DISAPPROVAL BY JOINT RESOLUTION

Although Congress has more than one legislative option it can use to block or modify an arms sale, one option explicitly set out in law for blocking a proposed arms sale is the use of a joint resolution of disapproval as provided for in Section 36(b) of the AECA. Under that law, the formal notification is legally required to be submitted to the chairman of the Senate Foreign Relations Committee and the Speaker of the House. The Speaker has routinely referred these notifications to the House Foreign Affairs Committee as the committee of jurisdiction. As a courtesy, the Defense Department has submitted a copy of the statutory notification to the House Foreign Affairs Committee when that notification is submitted to the Speaker of the House. Under this option, after receiving a statutory Section 36(b) notification from the executive branch, opponents of the arms sale would introduce joint resolutions in the House and Senate drafted so as to forbid by law the sale of the items specified in the formal sale notification(s) submitted to Congress. If no Member introduces such a measure, the AECA's provisions expediting congressional action, discussed below, do not take effect.

The next step would be committee hearings in both houses on the arms sale proposal. If a majority of either the House or the Senate committee supported the joint resolution of disapproval, they would report it to their respective chamber in accordance with its rules. Following this, efforts would be made to seek floor consideration of the resolution.

Senate Procedures

At this point, it is important to take note of procedures crafted to expedite the consideration of arms sales resolutions of disapproval. Since 1976, Section 36(b)(2) of the AECA has stipulated that consideration of any resolution of disapproval in the Senate under Section 36(b)(1) of the AECA shall be "in accordance with the provisions of Section 601(b) of the International Security Assistance and Arms Export Control Act of 1976" (P.L. 94-329, 90 Stat. 729). Since 1980, this stipulation has also applied to resolutions of disapproval in the Senate relating to commercially licensed arms sales under Section 36(c)(1) of the AECA. The purpose of Section 601(b) was to establish rules to facilitate timely consideration of any resolution of disapproval in the Senate. The rules set forth in Section 601(b) supersede the standing rules of the Senate and include the following:

- Give the committee with jurisdiction [the Senate Foreign Relations Committee] 10 calendar days from the date a resolution of disapproval is referred to it to report back to the Senate its recommendation on any such resolution (certain adjournment periods are excluded from computation of the 10 days);

- Make it in order for a Senator favoring a disapproval resolution to move to discharge the committee from further consideration of the matter if the committee fails to report back to the Senate by the end of the 10 calendar days it is entitled to review the resolution (the AECA expressly permits a discharge motion after 5 calendar days for sales to NATO, NATO countries, Japan, Australia, South Korea, Israel, and New Zealand);
- Make the discharge motion privileged, limit floor debate on the motion to one hour, and preclude efforts to amend or to reconsider the vote on such a motion;
- Make the motion to proceed to consider a resolution of disapproval privileged and preclude efforts to amend or to reconsider the vote on such motion;
- Limit the overall time for debate on the resolution of disapproval to 10 hours and preclude efforts to amend or recommit the resolution of disapproval;
- Limit the time (one hour) to be used in connection with any debatable motion or appeal; provide that a motion to further limit debate on a resolution of disapproval, debatable motion, or appeal is not debatable.

The Senate is constitutionally empowered to amend its rules or to effect a rule change at any time. The fact that an existing rule is in Section 601 of the International Security Assistance and Arms Export Control Act of 1976 is not an obstacle to changing it by Senate action alone should the Senate seek to do so.

House Floor Procedures[11]

The House of Representatives is directed by Sections 36(b)(3) and 36(c)(3)(B) of the AECA to consider a motion to proceed to the consideration of a joint resolution disapproving an arms sale reported to it by the appropriate House committee as "highly privileged." Generally, this means that the resolution will be given precedence over most other legislative business of the House, and may be called up on the floor without a special rule reported by the Rules Committee. Unlike for the Senate, however, the AECA contains no provision for discharge of the House committee if it does not report on the joint resolution. If reported and called up, the measure will be considered in the Committee of the Whole, meaning that amendments can be offered under the "five-minute rule." Nevertheless, amendments to joint resolutions disapproving arms sales have apparently never been offered in the House.

The Rules Committee usually sets the framework for floor consideration of major legislation in the House of Representatives, however, and could do so for a joint

[11] H.R. 7080, the Arms Sale Oversight Act, introduced on October 19, 2018, would change these conditions for considering joint resolutions of disapproval in the House.

resolution of disapproval. Upon receiving a request for a rule to govern consideration of such a resolution, the House Rules Committee could set a time limit for debate, exclude any amendments to, and waive any points of order against the resolution. If the House adopted the rule reported by the committee, it would govern the manner in which the legislation would be considered, superseding the statutory provision.

Final Congressional Action

After a joint resolution is passed by both the House and the Senate, the measure would next be sent to the President. Once this legislation reaches the President, presumably he would veto it in a timely manner. Congress would then face the task of obtaining a two-thirds majority in both houses to override the veto and impose its position on the President.

Congressional Use of Other Legislation

Congress can also block or modify a proposed sale of major defense equipment, or defense articles and services, if it uses the regular legislative process to pass legislation prohibiting or modifying the sale or prohibiting delivery of the equipment to the recipient country. While it is generally presumed that Congress will await formal notification under Section 36(b) or 36(c) of the AECA before acting in opposition to a prospective arms sale, it is clear that a properly drafted law could block or modify an arms sale transaction at any time—including before a formal AECA notification was submitted or after the 30-day AECA statutory notification period had expired— so long as the items have not been delivered to the recipient country.

Congressional use of its lawmaking power regarding arms sales is not constrained by the AECA reporting requirements. In order to prevail, however, Congress must be capable of overriding a presidential veto of this legislation, for the President would presumably veto a bill that blocked his wish to make the arms sale in question. This means, in practical terms, that to impose its view on the President, Congress must be capable of securing a two-thirds majority of those present and voting in both houses.

There are potentially important practical advantages, however, to prohibiting or modifying a sale, if Congress seeks to do so, prior to the date when the formal contract with the foreign government is signed—which could occur at any time after the statutory 30-day period. These likely advantages include (1) limiting political damage to bilateral relations that could result from signing a sales contract and later nullifying it with a new law; and (2) avoiding financial liabilities which the United States Government might face for breaking a valid sales contract. The legislative vehicle designed to prohibit or modify a specific arms sale can take a variety of forms, ranging from a rider to any appropriation or authorization bill to a freestanding bill or joint resolution. The only essential features that the vehicle must have are (1) that it is legislation passed by both houses of Congress and presented to the President for his signature or veto and, (2) that it contains an express restriction on the sale and/or the delivery of military equipment (whether it applies to specific items or general categories) to a specific country or countries.

PRESIDENTIAL WAIVER OF CONGRESSIONAL REVIEW

It is important to note that the President also has the legal authority to waive the AECA 30-day statutory review period. For example, if the President states in the formal notification to Congress under AECA Sections 36(b)(1) or 36(c)(1) that "an emergency exists" which requires the sale (or export license approval) to be made immediately "in the national security interests of the United States," he is free to proceed with the sale without further delay. He must provide Congress at the time of this notification a "detailed justification for his determination, including a description of the emergency circumstances" which necessitated his action and a "discussion of the national security interests involved."

Section 614(a) of the Foreign Assistance Act of 1961 (FAA), as amended,[12] also allows the President, among other things, to waive provisions of the AECA, the FAA, and any act authorizing or appropriating funds for use under either the AECA or FAA in order to make available, during each fiscal year, up to $750 million in cash arms sales and up to $250 million in funds. Not more than $50 million of the $250 million limitation on funds use may be made available to any single country in any fiscal year through this waiver authority unless the country is a "victim of active aggression." Not more than $500 million of cash sales (or cash sales and funds made available combined) may be provided under this waiver authority to any one country in any fiscal year. To waive the provisions of these acts related to arms sales, the President must determine and notify Congress in writing that it is "vital" to the "national security interests" of the United States to do so. Before exercising the authority granted in Section 614(a), the President must "consult with" and "provide a written policy justification to" the House Foreign Affairs and the Senate Foreign Relations Committees and House and Senate Appropriations Committees.

In summary, in the absence of a strong majority in both houses of Congress supporting legislation to block or modify a prospective arms sale, the practical and procedural obstacles to passing such a law—whether a freestanding measure or one within the AECA framework—are great. Even if Congress can pass the requisite

---

[12] 22 U.S.C. 2364(a).

legislation to work its will on an arms sale, the President need only veto it and secure the support of one-third plus one of the Members of either the Senate or the House to have the veto sustained and permit the sale.

It should be noted that Congress has never successfully blocked a proposed arms sale by use of a joint resolution of disapproval, although it has come close to doing so (see section below for selected examples). Nevertheless, Congress has—by expressing strong opposition to prospective arms sales, during consultations with the executive branch—affected the timing and the composition of some arms sales, and may have dissuaded the President from formally proposing certain arms sales.

Examples of AECA Resolutions of Disapproval

On October 14, 1981, the House adopted a resolution (H.Con.Res. 194) objecting to President Reagan's proposed sale to Saudi Arabia of E-3A airborne warning and control system (AWACS) aircraft, Sidewinder missiles, Boeing 707 refueling aircraft, and defense articles and services related to F-15 aircraft. An October 28, 1981, Senate vote on identical legislation failed, however, after President Reagan made a series of written commitments to Congress regarding the proposed sale. Congress later enacted legislation requiring the President to certify that the commitments made in 1981 regarding the proposed sale had been met prior to the delivery of the AWACS planes (Section 127 of the International Security and Development Cooperation Act of 1985; P.L. 99-83).

On April 8, 1986, President Ronald Reagan formally proposed the sale to Saudi Arabia of 1,700 Sidewinder missiles, 100 Harpoon missiles, 200 Stinger missile launchers, and 600 Stinger missile reloads. On May 6, 1986, the Senate passed legislation to block these sales (S.J.Res. 316) by a vote of 73-22. The House concurred with the Senate action on May 7, 1986, by passing H.J.Res. 589 by a vote of 356-62. The House then passed S.J.Res. 316 by a voice vote and (in lieu of H.J.Res. 589) sent it to the President. On May 21, 1986, President Reagan vetoed S.J.Res. 316. But, in a letter that day to then-Senate Majority Leader Robert Dole, President Reagan said he would not include the controversial Stinger missiles and launchers in the sales proposal. On June 5, 1986, the Senate, by a 66-34 vote, sustained the President's veto of S.J.Res. 316, and the sale of the Sidewinder and Harpoon missiles to Saudi Arabia proceeded.

More recently, on March 10, 2016, the Senate Foreign Relations Committee rejected a motion to discharge a joint resolution (S.J.Res. 31) prohibiting the sale of several defense articles, particularly eight F-16 Block 52 aircraft.[13] H.J.Res 82 was the House companion bill. On May 5, 2016, a State Department spokesperson, noting congressional objections to using Foreign Military Financing funds for the aircraft, told reporters that

---

[13] K. Alan Kronstadt, Specialist in South Asian Affairs, contributed to this paragraph.

the United States had "told the Pakistanis that they should put forward national funds for the purchase."[14] In late May, the U.S. offer expired after Islamabad failed to submit a letter of acceptance by the required deadline.

On June 13, 2017, the Senate voted to reject a motion to discharge from the Senate Foreign Relations Committee a joint resolution (S.J.Res. 42) prohibiting certain proposed defense exports to Saudi Arabia, such as "technical data, hardware, and defense services" to support the Royal Saudi Air Force's deployment of the Joint Direct Attack Munition and integration of the FMU-152A/B JPB Fuze System into several warhead types. The bill also would have prohibited the transfer of "defense articles, defense services, and technical data to support the assembly, modification, testing, training, operation, maintenance, and integration" of certain precision guided munitions for the certain Royal Saudi Air Force planes. H.J.Res 102 was the House companion bill.

[14] Mark Toner, State Department Regular News Briefing, May 5, 2016.

INDEX

A

B

C

D

E

I

J

K

L

M

N

S

T

U

V

W